华章IT

HZBOOKS | Information Technology

Spring Boot 开发实战

陈光剑 编著

机械工业出版社
China Machine Press

图书在版编目（CIP）数据

Spring Boot 开发实战 / 陈光剑编著．—北京：机械工业出版社，2018.7
（华章程序员书库）

ISBN 978-7-111-60333-7

I. S…　II. 陈…　III. JAVA 语言－程序设计　IV. TP312.8

中国版本图书馆 CIP 数据核字（2018）第 145768 号

Spring Boot 开发实战

出版发行：机械工业出版社（北京市西城区百万庄大街 22 号　邮政编码：100037）
责任编辑：吴　怡　　　　责任校对：李秋荣
印　　刷：北京文昌阁彩色印刷有限责任公司　　　　版　　次：2018 年 8 月第 1 版第 1 次印刷
开　　本：186mm×240mm　1/16　　　　印　　张：23.5
书　　号：ISBN 978-7-111-60333-7　　　　定　　价：89.00 元

凡购本书，如有缺页、倒页、脱页，由本社发行部调换
客服热线：（010）88379426　88361066　　　　投稿热线：（010）88379604
购书热线：（010）68326294　88379649　68995259　　　　读者信箱：hzit@hzbook.com

Preface 前　　言

Spring Boot 是由 Pivotal 团队提供的全新框架，其设计目的是简化新 Spring 应用的初始搭建以及开发过程。在 Java 开发领域中，有很多著名框架都是 Pivotal 团队的产品，如：Spring 框架及其衍生框架、缓存 Redis、消息队列框架 RabbitMQ、Greenplum 数据库等。还有 Tomcat、Apache Http Server、Groovy 里的一些顶级开发者、DevOps 理论的提出者都属于 Pivotal 团队。Spring 团队在现有 Spring 框架的基础上，开发了一个新框架 :Spring Boot，用来简化配置和部署 Spring 应用程序的过程，去除了那些烦琐的开发步骤和样板代码及其配置，使得基于 Spring 框架的 Java 企业级应用开发“极简化”。相比于传统的 Spring/Spring MVC 框架的企业级应用开发 (Spring 的各种配置太复杂了，我们之前是用“生命”在搞这些配置)，Spring Boot 用简单的注解和 application.properties 配置文件，避免了烦琐而且容易出错的 XML 配置文件，极大地简化了基于 Spring 框架的企业级应用开发的配置。

Kotlin 是由 JetBrains 团队开发的多平台、静态类型、强工程实用性的编程语言，Kotlin 100% 兼容 Java，比 Java 更强大、更安全、更简洁、更优雅。Kotlin 是 Google 公司的 Android 官方支持的开发语言。Spring 官方也正式支持 Kotlin 语言，Spring Boot 2.0 版本中为 Kotlin 提供了一流的支持。其实，在 Spring Boot 2.0 和 Spring 5.0 框架源代码中，已经可以看到 Kotlin 代码。

本书可以说是我对使用 Spring Boot + Kotlin 进行服务端开发的实战和思考过程的粗浅总结。通过本书的写作，加深了我对 Spring Boot 框架和 Kotlin 编程语言的理解，我深刻体会到了学无止境的含义。写书的过程也是我系统学习与思考的过程，如果本书能够对你有所帮助，将不胜欣慰。

如何阅读本书

本书系统介绍了使用 Spring Boot 2.0 框架，并基于 Gradle + Kotlin 来开发企业级应用。希望通过简练的表述，系统全面地介绍如何使用 Spring Boot 2.0 框架开发项目，每章的关联度不大，读者可根据自己的需求阅读本书。

全书共分三大部分：

- 第Ⅰ部分　Spring Boot 框架基础（第 1 ～ 3 章）
- 第Ⅱ部分　Spring Boot 项目综合实战（第 4 ～ 17 章）
- 第Ⅲ部分　Spring Boot 系统监控、测试与运维（第 18 ～ 20 章）

建议初学者最好按照章节顺序来阅读本书。如果想直接使用 Spring Boot 框架进行项目的实战，可以直接进入第Ⅱ部分，如果对 Spring Boot 应用的监控、测试与运维感兴趣，那么可以从第Ⅲ部分直接开始阅读。

本书共 20 章，各个章节内容简介如下。

第 1 章：简单介绍了 Spring Boot 框架的历史、组成、特性等。

第 2 章：使用 Spring Boot 2.0 快速实现一个基于 Kotlin 和 Gradle 的 HelloWorld 应用。

第 3 章：介绍 Spring Boot 是怎样通过自动配置实现“极简化配置”的应用开发。

第 4 章：介绍如何使用 Spring Boot 集成 MyBatis 来进行数据库层开发。

第 5 章：介绍如何使用 Spring Boot 集成 Spring Data JPA 来进行数据库层开发。

第 6 章：介绍如何开发一个 Gradle 插件，以及如何简化 开发过程中样板代码的编写。

第 7 章：介绍 Kotlin 编程语言，以及如何集成 Spring Boot 和 Spring MVC 进行服务端开发。

第 8 章：介绍在 Spring Boot 项目中怎样自定义 Web MVC 配置。

第 9 章：介绍基于 Spring Boot + Spring MVC，使用 AOP + Filter 如何实现一个简单的用户登录鉴权与权限控制系统。

第 10 章：介绍如何使用 Spring Boot 集成 Spring Security 开发一个自动化测试平台。

第 11 章：介绍 Spring Boot 集成 React.js 开发前后端分离项目的实战案例。

第 12 章：介绍如何开发任务调度、邮件服务等系统功能。

第 13 章：介绍如何用 Spring Boot 集成 WebFlux 开发响应式 Web 应用。

第 14 章：介绍在 Spring Boot 项目开发中怎样使用 Spring Cache 实现数据的缓存。

第 15 章：介绍如何使用 Spring Session 集成 Redis 实现 Session 共享，从而实现水平扩展。

第 16 章：介绍如何使用 Netflix Zuul 实现一个微服务 API Gateway 来完成简单代理转发和过滤器功能。

第 17 章：详细介绍 Spring Boot 应用的日志配置与使用，主要介绍 Logback 日志框架。

第 18 章：介绍如何使用 Spring Boot Actuator 和 Spring Boot Admin 实现监控与管理。

第 19 章：介绍 Spring Boot 应用的测试，以及如何在实际项目中进行分层测试。

第 20 章：介绍如何使用 Docker 来构建部署运行 Spring Boot 应用。

谁适合阅读本书

本书适合于所有 Java、Kotlin 程序员，以及任何对编程感兴趣的朋友。如果你目前还不是程序员，但想进入企业级应用开发的编程世界，那么你也可以尝试从本书开始学习。

虽然书中的部分内容需要一定的 Java 和 Kotlin 编程基础，还需要了解 Spring 框架，但是如果你想快速开始企业级应用开发，不妨从这里开始——Spring Boot 2.0 + Kotlin ，这种方式的极简特性定能激发你对编程的兴趣。

代码下载

每章末尾基本上都附了该章示例工程源代码地址。这些源码都在 https://github.com/Easy-SpringBoot 。可以根据需要，自由克隆下载学习。

致谢

在本书的写作出版过程中，得到了很多人的帮助和陪伴。首先要感谢的是我的妻子和两个可爱的孩子。正是有了你们的陪伴，我的生活才更加有意义。我始终感谢我的父母，虽然你们可能不知道我写的东西是什么，但是因为有了你们的辛勤养育，我才能长成今天的我。我要衷心地感谢吴怡编辑。在本书的写作修改过程中，她耐心细致地对稿件进行了详尽、细致的审阅和批注，还提出了很多宝贵的修改建议。感谢本书出版过程中所有付出辛勤劳动的工作人员。我还要感谢在我的工作学习生活中认识的，所有朋友和同事们，能够认识你们并跟你们一起学习共事，是我的荣幸。

请联系我

虽然在本书写作与修改的过程中，我竭尽全力追求简单正确、清晰流畅地表达内容，但是限于自身水平和有限的时间，也许仍有错误与疏漏之处，还望各位读者不吝指正。

关于本书的任何问题、意见或者建议都可以通过邮件 universsky@163.com 与我交流。

快乐生活，快乐学习，快乐分享，快乐实践出真知。

最后，祝大家阅读愉快！

陈光剑

2018 年 4 月于杭州

目　录 *Contents*

第Ⅱ部分 Spring Boot 项目综合实战

第 I 部分 *Part 1*

Spring Boot 框架基础

- 第 1 章　Spring Boot 简介
- 第 2 章　快速开始 HelloWorld
- 第 3 章　深入理解 Spring Boot 自动配置

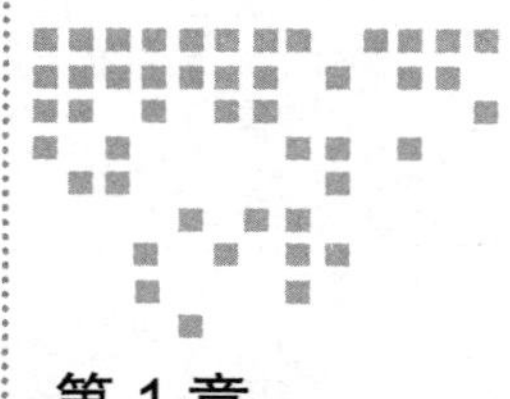

Chapter 1 第1章

Spring Boot简介

认识一个事物最好的方式就是首先去了解它的历史。

Spring框架是由Rod Johnson在2001年开始开发的一个开源框架，主要为了解决企业级应用程序开发的复杂性。Spring提倡“零”侵入设计原则，颠覆了传统的编程模式。Spring引入控制反转（Inversion of Control，IoC）的核心编程思想，控制反转还有一个名字叫作依赖注入（Dependency Injection，DI），就是由容器来管理协同Bean之间的关系，而非传统实现中，由程序代码直接操控。同时，Spring还把面向切面编程（AOP）集成进来，使得AOP的编程范式发扬光大。

Spring从IoC容器发展而来，通过不断集成AOP、MVC、OR/Mapping以及几乎你能想到的各项服务而提供完善的企业应用框架。目前大多数J2EE项目都已经采用Spring框架。

随着Spring功能的不断丰富，版本的不断迭代发展，Spring框架渐渐暴露出了一些问题和弊端。例如太多样板化的配置、烦琐复杂的使用过程等，我们不仅需要维护程序代码，还需要额外去维护相关的配置文件。Spring项目的配置越来越复杂，让人难以承受。大量的XML配置以及复杂的依赖管理使得人们不得不去解决这个问题——Spring Boot由此应运而生。

在本章中，我们先来简单了解一下Spring Boot框架的历史、组成、特性等。

1.1 从Spring到Spring Boot

本节将介绍Spring Boot的产生背景。我们先来回顾一下Spring框架的前世今生。

1.1.1　从 EJB 到 Spring

EJB（Enterprise Java Bean）最初的设计思想是为分布式应用服务的。分布式是针对大型应用构造的跨平台的协作计算，EJB 最初的目的就是为这种计算服务的。使用 EJB 技术的系统整体架构如图 1-1 所示。

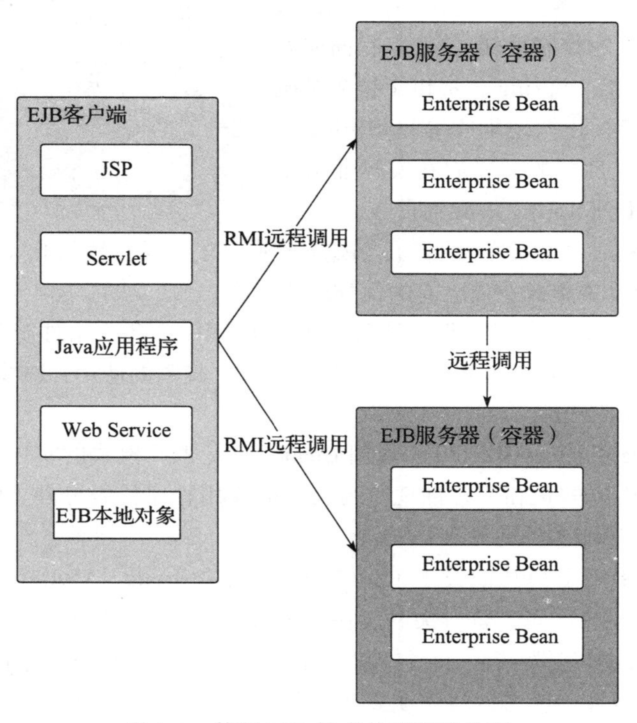

图 1-1　使用 EJB 技术的系统架构图

EJB 的基础是 RMI（Remote Method Invocation，远程方法调用），RMI 利用 Java 对象序列化的机制实现分布式计算，实现远程类对象的实例化以及调用。通过 RMI，J2EE 将 EJB 组件创建为远程对象。RMI 将各种任务与功能的类放到不同的服务器上，然后通过各个服务器间建立的调用规则实现分布式的运算。通过 RMI 的通信（底层仍然是 Socket），连接不同功能模块的服务器，以实现一个完整的功能。

EJB 规范定义了 EJB 组件在何时如何与它们的容器进行交互作用，容器负责提供公用的服务，例如目录服务、事务管理、安全性、资源缓冲池以及容错性。但这里值得注意的是，EJB 并不是实现 J2EE 的唯一途径。但是软件发展到目前为止，大多数应用不需要采用这么重的解决方案，因此用 EJB 显得太臃肿了。

提示　更多关于 J2EE 的内容，可以参考：https://github.com/javaee。

对于中小型的应用项目而言，基本不采用分布式的解决方案，那么为什么要采取一个为分布式设计的方案来解决非分布式的问题呢？ Spring 就是为了解决这个问题而诞生的。

Spring 的目的是为了解决企业应用开发的复杂性，它的主要功能是使用基本的 Java Bean 代替 EJB，并提供了更多的企业应用功能。Spring 使得已存在的技术更加易用。简单来说，Spring 是一个轻量级的控制反转（IoC）和面向切面（AOP）的容器框架。Spring 也提供了很多基础功能（事务管理、持久化框架集成等）。Spring 的设计原则是“非侵入性”的，我们在实际业务逻辑代码中几乎感觉不到 Spring 框架的存在。

Spring 框架的核心功能简单概括为：解耦依赖（DI）、系统模块化（AOP）。Spring“不重复发明轮子”，而是去集成业内已有的优秀解决方案。

Spring 容器以 Bean 的方式来组织和管理 Java 应用中的各个组件及其组件之间的关系。基于 Java Beans 的配置管理，特别是对依赖注入（DI）技术的使用，减少了各组件间对业务逻辑具体实现的相互依赖性。

Spring 使用 BeanFactory 来产生和管理 Bean，它是工厂模式的实现。BeanFactory 使用控制反转模式将应用的配置和依赖性规范与实际的应用程序代码分开。BeanFactory 使用依赖注入的方式给组件提供依赖。

Spring 框架主要用于与其他技术（例如 Struts，Hibernate，MyBatis 等）进行整合，将应用程序中的 Bean 组件实现低耦合关联，提高了系统的可扩展性和维护性。

Spring 集成的 AOP 框架提供了诸如数据库声明式事务等服务。通过使用 Spring AOP，我们无须依赖 EJB 组件，就可以将声明式事务管理集成到应用程序中。AOP 的目的是提高系统的模块化程度。

当然，作为一个完整的 J2EE 框架，Spring 生态中也给出了完整的分布式系统架构的解决方案，那就是 Spring Boot + Spring Cloud，这个解决方案中包含了服务发现（Service Discovery）、断路器（Circuit Breaker）、OAuth2（实现 SSO、登录 token 的管理）、服务配置（Configuration Server）、消费者驱动契约（Consumer-Driven Contracts）、API Gateway 等。

Spring 的微服务系统架构如图 1-2 所示。

1.1.2 Spring 框架发展简史

Spring 框架首次在 2003 年 6 月的 Apache2.0 使用许可中发布。第一个具有里程碑意义的版本是 2004 年 3 月发布的 1.0。

下面是 Spring 框架的发展简史：

- 2003 年，Spring0.9 发布。2003 年 11 月，Ben Alex 将 Acegi Security 的代码贡献给 Rod 和 Juergen，2006 年 5 月发布 Acegi Security。

- 2006 年 6 月发布 Spring Webflow 1.0、2006 年 8 月发布 Spring LDAP。2006 年 10 月发布 Spring 2.0。
- 2007 年 5 月发布 Spring Batch。2007 年 11 月发布 Spring 2.5。Spring 2.5 是 Spring 2.1 各个里程碑版本的终结。
- 2011 年 6 月发布 Spring Data JPA 1.0。2011 年 12 月发布 Spring 3.1
- 2014 年 4 月发布 Spring Boot 1.0。2014 年 12 月发布 Spring 4.1.3
- 2015 年 7 月发布 Spring 4.2
- 2016 年 6 月发布 Spring 4.3
- 2017 年 9 月发布 Spring 5.0。2017 年 11 月发布 Spring Boot v2. 0.0.M7
- 2018 年 3 月 1 日发布 Spring Boot v2.0.0.Release；2018 年 4 月 5 日发布 Spring Boot 2.0.1.Release 版本，是目前最新版本。

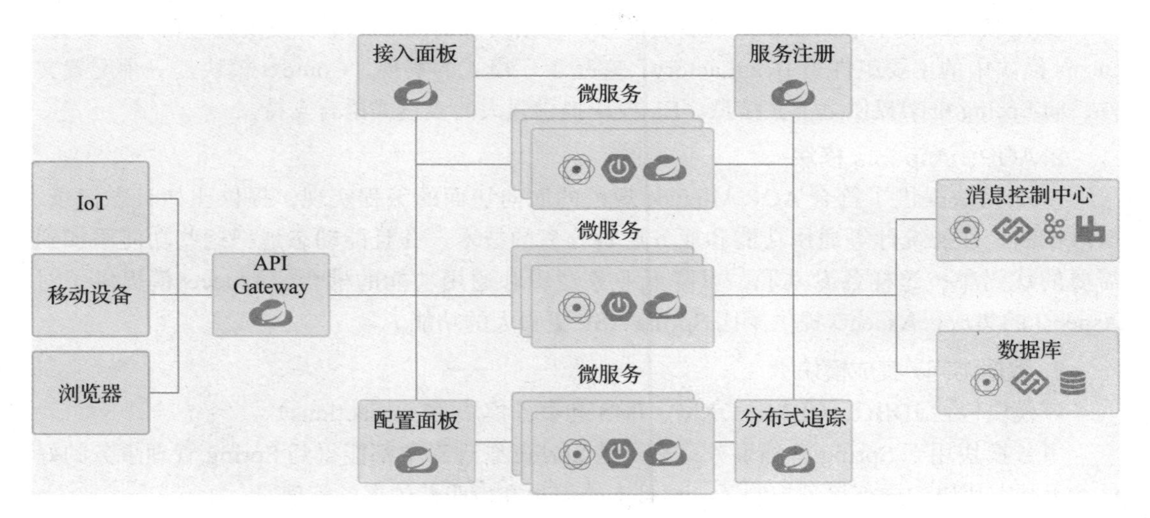

图 1-2　Spring 的微服务系统架构图

提示　详细的发布日志参考 https://github.com/spring-projects/spring-boot/releases。

1.1.3　Spring 框架的核心模块

Spring 框架如图 1-3 所示。组成 Spring 框架的每个模块（或组件）都可以单独存在，或者与其他一个或多个模块联合实现。下面我们分别介绍。

1. 核心容器模块

核心容器提供 Spring 框架的基本功能，包括 Core、Beans、Context、EL 模块。

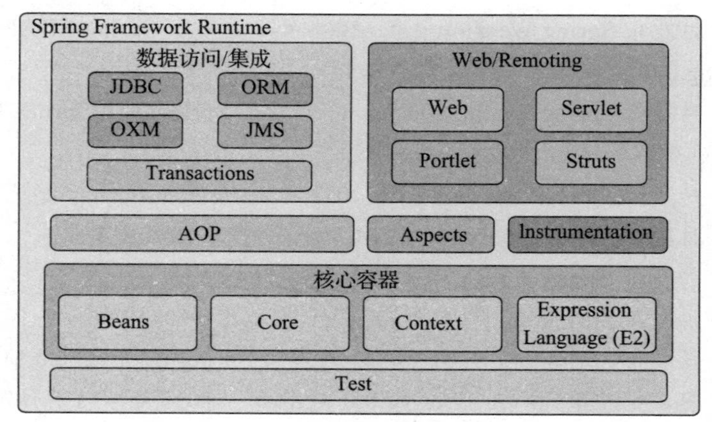

图 1-3 Spring 架构图

Core 模块封装了框架依赖的最底层部分，包括资源访问、类型转换及一些常用工具类。Beans 模块中的主要组件是 BeanFactory，它是工厂模式的实现。Context 模块是一个配置文件，向 Spring 框架提供上下文信息。EL 模块提供强大的表达式语言支持。

2. AOP、Aspects 模块

AOP 模块提供了符合 AOP Alliance 规范的面向切面的编程实现，提供比如日志记录、权限控制、性能统计等通用功能和业务逻辑分离的技术，并且能动态地把这些功能添加到需要的代码中；这样各专其职，可降低业务逻辑和通用功能的耦合。Aspects 模提供了对 AspectJ 的集成，AspectJ 提供了比 Spring ASP 更强大的功能。

3. 数据访问 / 集成模块

该模块包括 JDBC、ORM、OXM、JMS 和事务模块（Transactions）。

事务模块用于 Spring 管理事务，只要是 Spring 管理对象都能得到 Spring 管理事务的好处，无须在代码中进行事务控制了，而且支持编程和声明性的事务管理。

JDBC 模块提供了一个 JBDC 的样例模板，使用这些模板能消除传统冗长的 JDBC 编码还有必须的事务控制，而且能享受到 Spring 管理事务的好处。

ORM 模块提供与流行的“对象 – 关系映射”ORM 框架的无缝集成，包括 Hibernate、JPA、MyBatis 等。

OXM 模块提供了一个对 Object/XML 映射实现，将 Java 对象映射成 XML 数据，或者将 XML 数据映射成 Java 对象，Object/XML 映射实现包括 JAXB、Castor、XMLBeans 和 XStream。

JMS（Java Messaging Service）模块提供一套“消息生产者、消息消费者”模板以便更加简单地使用 JMS，JMS 用于在两个应用程序之间，或分布式系统中发送消息，进行异步通信。

4. Web/Remoting 模块

Web/Remoting 模块包含 Web、Web-Servlet、Web-Struts、Web-Porlet 模块。

Web 模块提供了基础的 Web 功能。例如多文件上传、集成 IoC 容器、远程过程访问

（RMI、Hessian、Burlap）以及 Web Service 支持，并提供一个 RestTemplate 类来提供方便的 Restful services 访问。

Web Servlet 模块提供了一个 Spring MVC Web 框架实现。

Web Struts 模块提供了与 Struts 无缝集成，Struts1.x 和 Struts2.x 都支持。

5. Test 模块

Test 模块支持 Junit 和 TestNG 测试框架，而且还额外提供了一些基于 Spring 的测试功能，比如在测试 Web 框架时，模拟 Http 请求的功能。

当下 Spring 生态中，Spring Boot、Spring Cloud 和 Data Flow 三驾马车带领使用 Spring 进行应用开发勇往直前，如图所示：

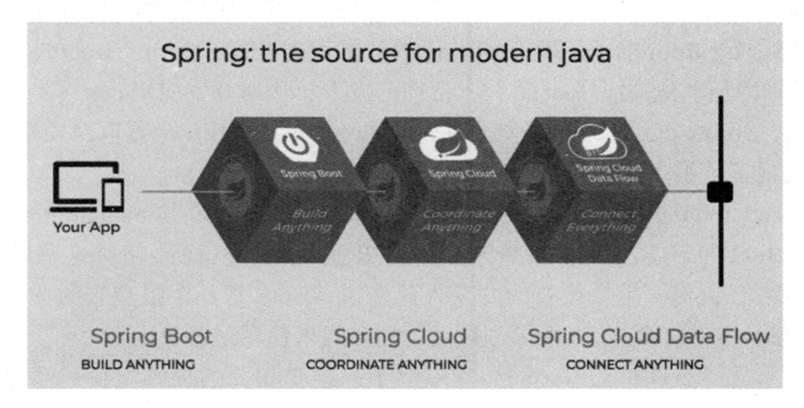

使用 Spring Boot 构建一切服务。Spring Boot 旨在让你尽可能快地启动和运行，并极简化 Spring 配置。

使用 Spring Cloud 协调一切服务。Spring Cloud 使得实现分布式的、微服务风格的架构更加简单。

使用 Spring Cloud Data Flow 连接一切服务。Data Flow 将企业服务连接到任何移动设备、传感器、可穿戴设备、汽车等的互联网上。Spring Cloud 数据流提供了一个统一的服务，用于创建地址流和基于 ETL 的数据处理模式、可组合的数据微服务。

1.2 Spring Boot 简介

在本节中，我们从整体上简要介绍一下 Spring Boot 框架。

1.2.1 Spring Boot 是什么

Java Web 开发涉及的技术比较繁杂，有很多开发框架和工具（Java、Scala、Kotlin、Clojure、Groovy、Grails、Gradle、Maven、JDBC、MySQL、Oracle、

MongoDB、Tomcat、Jetty、Spring、Struts、Hibernate、MyBatis、JPA、JSP、Velocity、FreeMarker、Thymeleaf、Redis 等)，而且它们各有所长，并不是一个完整的体系。这提高了程序员进行 Jave Web 开发的技术门槛和学习成本。

有没有一个像“航空母舰”式的威力强大的武器，可以整合这一切呢？答案就是：Spring Boot。

Spring Boot 由 Pivotal 团队提供的全新框架，其设计目的是用来简化新 Spring 应用的初始搭建以及开发过程。Spring Boot 是伴随着 Spring 4.0 诞生的。从字面理解，Boot 是引导的意思，因此 Spring Boot 极大地帮助了开发者快速搭建使用 Spring 框架开发应用程序的过程。例如，Spring Boot 可以直接快速启动一个内嵌的 Web 容器，而无须单独安装和配置 Web 服务器。

Spring Boot 框架遵循“约定优于配置”的思想。清除了原先使用 Spring 框架的那些样板化的配置。Spring Boot 继承了原有 Spring 框架的优秀基因；Spring Boot 使得基于 Spring 的开发过程更加简易。Spring Boot 致力于帮助开发人员快速开发应用。

多年以来，Spring IO 平台饱受非议的一点就是大量的 XML 配置以及复杂的依赖管理。Spring Boot 实现了“零 XML 配置”的极简开发体验。

然而，Spring Boot 并不是要成为 Spring IO 平台里面众多“ Foundation ”层项目的替代者。Spring Boot 的目标是为平台带来另一种开发体验，从而简化对这些已有技术的使用。对于已经熟悉 Spring 生态系统的开发人员来说，Boot 是一个很理想的选择；对于采用 Spring 技术的新人来说，Boot 提供了一种极简的方式来使用这些技术。

作为当前主流的企业框架 Spring，它提供了一整套相关的顶级项目，能让开发者快速上手实现自己的应用。Spring Boot 在整个 Spring 生态中的位置如图 1-4 所示。

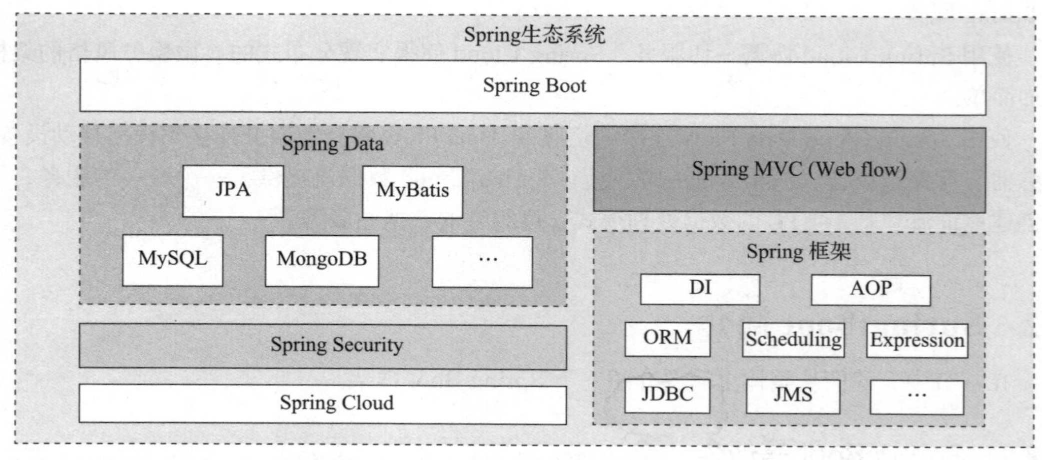

图 1-4 Spring Boot 在整个 Spring 生态中的位置

Spring Boot 是构建基于 Spring 的应用程序的起点。Spring Boot 旨在让你尽可能快地启动和运行，并以最小的预先配置的 Spring 配置。使用 Spring Boot 我们可以体验到下面的这些（但不仅限于）特性：

- 使用 Spring Initializr 在数秒内创建 Spring 应用程序。
- 构建任何东西——REST API、WebSocket、Web、流媒体、任务等。
- 简化了安全（Security）权限的开发。
- 丰富的 SQL 和 NoSQL 支持。
- 嵌入式运行时支持——Tomcat、Jetty 和 Undertow。
- 开发人员的生产力工具，例如实时重载（reload）和自动重启（restart）。
- 开箱即用的模块化依赖。
- 供生产环境直接使用的特性，如跟踪、度量和健康状态的监控。
- 丰富的 IDE 支持：Spring Tool Suite、IntelliJ IDEA 和 NetBeans。

Spring Boot 的核心特性如下：

- 创建一键运行的 Spring 应用。
- 能够使用内嵌的 Tomcat、Jetty 或 Undertow，不需要部署 war。
- 提供定制化的启动器 starters 简化第三方依赖配置。
- 追求极致的自动配置 Spring。
- 提供一些生产环境的特性，比如特征指标、健康检查和外部配置。
- 零代码生成和零 XML 配置

Java EE 原来开发应用的步骤是：

- 应用打成 war 包。
- 启动应用服务器。
- 在应用服务器中进行部署。

微服务时代，从部署到服务器中改造为直接启动应用进程，内嵌一个 Web 容器。把所需要的 jar 和应用代码全部打包到一个 jar 或者 war 中。如果打成可执行 jar 包，我们可以直接通过 java -jar example.war 的方式来启动服务。

嵌入式 Tomcat 早就存在，Spring Boot 支持内嵌 Tomcat、Jetty 和 Undertow 等 Web 服务器。测试表明 Undertow 比 Tomcat 性能更好。类似于 Wildfly-swarm 等微服务框架，Spring Boot 拥有相似的架构和开发 / 构建方法。例如，Wildfly-swarm 和 Spring Boot 的基础组件对比见表 1-1。

表 1-1　Spring Boot 和 Wildfly-swarm 基础组件对比

	Spring Boot	Wildfly-swarm
注入服务（Bean 管理）	SpringFramework 容器	Weld CDI 容器
Web 容器	嵌入式的 Tomcat 和嵌入式的 Undertow 等	嵌入式的 Undertow 等
Rest 数据	SpringMVC	JaxRS 实现的 RestEasy
持久层	采用 JPA 和 Hibernate 作为实现	采用 JPA 和 Hibernate 作为实现
嵌入式的数据库	HsqlDB 和 H2 数据库	HsqlDB 和 H2 数据库
构建	Maven、Gradle	Maven、Gradle

1.2.2 Spring Boot 核心模块

Spring Boot 核心模块如图 1-5 所示。

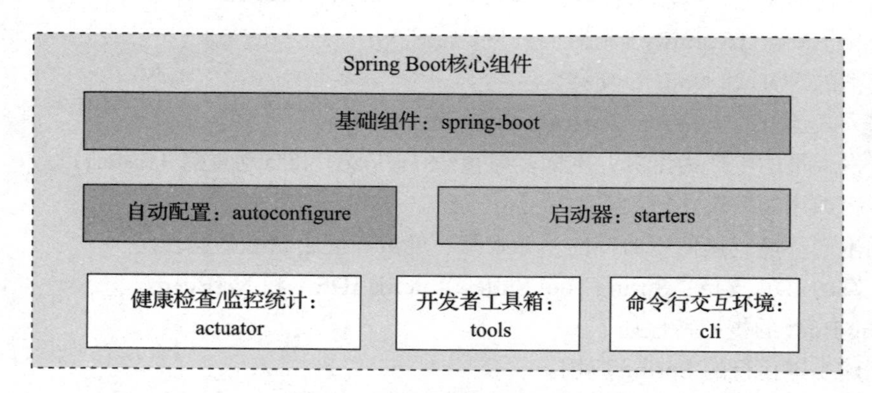

图 1-5 Spring Boot 核心模块

下面我们简要介绍一下 Spring Boot 的核心模块。

1. spring-boot

Spring Boot 核心工程。

2. starters

是 Spring Boot 的启动服务工程。spring-boot 中内置提供的 starter 列表可以在 Spring Boot 项目源代码工程 spring-boot/spring-boot-starters 中看到。这些 starters 的使用例子，在源码中的 spring-boot/spring-boot-samples 工程中。

3. autoconfigure

是 Spring Boot 实现自动配置的核心工程。

4. actuator

提供 Spring Boot 应用的外围支撑性功能。比如：应用状态监控管理、应用健康指示表、远程 shell 支持、metrics 支持等。

5. tools

提供了 Spring Boot 开发者的常用工具集。诸如，spring-boot-gradle-plugin、spring-boot-maven-plugin 就在这个模块里面。

6. cli

是 Spring Boot 命令行交互工具，可用于使用 Spring 进行快速原型搭建。可以用它直接运行 Groovy 脚本。如果你不喜欢 Maven 或 Gradle，可用 CLI（Command Line Interface）来开发运行 Spring 应用程序。可以使用它来运行 Groovy 脚本，甚至编写自定义命令。

1.3　约定优于配置极简化理念

Spring Boot 充分利用了 JavaConfig 的配置模式以及“约定优于配置”（Convention Over Configuration, COC）的理念，极大地简化了基于 Spring MVC 的 Web 应用和 REST 服务的开发。

不用看那一堆带着无数尖括号的 XML 真的让人很神清气爽。用 JavaConfig 注解方式可以让人很容易明白配置代码中的关键信息。

例如，一个标准的基于 Gradle 构建的 Spring Boot 应用程序目录结构约定如下：

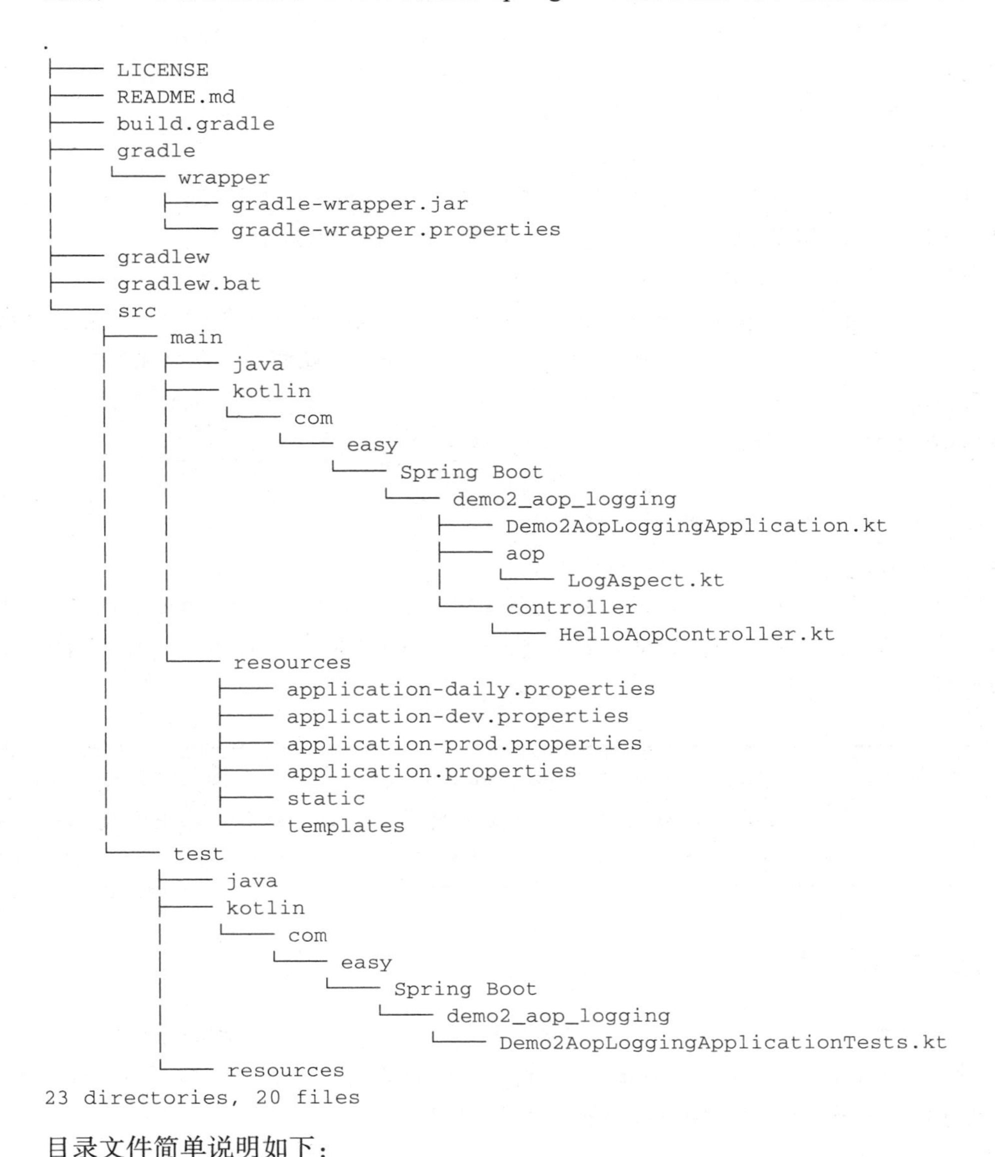

```
.
├── LICENSE
├── README.md
├── build.gradle
├── gradle
│   └── wrapper
│       ├── gradle-wrapper.jar
│       └── gradle-wrapper.properties
├── gradlew
├── gradlew.bat
└── src
    ├── main
    │   ├── java
    │   ├── kotlin
    │   │   └── com
    │   │       └── easy
    │   │           └── Spring Boot
    │   │               └── demo2_aop_logging
    │   │                   ├── Demo2AopLoggingApplication.kt
    │   │                   ├── aop
    │   │                   │   └── LogAspect.kt
    │   │                   └── controller
    │   │                       └── HelloAopController.kt
    │   └── resources
    │       ├── application-daily.properties
    │       ├── application-dev.properties
    │       ├── application-prod.properties
    │       ├── application.properties
    │       ├── static
    │       └── templates
    └── test
        ├── java
        ├── kotlin
        │   └── com
        │       └── easy
        │           └── Spring Boot
        │               └── demo2_aop_logging
        │                   └── Demo2AopLoggingApplicationTests.kt
        └── resources
23 directories, 20 files
```

目录文件简单说明如下：

- build.gradle——Gradle 工程项目配置文件。
- src/main/java——项目 Java 源代码目录。
- src/main/kotlin——项目 Kotlin 源代码目录。
- src/main/resources——项目资源文件目录。
- src/test/java——测试 Java 源代码目录。
- src/test/kotlin——测试 Kotlin 源代码目录。
- src/test/resources——测试资源文件目录。

许多框架使用了 COC 的思想，包括：Spring、Ruby on Rails、Kohana PHP、Grails、Grok、Zend Framework、CakePHP、Symfony、Maven、ASP.NET MVC、Web2py（MVC）、Apache Wicket 等。COC 是一个古老的思想理念，甚至在 Java 类库中也可以找出这一概念的踪迹。JavaBean 规范中很多就是依赖这个理念。

例如，在知名的 Java 对象关系映射（ORM）框架 Hibernate 的早期版本中，将类及其属性映射到数据库上在 XML 文件中进行配置，而其中大部分信息都应能够按照约定得到，如将类映射到对应的数据库表，将类属性一一映射到表上的字段。在后续的版本中抛弃了这样的 XML 配置文件，而是采用 Java 类属性使用驼峰式命名对应数据库表中的下划线命名这个恰当的约定，大大简化了配置。而对于不符合这些约定的特殊情形，就使用 Java 注解来标注说明。

例如，Spring 通过使用约定好的注解来标注 Spring 应用中各层中的 Bean 类：

- @Component——标注一个普通的 Spring Bean 类。
- @Controller——标注一个控制器组件类。
- @Service——标注一个业务逻辑组件类。
- @Repository——标注一个 DAO 组件类。

其实 Java 的成功，Spring 的成功，XML 的成功，Maven 的成功等，都有其必然性，因为它们的设计理念都包含一个很简单但很深刻的道理——那就是“通用”。为什么通用？因为遵循约定。

约定优于配置（Convention Over Configuration，COC）也称为按约定编程，是一种软件设计范式，旨在减少软件开发人员需做决定的数量，获得简单的好处，而又不失灵活性。

1.4 本章小结

Spring Boot 是一个名词，反过来念就是“ Boot Spring”，是一个动宾结构的词语，意即：“起飞吧，Spring！”。这正是 Spring Boot 框架设计的初心所在。自始至终，Spring 都在努力使开发者能够“极简”“快速”地创建并开发应用。

第 2 章 Chapter 2

快速开始 HelloWorld

在本章中，我们使用 Spring Boot2.0 快速实现一个基于 Kotlin 和 Gradle 的 HelloWorld 应用。下面我们直接开始吧。

2.1 创建 Spring Boot 项目

本节我们使用 IDEA 集成开发环境来快速创建实现一个 Spring Boot 版本的 Hello World 项目。

基本的 JDK 运行环境、Gradle 环境配置就不在这里赘述。如果是初次进入 Java 企业级应用的开发者，可以先把基本的开发环境配置好。

首先打开 IDEA，依次点击 File → New → Project，然后，我们进入 New Project 界面，如图 2-1 所示。

选择 Spring Initializr，Project SDK 设置为 JDK1.8，采用 Spring 官方的 Initializr 服务 URL https://start.spring.io/，点击 Next 进入 Project Metadata 设置界面，如图 2-2 所示。如图中设置项目元数据。其中：

- Type：选择 Gradle Project 表示我们创建的是一个基于 Gradle 构建的项目。
- Language：选择 Kotlin 表示我们采用 Kotlin 编程语言。
- Packag：表示项目的主包路径是 com.easy.Spring Boot.demo0_hello_world。

点击 Next 进入 Spring Initializr 初始化项目界面，如图 2-3 所示。

选择 Spring Boot 版本为 2.0.0.M7，选择“Web”启动器依赖（即 Full-stack web development with Tomcat and Spring MVC）。点击 Next 进入项目名称、存放路径等信息的配置，如图 2-4 所示。

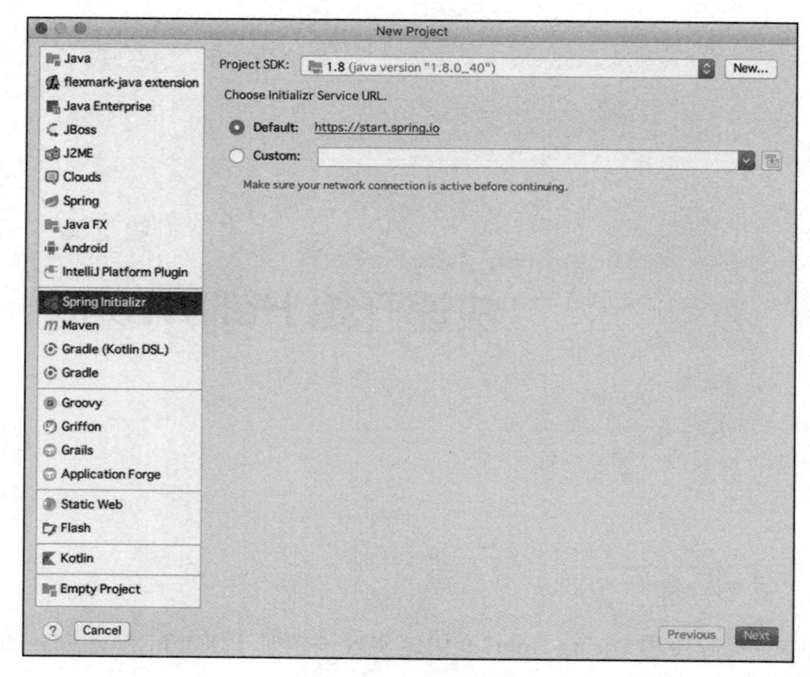

图 2-1 New Project 界面

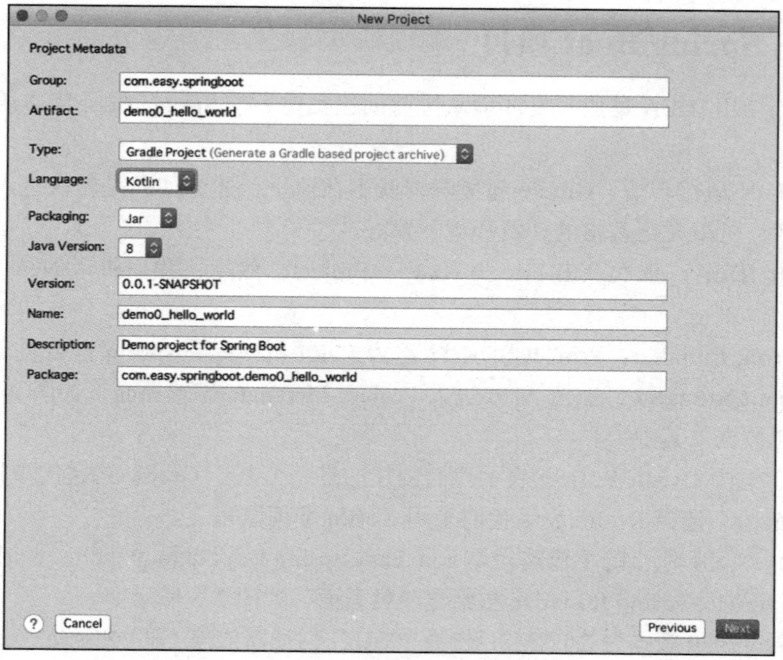

New Project
Project Metadata
Group: com.easy.springboot
Artifact: demo0_hello_world
Type: Gradle Project (Generate a Gradle based project archive)
Language: Kotlin
Packaging: Jar
Java Version: 8
Version: 0.0.1-SNAPSHOT
Name: demo0_hello_world
Description: Demo project for Spring Boot
Package: com.easy.springboot.demo0_hello_world
Cancel
Previous
Next

图 2-2 Project Metadata 设置界面

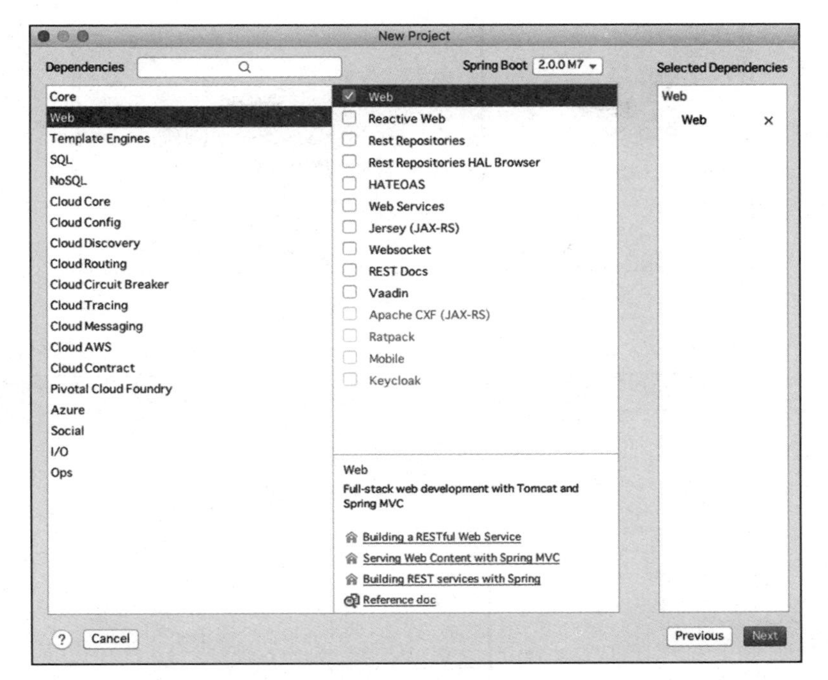

图 2-3　Spring Initializr 初始化项目界面

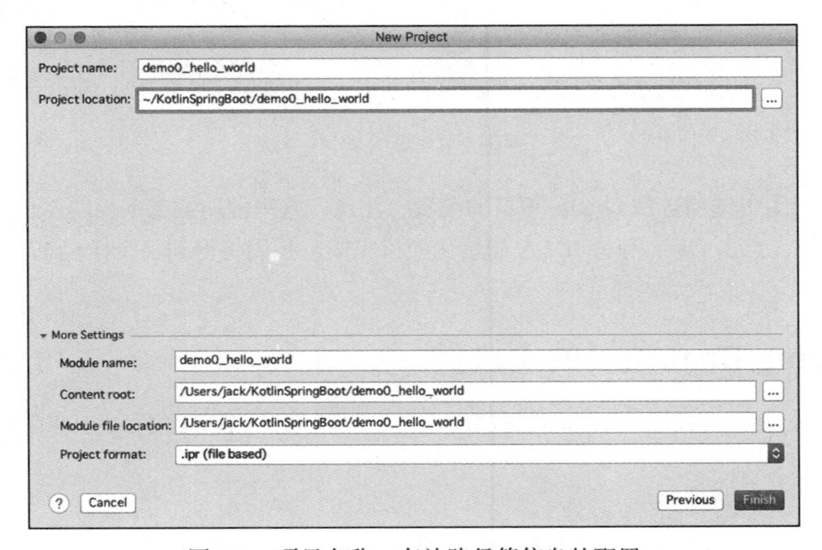

图 2-4　项目名称、存放路径等信息的配置

点击 Finish 会弹出创建目录的对话框，如图 2-5 所示。直接点击 OK，IDEA 将会为我们完成剩下的一切：创建项目标准目录，下载 Gradle 项目依赖等。当完成项目的初始化创建后，IDEA 会自动导入该项目，如图 2-6 所示。

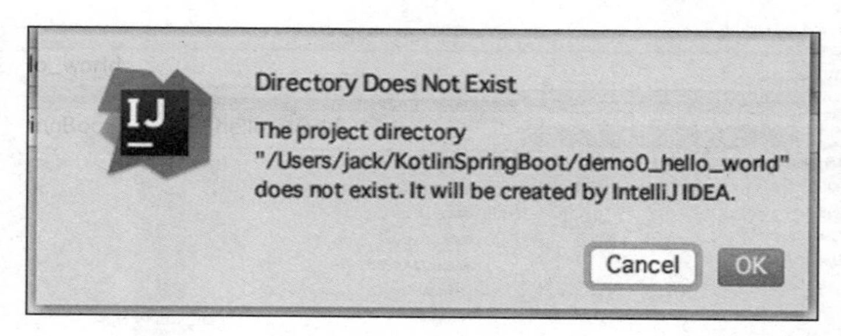

图 2-5　创建目录的对话框

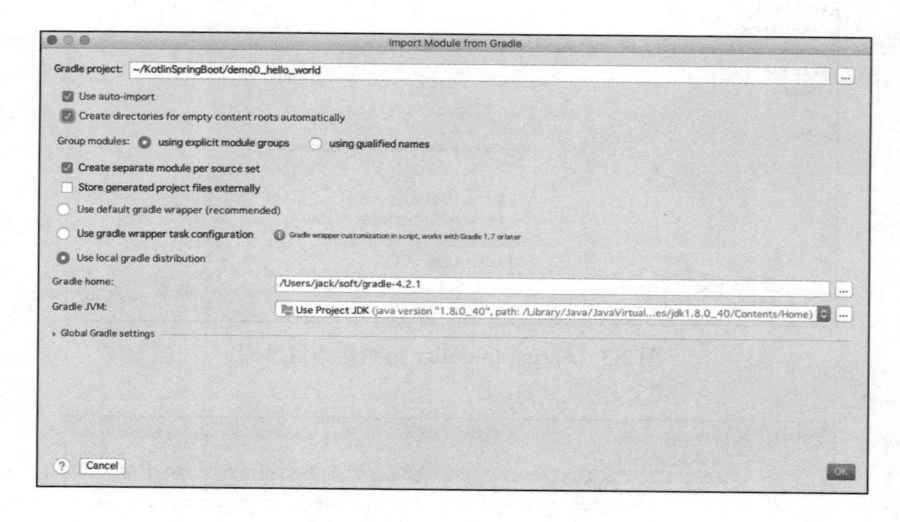

图 2-6　导入项目

我们按照图中选项选择 Gradle 项目的配置，注意，这里使用的是 local gradle distribution，这样会比较快。点击 OK，等待 IDEA 初始化项目完毕，我们将得到一个样板工程。

2.2　Spring Boot 项目的入口类

在样板工程中，Demo0HelloWorldApplication 是 Spring Boot 项目的入口类，它的关键源代码如下：

```
@Spring BootApplication
class Demo0HelloWorldApplication

fun main(args: Array<String>) {
    runApplication<Demo0HelloWorldApplication>(*args)
}
```

其中，org.springframework.boot.runApplication 是 Spring Boot2.0 中针对 Kotlin 扩展的

功能类 SpringApplicationExtensions.kt 中提供的内联函数。关键代码如下：

```
inline fun <reified T : Any> runApplication(vararg args: String): Configurable
        ApplicationContext = SpringApplication.run(T::class.java, *args)
```

我们在 main 函数里面打印一行日志：

```
fun main(args: Array<String>) {
    println("Spring Boot 2.0 极简教程 ")
    runApplication<Demo0HelloWorldApplication>(*args)
}
```

点击 IDEA 的运行按钮，如图 2-7 所示。

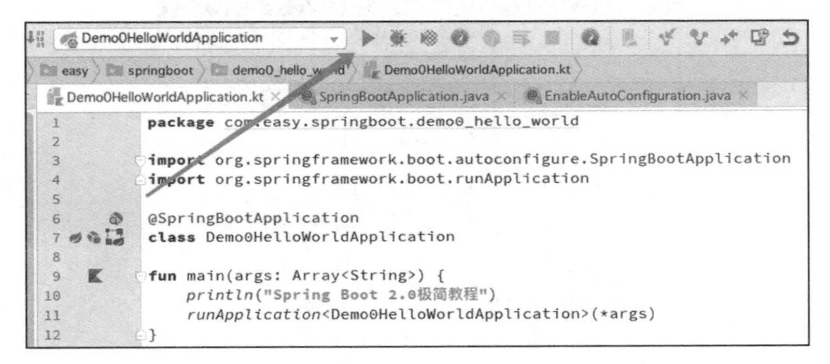

图 2-7　运行按钮

我们可以看到后台日志中打印出了我们代码中的内容，如图 2-8 所示。

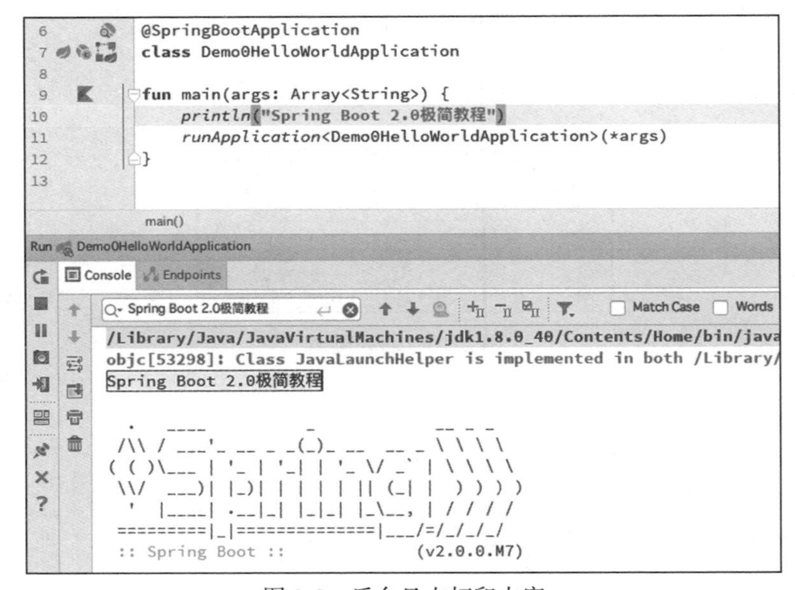

图 2-8　后台日志打印内容

2.3 添加 HelloWorldController

接着上节的工程，在目标 package 上单击，按下快捷键 Command + N 创建 Kotlin Class，如图 2-9 所示。

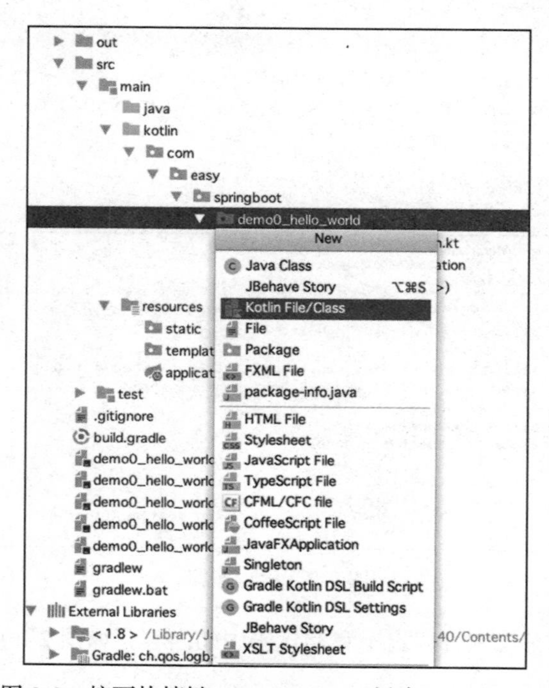

图 2-9 按下快捷键 Command + N 创建 Kotlin Class

输入类名，如图 2-10 所示。

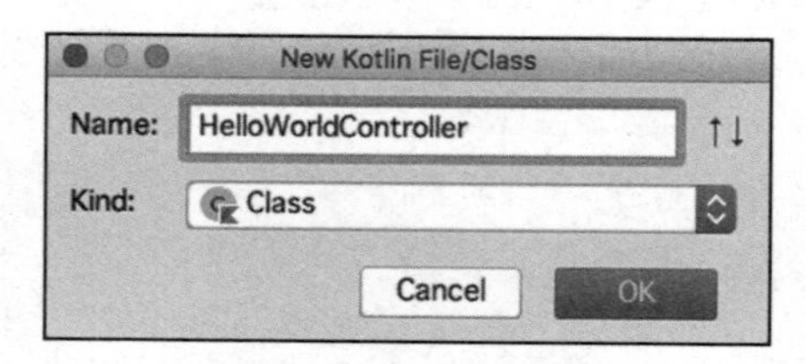

图 2-10 输入类名

实现代码如下：

```
@RestController
class HelloWorldController {
    @GetMapping(value = ["", "/", "/hello"]) // 匹配请求的 URL 列表: "", "/", "/hello"
    fun hello(): Greeting {
        return Greeting(name = "World", greeting = "Hello")
    }
```

```
    data class Greeting(var name: String, var greeting: String)
}
```

其中，data class Greeting 是 Kotlin 中的数据类。

提示　关于 Kotlin 编程语言的相关内容你可以参考《Kotlin 极简教程》(机械工业出版社，2017 年 9 月出版)。

启动应用，在浏览器中打开 http://127.0.0.1:8080，可以看到输出结果，如图 2-11 所示。

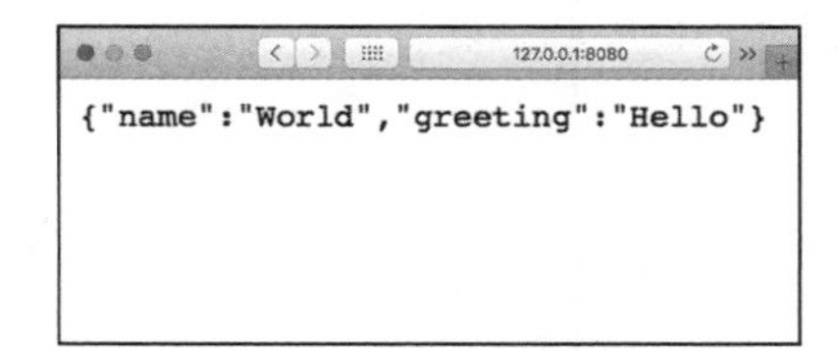

图 2-11　浏览器中的输出结果

到这里，我们已经完成了一个 RESTful Web HTTP Service。

提示　本节示例工程源代码位于 https://github.com/KotlinSpringBoot/demo0_hello_world。

下面我们重点讲解一下示例工程中用到的几个核心注解。

2.4　Spring Boot 应用注解 @Spring BootApplication

我们在上面看到在 Spring Boot 入口类上面添加了注解 @Spring BootApplication，这个注解的定义如下：

```
@Target(ElementType.TYPE)
@Retention(RetentionPolicy.RUNTIME)
@Documented
@Inherited
@Spring BootConfiguration
@EnableAutoConfiguration
@ComponentScan(excludeFilters = {
        @Filter(type = FilterType.CUSTOM, classes = TypeExcludeFilter.class),
        @Filter(type = FilterType.CUSTOM, classes = AutoConfigurationExclu
            deFilter.class) })
public @interface Spring BootApplication {...}
```

@Spring BootApplication 注解实际上封装了以下三个注解：

- @Spring BootConfiguration：配置类注解。

- @EnableAutoConfiguration：启用自动配置注解。
- @ComponentScan：组件扫描注解。

下面我们分别来介绍。

2.4.1 Spring Boot 配置类注解

@SpringBootConfiguration 与 @Component 注解是一样的。@SpringBootConfiguration 其实是 Spring Boot 包装的 @Configuration 注解：

```
@Target(ElementType.TYPE)
@Retention(RetentionPolicy.RUNTIME)
@Documented
@Configuration
public @interface Spring BootConfiguration {

}
```

而 @Configuration 注解使用的又是 @Component 注解：

```
@Target(ElementType.TYPE)
@Retention(RetentionPolicy.RUNTIME)
@Documented
@Component
public @interface Configuration {
    @AliasFor(annotation = Component.class)
    String value() default "";

}
```

我们知道，@Component 注解的功能是把普通 POJO 实例化到 Spring 容器中，相当于配置文件中的 <bean id = ""class = ""/>。

在类上添加注解 @Configuration，表明这个类代表一个 Spring 配置文件，与原来 XML 配置是等效的。只不过现在用 Java 类加上一个 @Configuration 注解进行配置了，这种方式与 XML 相比可以称得上是极简风格了。同时基于注解的配置风格，使得代码的可读性也大大增高了。

Spring 容器可以扫描出任何我们添加了 @Component 注解的类，Bean 的注册逻辑在 ClassPathScanningCandidateComponentProvider 这个类的 registerDefaultFilters 方法里。

提示　注解（Annotation）是 JDK1.5 中引入的一个新特性。从 Spring2.0 以后的版本中，Spring 引入了基于注解方式的配置，用于取代 XML 配置文件，从而极简化了 Bean 的配置，Spring 后来的新版本。在 Spring Boot 中完全采用基于注解（Spring4.x 引入了更加智能的 @Condition 系列注解，我们将会在后面的章节中详细介绍）的配置，实现“零 XML 的配置”（当然，同时也支持之前的 XML 配置文件方式）。

2.4.2 启用自动配置注解

@EnableAutoConfiguration 这个注解是 Spring Boot 的最核心注解。首先我们看它的定义：

```
@Target(ElementType.TYPE)
@Retention(RetentionPolicy.RUNTIME)
@Documented
@Inherited
@AutoConfigurationPackage
@Import(AutoConfigurationImportSelector.class)
public @interface EnableAutoConfiguration {...}
```

其中，导入配置类注解 @Import 标识导入 @Configuration 标注的配置类。@Import 用来整合所有在 @Configuration 注解中定义的 Bean 配置。这与我们将多个 XML 配置文件导入到单个文件的场景一致。@Import 注解实现了相同的功能。

使用 @EnableAutoConfiguration 注解可以启用 Spring 应用程序上下文的自动配置，Spring Boot 会去尝试猜测和配置你可能需要的 Bean。自动配置类通常是根据类路径中你定义的 Bean 来推断可能需要怎样的配置。

例如，如果在你的类路径中有 tomcat-embedded.jar 这个类库，那么 Spring Boot 会根据此信息来判断你可能需要一个 TomcatServletWebServerFactory（除非你已经定义了你自己的 ServletWebServerFactory Bean）。当然我们还可以通过设置 exclude 或者 excludeName 变量的值来手动排除你不想要的自动配置。

Spring Boot 默认扫描的包路径是入口类 Demo0HelloWorldApplication 所在的根包中，及其所有的子包。通常，Spring Boot 自动配置 Bean 是根据 Conditional Bean（条件 Bean）中注解的类信息来推断的。例如 @ConditionalOnClass、@ConditionalOnMissingBean 注解。关于 Spring Boot 自动配置的相关内容我们将在后面的章节中详细介绍。

2.4.3 组件扫描注解

组件扫描注解 @ComponentScan 提供的功能与 Spring XML 配置文件中的 <context:component-scan> 元素等价。对应 @ComponentScan 注解的处理类是 ConfigurationClassParser。@ComponentScan 告诉 Spring 去哪个 package 下面扫描 Spring 注解。Spring 会去自动扫描这些被 Spring 注解标注的类，并且将其注册到 Bean 容器中。例如下面的 XML 配置：

```
<beans>
    <context:component-scan base-package="com.easy.SpringBoot"
        name-generator="com.easy.SpringBoot.MyApp" />
</beans>
```

对应到 Java Config 风格如下：

```
@Configuration
@ComponentScan(basePackages = "com.easy.SpringBoot", nameGenerator = MyApp.class)
```

```
public class AppConfig {
    ...
}
```

如果你有个类用 @Controller 注解标识了，但是没有加上 @ComponentScan 告诉 Spring 去扫描这个类所在的包，那么该 Controller 就不会被注册到 Spring 容器中。

不过，Spring Boot 中如果不显式地使用 @ComponentScan 指明对象扫描的包，那么默认只扫描当前启动类所在的包里的类。

我们可以设置 basePackageClasses 的值来指定要扫描哪个类所在的包，代码示例如下：

```
@Spring BootApplication
@ComponentScan(basePackageClasses = MyApplication.class) //指定扫描 MyApplication
                                                         //类所在的包
public class MyApplication {

    public static void main(String[] args){
        SpringApplication.run(MyApplication.class, args);
    }
}
```

2.5 XML 配置与注解配置

本节简单对比一下传统的 Spring XML 配置与基于 JavaConfig 的注解配置 Bean 的两种方式。

Spring 对于 Bean 的配置有两种方式：XML 配置，注解配置。

1. XML 配置

优点：可以在后期维护的时候适当地调整 Bean 管理模式，并且只要遵循一定的命名规范，可以让程序员不必关心 Bean 之间的依赖关系。

缺点：系统越庞大，XML 配置文件就越大；关系错综复杂，容易导致错误。

2. 注解配置

优点：配置比较方便，程序员只要在 service 层代码设置即可实现，不需要知道系统需要多少个 Bean，交给容器来注入就好了。

缺点：当你要修改或删除一个 Bean 的时候，你无法确定到底有多少个其他的 Bean 依赖于这个 Bean。(解决方法：需要有严格的开发文档，在修改实现时尽可能继续遵守相应的接口规则，避免使其他依赖于此的 Bean 不可用。)

2.6 本章小结

Spring Boot 可以说是 Spring 践行“约定优于配置”（Convention Over Configuration）理念的极佳范例。“Spring Boot 无它，唯 Spring”是也。

第3章 Chapter 3

深入理解 Spring Boot 自动配置

Spring 框架提供了以多种方式配置 Bean 的灵活性，如 XML、注释和 JavaConfig。随着特性的增加，复杂性也增加了，而配置 Spring 应用程序变得单调乏味且容易出错。Spring 团队于是创建了 Spring Boot 以解决配置的复杂性。本章先来快速回顾一下使用 Spring 框架开发的复杂性，看看 Spring Boot 试图解决的问题是什么，然后介绍 Spring Boot 自动配置原理，以及实测分析。

3.1 传统的 SSM 开发过程

传统的 SSM（Spring + SpringMVC + MyBatis），曾经是主流的企业级架构方案：标准的 MVC 分层架构设计模式，将整个系统划分为模板视图（View）层、控制器（Controller）层、业务逻辑 Service 层、数据库访问的 Dao 层。我们使用 Spring MVC 负责请求的转发和视图管理，使用 Spring 核心容器实现业务对象的协作和生命周期的管理，MyBatis 作为数据库 ORM 层的对象持久化引擎。

我们需要小心翼翼地配置 pom.xml 中的各种项目依赖及其版本以保证 jar 包不冲突。这个 pom.xml 将是一个很庞大的依赖配置，动辄上百行。这么多的依赖，各种版本号也都必须要对得上，不能发生版本不兼容的情况。然后，我们还需要仔细配置 Spring 上下文 spring.xml 文件。这个 Spring 配置文件是 Spring 的 BeanFactory 工厂进行 Bean 生产、依赖关系注入（装配）及 Bean 实例分发的“图纸总纲”。Java EE 程序员必须学会并灵活应用这份“图纸”来准确地表达自己的“生产意图”。

Spring 配置文件是一个或多个标准的 XML 文档，如果在 web.xml 中没有显式指定

contextConfigLocation，将会使用 XmlWebApplicationContext 的默认的配置 /WEB-INF/applicationContext.xml。applicationContext.xml 是 Spring 的默认配置文件，当容器启动时找不到指定的配置文档时，将会尝试加载这个默认的配置文件。

如果我们使用自定义名称的 spring.xml 文件，就需要在 web.xml 中通过配置 contexConfigLocation 参数来指定 Spring 的配置文件。代码示例如下：

```
<?xml version="1.0" encoding="UTF-8"?>
<web-app xmlns:xsi="http://www.w3.org/2001/XMLSchema-instance"
         xmlns="http://java.sun.com/xml/ns/j2ee"
         xsi:schemaLocation="http://java.sun.com/xml/ns/j2ee http://java.sun.com/
             xml/ns/j2ee/web-app_2_4.xsd"
         version="2.4">

    <!-- 配置 Spring -->
    <context-param>
        <param-name>contextConfigLocation</param-name>
        <param-value>classpath:spring.xml</param-value>
    </context-param>
    …

</web-app>
```

完整的 Spring 配置文件 spring.xml 内容通常也非常庞大。

如果是 Java Web 项目的开发，通常还需要配置 SpringMVC 的上下文 spring-mvc.xml 这个 XML。SpringMVC 的上下文配置文件主要是：

```
org.springframework.web.servlet.DispatcherServlet
```

这个 DispatcherServlet 类在初始化过程中使用。DispatcherServlet 提供 Spring Web MVC 的集中访问点，负责职责的分派，而且与 Spring IoC 容器无缝集成，从而可以获得 Spring 的所有好处。我们需要在 spring-mvc.xml 中配置默认的注解映射的支持、自动扫描包路径、视图模板引擎等等一系列配置，完整的 spring-mvc.xml 配置文件内容参考示例工程源代码。

在 Web 应用中的 web.xml 中通常还需要配置 DispatcherServlet，在应用程序目录中的 classpath: spring-mvc.xml 配置 springMVC 的配置文件位置。一个配置实例如下：

```
    <!-- 配置 springmvc -->
    <servlet>
        <servlet-name>springMVC</servlet-name>
        <servlet-class>org.springframework.web.servlet.DispatcherServlet</
            servlet-class>
        <init-param>
            <param-name>contextConfigLocation</param-name>
            <param-value>classpath:spring-mvc.xml</param-value>
        </init-param>
        <load-on-startup>1</load-on-startup>
    </servlet>
```

```
    <servlet-mapping>
        <servlet-name>springMVC</servlet-name>
        <url-pattern>/</url-pattern>
    </servlet-mapping>
```

这样，Spring Web MVC 框架将加载“ classpath: spring-mvc.xml”来进行初始化上下文而不是约定的默认文件路径“/WEB-INF/[servlet 名字]-servlet.xml”。

项目完整的 web.xml 配置文件的内容参考示例工程。

提示　本节介绍的传统 SSM 实例工程源代码参考：https://github.com/KotlinSpring Boot/spring_mybatis_demo

web.xml 文件是用来初始化整个项目的配置信息的。比如 Welcome 页面、servlet、servlet-mapping、filter、listener、启动加载级别等。web.xml 又叫部署描述符文件，是在 Servlet 规范中定义的，是 web 应用用的配置文件。部署描述符文件就像所有 XML 文件一样，必须以一个 XML 头开始。这个头声明可以使用的 XML 版本并给出文件的字符编码。DOCYTPE 声明必须立即出现在此头之后。这个声明告诉服务器适用的 servlet 规范的版本（如 2.2 或 2.3）并指定管理此文件其余部分内容的语法的 DTD（Document Type Definition，文档类型定义）。所有部署描述符文件的顶层（根）元素为 web-app。请注意，XML 元素是大小写敏感的。因此，web-App 和 WEB-APP 都是不合法的，web-app 必须用小写。

web.xml 的加载顺序是：

```
<context-param>→<listener>→<filter>→<servlet>
```

其中，如果 web.xml 中出现了相同的元素，则按照在配置文件中出现的先后顺序来加载。另外，当我们使用 Spring 的 @Service、@Controller 等注解的时候，需要告诉 Spring 去哪里扫描并注册这些 Bean，这个配置在 spring.xml 中，例如：

```
<!-- 扫描 service、dao 组件 --><context:component-scan base-package="com.easy.
    Spring Boot"/>
```

我们在上一章中已经知道了：

```
<context:component-scan base-package="com.easy.Spring Boot"/>
```

这个配置等价于下面这段使用注解配置的代码：

```
@ComponentScan(basePackage="com.easy.Spring Boot")
```

在 Spring Boot 中就是大量使用基于注解的配置，从而去除 XML 配置。

传统的 Java Web 项目的开发过程中，通常还需要单独去配置 Tomcat 服务器，然后在 IDE 中配置集成。这个过程也比较费时。

3.2 Spring Boot 自动配置原理

那么，有没有一种方案，可以把上面这些繁杂费时费力的重复性劳动“一键打包、开箱即用”？

接下来，我们就逐步展示 Spring Boot 是怎样通过自动配置和提供一系列开箱即用的启动器 starter 来封装上面的复杂性使其简单化的。

3.2.1 Java 配置

在整个 Spring Boot 应用程序中，我们将看不到一个传统意义上的 Spring XMI 配置文件。其实，在 Spring3.x 和 Spring4.x 中就出现了大量简化 XML 配置的解决方案。例如：

- 组件扫描（Component Scan）：Spring 去自动发现应用上下文中创建的 Bean。
- 自动装配（Autowired）：Spring 自动创建 Bean 之间的依赖。
- 通过 JavaConfig 方式实现 Java 代码配置 Bean。

下面是一个使用 Java Config 方式配置 Thymeleaf 视图模板引擎的代码示例：

```
@Configuration
@ComponentScan(basePackages = { "com.easy.Spring Boot"})
@EnableWebMvc // 启用 WebMVC 配置（关于 WebMVC 的自定义配置我们将在后面章节中介绍）
public class WebMvcConfig extends WebMvcConfigurerAdapter
{
    @Bean
    public TemplateResolver templateResolver()   {// 配置模板解析器
        TemplateResolver templateResolver = new ServletContextTemplateResolver();
        templateResolver.setPrefix("/WEB-INF/views/");
        templateResolver.setSuffix(".html");
        templateResolver.setTemplateMode("HTML5");
        templateResolver.setCacheable(false);
        return templateResolver;
    }
    @Bean
    public SpringTemplateEngine templateEngine() {// 配置模板引擎
        SpringTemplateEngine templateEngine = new SpringTemplateEngine();
        templateEngine.setTemplateResolver(templateResolver());
        return templateEngine;
    }
    @Bean
    public ThymeleafViewResolver viewResolver()  {// 配置视图解析器
        ThymeleafViewResolver thymeleafViewResolver = new ThymeleafView
            Resolver();
        thymeleafViewResolver.setTemplateEngine(templateEngine());
        thymeleafViewResolver.setCharacterEncoding("UTF-8");
        return thymeleafViewResolver;
    }
    @Override
    public void addResourceHandlers(ResourceHandlerRegistry registry)
```

```
                                                {// 静态资源处理器配置
        registry.addResourceHandler("/resources/**").addResourceLocations
            ("/resources/");
    }
    …
    @Bean(name = "messageSource")
    public MessageSource configureMessageSource(){// 消息源配置
        ReloadableResourceBundleMessageSource messageSource = new Reloada
            bleResourceBundleMessageSource();
        messageSource.setBasename("classpath:messages");
        messageSource.setCacheSeconds(5);
        messageSource.setDefaultEncoding("UTF-8");
        return messageSource;
    }
}
```

在 WebMvcConfig.java 配置类中，我们做了如下的配置：

- 将它标记为使用 @Configuration 注释的 Spring 配置类。
- 启用基于注释的 Spring MVC 配置，使用 @EnableWebMvc。
- 通过注册 TemplateResolver、SpringTemplateEngine、ThymeleafViewResolver Bean 来配置 Thymeleaf ViewResolver。
- 注册的 ResourceHandlers Bean 用来配置 URI/resources/** 静态资源的请求映射到 /resources/ 目录下。
- 配置的 MessageSource Bean 从 classpath 路径下的 ResourceBundle 中的 messages-{country-code}.properties 消息配置文件中加载 i18n 消息。

这些样板化的 Java 配置代码比 XML 要更加简单些，同时易于管理。而 Spring Boot 则是引入了一系列的约定规则，将上面的样板化配置抽象内置到框架中去，用户连上面的 Java 配置代码也将省去。

3.2.2　条件化 Bean

Spring Boot 除了采用 Java、Config 方式实现“零 XML”配置外，还大量采用了条件化 Bean 方式来实现自动化配置，本节就介绍这个内容。

1. 条件注解 @Conditional

假如你想一个或多个 Bean 只有在应用的路径下包含特定的库时才创建，那么使用这节我们所要介绍的 @Conditional 注解定义条件化的 Bean 就再适合不过了。

Spring4.0 中引入了条件化配置特性。条件化配置通过条件注解 @Conditional 来标注。条件注解是根据特定的条件来选择 Bean 对象的创建。条件注解根据不同的条件来做出不同的事情（简单说就是 if else 逻辑）。在 Spring 中条件注解可以说是设计模式中状态模式的一种体现方式，同时也是面向对象编程中多态的应用部分。

常用的条件注解如表 3-1 所示。

表 3-1 常用的条件注解

条件注解	条件说明
@ConditionalOnBean	仅在当前上下文中存在某个对象时，才会实例化一个 Bean
@ConditionalOnClass	当 class 位于类路径上，才会实例化一个 Bean
@ConditionalOnExpression	当表达式为 true 的时候，才会实例化一个 Bean
@ConditionalOnMissingBean	仅在当前上下文中不存在某个对象时，才会实例化一个 Bean
@ConditionalOnMissingClass	当类路径上不存在某个 class 的时候，才会实例化一个 Bean
@ConditionalOnNotWebApplication	当不是一个 Web 应用时

2. 条件注解使用实例

下面我们通过实例来说明条件注解 @Conditional 的具体工作原理。

1）创建示例工程。

为了精简篇幅，这里只给出关键步骤。首先使用 Spring Initializr 创建一个 Spring Boot 工程，选择 Web Starter 依赖，配置项目名称和存放路径，配置 Gradle 环境，最后导入到 IDEA 中，完成工程的创建工作。

2）实现 Condition 接口。

下面我们来实现 org.springframework.context.annotation.Condition 接口，实现类是 MagicCondition。

实现类的“条件”逻辑是：当 application.properties 配置文件中存在“magic”配置项，同时当值是 true 的时候：

```
magic=true
#magic=false
```

就表示条件匹配。

新建 MagicCondition 类，实现 Condition 接口。在 IDEA 中会自动提示我们实现其中的方法，如图 3-1 所示。

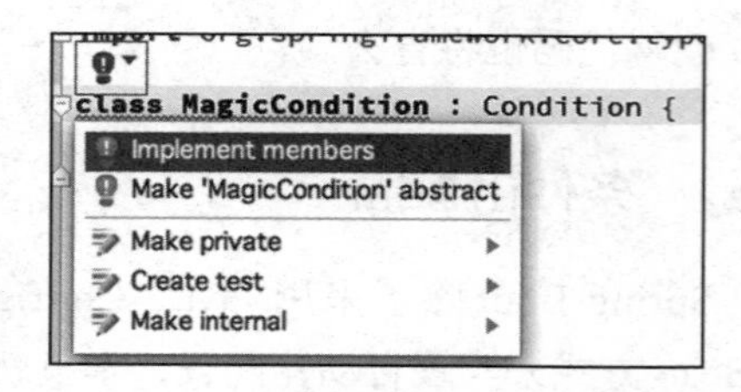

图 3-1 IDEA 会自动提示我们实现其中的方法

选择要实现的 matches 函数，如图 3-2 所示。

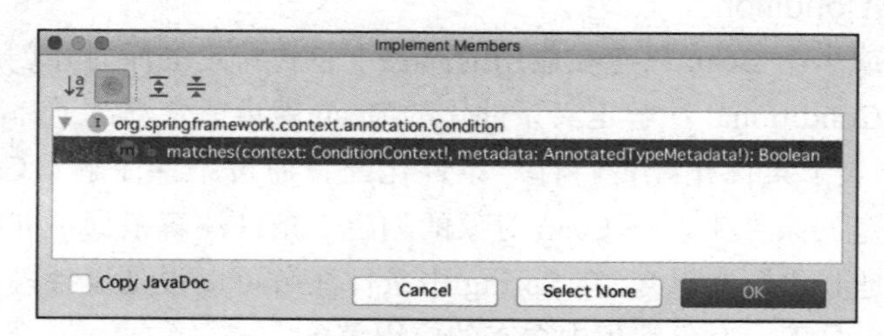

图 3-2 选择要实现的 matches 函数

完整的实现代码如下：

```
class MagicCondition : Condition {
    override fun matches(context: ConditionContext, metadata: AnnotatedType
        Metadata): Boolean {
        val env = context.getEnvironment()
        if (env.containsProperty("magic")) // 检查 application.properties 配置
            文件中是否存在 magic 属性 key
        {
            val b = env["magic"]            //获取 magic 属性 key 的值
            return b == "true"              //如果是 true，返回 true
        }
        return false                        // 返回 false
    }
}
```

实现这个 Condition 接口只需要实现 matches 方法。如果 matches 方法返回 true 就创建该 Bean，如果返回 false 则不创建 Bean。这就是否创建 MagicService Bean 的条件。

matches 方法中的第 1 个参数类型 ConditionContext 是一个接口，它的定义如下：

```
public interface ConditionContext {
    BeanDefinitionRegistry getRegistry();
    ConfigurableListableBeanFactory getBeanFactory();
    Environment getEnvironment();
    ResourceLoader getResourceLoader();
    ClassLoader getClassLoader();
}
```

ConditionContext 中的方法 API 说明如表 3-2 所示。

表 3-2　ConditionContext 中的方法 API

方　法	作　用
getRegistry()	返回保存 Bean 定义的 BeanDefinitionRegistry 对象，可以使用 BeanDefinitionRegistry 检查 Bean 的定义
getBeanFactory()	返回 ConfigrableListableBeanFactory 对象，用来检查 Bean 是否存在，以及检查 Bean 的属性
getEnvironment()	使用返回 Environment 检查环境变量是否存在以及读取它的值
getResourceLoader()	读取并检查它返回的 ResourceLoader 所加载的资源
getClassLoader()	使用返回的类加载器对象 ClassLoader 来加载类并检查类是否存在

matches 方法中的第 2 个参数类型 AnnotatedTypeMetadata，则能够让我们检查带有 @Bean 注解的方法上是否有其他注解。AnnotatedTypeMetadata 接口的定义如下：

```
public interface AnnotatedTypeMetadata {

    boolean isAnnotated(String annotationType);

    Map<String, Object> getAnnotationAttributes(String annotationType);
```

```
    Map<String, Object> getAnnotationAttributes(String annotationType, boolean
        classValuesAsString);

    MultiValueMap<String, Object> getAllAnnotationAttributes(String annota
        tionType);

    MultiValueMap<String, Object> getAllAnnotationAttributes(String annota
        tionType, boolean classValuesAsString);

}
```

使用 isAnnotated() 方法，能够判断带有 @Bean 注解的方法是不是还有其他特定的注解。使用另外的几个方法，我们能够检查 @Bean 注解的方法上，所标注的其他注解的属性。

例如 Spring4 使用 @Conditional 对多环境部署配置文件功能实现的 ProfileCondition 类的代码如下：

```
class ProfileCondition implements Condition {

    @Override
    public boolean matches(ConditionContext context, AnnotatedTypeMetadata
        metadata) {
        MultiValueMap<String, Object> attrs = metadata.getAllAnnotationA
            ttributes(Profile.class.getName());
        if (attrs != null) {
            for (Object value : attrs.get("value")) {
                if (context.getEnvironment().acceptsProfiles((String[])
                    value)) {
                return true;
                }
            }
            return false;
        }
        return true;
    }

}
```

我们可以看到，ProfileCondition 通过 AnnotatedTypeMetadata 得到了用于 @Profile 注解的所有属性：

```
MultiValueMap<String, Object> attrs = metadata.getAllAnnotationAttributes
    (Profile.class.getName());
```

然后循环遍历 attrs 这个 Map 中的属性“value”的值（包含了 Bean 的 profile 名称），使用 ConditionContext 中的 Environment 来检查这个 value 进而决定使用哪个 Profile 处于激活状态。

3）条件配置类 ConditionalConfig。

Spring4 提供了一个通用的基于特定条件创建 Bean 的方式：@Conditional 注解。编写条件配置类 ConditionalConfig 代码如下：

```
@Configuration
@ComponentScan(basePackages = ["com.easy.Spring Boot.demo_conditional_bean"])
class ConditionalConfig {
    @Bean
    @Conditional(MagicCondition::class) //指定条件类
    fun magicService(): MagicServiceImpl {
        return MagicServiceImpl()
    }
}
```

逻辑是当 Spring 容器中存在 MagicCondition Bean，并满足 MagicCondition 类的条件时，去实例化 magicService 这个 Bean。否则不注册这个 Bean。

4）MagicServiceImpl 逻辑实现。

MagicServiceImpl 业务 Bean 的逻辑很简单，就是打印一个标识信息。实现代码如下：

```
class MagicServiceImpl : MagicService {
    override fun info(): String {
        return "THIS IS MAGIC"          // 打印一个标识信息
    }
}
interface MagicService {
    fun info(): String
}
```

5）测试 MagicController。

我们使用一个 HTTP 接口来测试条件化 Bean 的注册结果：

```
@RestController
class MagicController {

    @GetMapping("magic")
    fun magic(): String {
        try {
            val magicService = SpringContextUtil.getBean("magicService") as
                              MagicService // 从 Spring 容器中获取 magicService Bean
            return magicService.info() //调用 info() 方法
        } catch (e: Exception) {
            e.printStackTrace()
        }
        return "null"
    }
}
```

其中 SpringContextUtil 实现代码如下：

```
object SpringContextUtil {
    lateinit var applicationContext: ApplicationContext

    fun setGlobalApplicationContext(context: ApplicationContext) {
        applicationContext = context
    }
```

```
    fun getBean(beanId: String): Any {
        return applicationContext.getBean(beanId)
    }
}
```

在 Spring Boot 启动入口类中，我们把 Spring Boot 应用的上下文对象放到 Spring Context-Util 中的这个 applicationContext 成员变量中：

```
@Spring BootApplication
class DemoConditionalBeanApplication

fun main(args: Array<String>) {
    val context = runApplication<DemoConditionalBeanApplication>(*args)
    SpringContextUtil.setGlobalApplicationContext(context)
}
```

完整的项目代码参考示例工程源代码。

提示 本小节的实例工程源码：https://github.com/KotlinSpringBoot/demo_conditional_bean

6）运行测试。

我们先来测试 magic = true。在 application.properties 中配置：

```
magic=true
```

重新启动应用程序，浏览器输入：http://127.0.0.1:8080/magic，输出“THIS IS MAGIC”。

再来测试 magic = false 。在 application.properties 中配置：

```
magic=false
```

重新启动应用程序，浏览器输入：http://127.0.0.1:8080/magic，输出：“null”。

这个时候，我们看到应用程序后台日志有如下输出：

```
org.springframework.beans.factory.NoSuchBeanDefinitionException: No bean
    named 'magicService' available ……
com.easy.Spring Boot.demo_conditional_bean.controller.MagicController.magic
    (MagicController.kt:14)
```

表明 magicService 这个 Bean 没有注册到 Spring 容器中。条件化注册 Bean 验证 OK。

3.2.3 组合注解

组合注解就是将现有的注解进行组合，生成一个新的注解。使用这个新的注解就相当于使用了该组合注解中所有的注解。这个特性还是蛮有用的，例如 Spring Boot 应用程序的入口类注解 @Spring BootApplication 就是典型的例子：

```
@Target(ElementType.TYPE)
```

```
@Retention(RetentionPolicy.RUNTIME)
@Documented
@Inherited
@Spring BootConfiguration
@EnableAutoConfiguration
@ComponentScan(excludeFilters = {
        @Filter(type = FilterType.CUSTOM, classes = TypeExcludeFilter.class),
        @Filter(type = FilterType.CUSTOM, classes = AutoConfigurationExclude
          Filter.class) })
public @interface Spring BootApplication
```

早期版本的 Spring Boot 中，用户需要使用如下三个注解来标注应用入口 main 类：

- @Configuration
- @EnableAutoConfiguration
- @ComponentScan

在 Spring Boot1.2.0 中只需用一个统一的注解 @Spring BootApplication。

3.3 Spring Boot 自动配置过程

Spring Boot 内置自动配置原理是怎样的呢？这一切都在 @EnableAutoConfiguration 这个注解里：

```
@Target(ElementType.TYPE)
@Retention(RetentionPolicy.RUNTIME)
@Documented
@Inherited
@AutoConfigurationPackage
@Import(AutoConfigurationImportSelector.class)
public @interface EnableAutoConfiguration
```

其中的核心注解是 @Import(EnableAutoConfigurationImportSelector.class)，借助 EnableAutoConfigurationImportSelector、@EnableAutoConfiguration、Spring Boot 应用将所有符合条件的 @Configuration 配置类都加载到当前 Spring 容器中——就像一只“八爪鱼”一样。具体的实现是使用了 Spring 框架中原有的一个工具类 SpringFactoriesLoader。这样，@EnableAutoConfiguration 就可以智能实现 Bean 的自动配置。

3.3.1 @EnableAutoConfiguration 注解

Spring Boot 中通过 @EnableAutoConfiguration 启用 Spring 应用程序上下文的自动配置，这个注解会导入一个 EnableAutoConfigurationImportSelector 的类，而 AutoConfigurationImportSelector 这个类会去读取一个 spring.factories 下 key 为 EnableAutoConfiguration 对应的类全限定名的值。其中的关键代码如下：

```
protected List<String> getCandidateConfigurations(AnnotationMetadata metadata,
```

```
        AnnotationAttributes attributes) {
    List<String> configurations = SpringFactoriesLoader.loadFactoryNames(
            getSpringFactoriesLoaderFactoryClass(), getBeanClassLoader());
    Assert.notEmpty(configurations,
            "No auto configuration classes found in META-INF/spring.factories.
                If you "
                    + "are using a custom packaging, make sure that file
                        is correct.");
    return configurations;
}
```

这个 spring.factories 里面配置的那些类，主要作用是告诉 Spring Boot 这个 stareter 所需要加载的那些 *AutoConfiguration 类，也就是你真正的要自动注册的那些 Bean 或功能。然后，再实现一个 spring.factories 指定的类，标上 @Configuration 注解，一个 starter 就定义完了。通过 org.springframework.boot.autoconfigure.AutoConfigurationImportSelector 里面的 getCandidateConfigurations 方法，获取到候选类的名字列表 List<String>。

其中，loadFactoryNames 的第 1 个参数是 getSpringFactoriesLoaderFactoryClass() 方法直接返回的是 EnableAutoConfiguration.class，代码如下：

```
protected Class<?> getSpringFactoriesLoaderFactoryClass() {
    return EnableAutoConfiguration.class;
}
```

所以，getCandidateConfigurations 方法里面的这段代码：

```
List<String> configurations = SpringFactoriesLoader.loadFactoryNames(
                getSpringFactoriesLoaderFactoryClass(), getBeanClassLoader());
```

会过滤出 key 为

```
org.springframework.boot.autoconfigure.EnableAutoConfiguration
```

的全限定名对应的值。其中，SpringFactoriesLoader 主要用来查询 META-INF/spring.factories 的 properties 配置中指定 class 对应的所有实现类。下节介绍这个文件。

全限定名都使用如下命名方法：

```
包名 . 外部类名
包名 . 外部类名 $ 内部类名
```

例如：

```
org.springframework.boot.autoconfigure.context.PropertyPlaceholderAutoConfi
    guration
```

3.3.2 spring.factories 文件

Spring Boot 中 的 META-INF/spring.factories（spring-boot/spring-boot-autoconfigure/src/

main/resources/META-INF/spring.factories）配置文件的完整内容可参考Spring Boot源代码工程，其中关于EnableAutoConfiguration的配置是：

```
# Auto Configure
org.springframework.boot.autoconfigure.EnableAutoConfiguration=\
org.springframework.boot.autoconfigure.admin.SpringApplicationAdminJmxAut
    oConfiguration,\
......
org.springframework.boot.autoconfigure.webservices.WebServicesAutoConfiguration
```

当然了，这些AutoConfiguration不是所有都会加载的，会根据AutoConfiguration上的@ConditionalOnClass等条件，再进一步判断是否加载。

3.3.3　获取候选配置类

在上面的getCandidateConfigurations方法中，我们可以看到读取spring.factories文件由SpringFactoriesLoader来完成的。SpringFactoriesLoader的实现类似于SPI(Service Provider Interface，在java.util.ServiceLoader的文档里有比较详细的介绍。Java SPI提供一种服务发现机制，为某个接口寻找服务实现的机制。有点类似IOC的思想，就是将装配的控制权移到程序之外，在模块化设计中这个机制尤其重要。

SpringFactoriesLoader会加载classpath下所有JAR文件里面的META-INF/spring.factories文件。

其中加载spring.factories文件的代码在loadFactoryNames()方法里。Spring Boot自动配置的过程可以用一张图说明，如图3-3所示。

3.4　FreeMarkerAutoConfiguration实例分析

本节通过FreeMarkerAutoConfiguration实例来分析Spring Boot中集成Freemarker模板引擎的整个自动配置的过程。

3.4.1　spring-boot-starter-freemarker工程

spring-boot-starter-freemarker工程是实现Free Marker模板引擎自动配置的启动工程，其目录结构如下：

```
.
├── pom.xml
├── spring-boot-starter-freemarker.iml
└── src
    └── main
        └── resources
            └── META-INF
```

```
└── spring.provides

4 directories, 3 files
```

我们可以看出，这个工程没有任何 Java 代码，只有两个文件：pom.xml 与 spring.provides，其中，spring.provides 文件如下：

```
provides: freemarker,spring-context-support
```

主要是给这个 starter 起个好区分的名字。

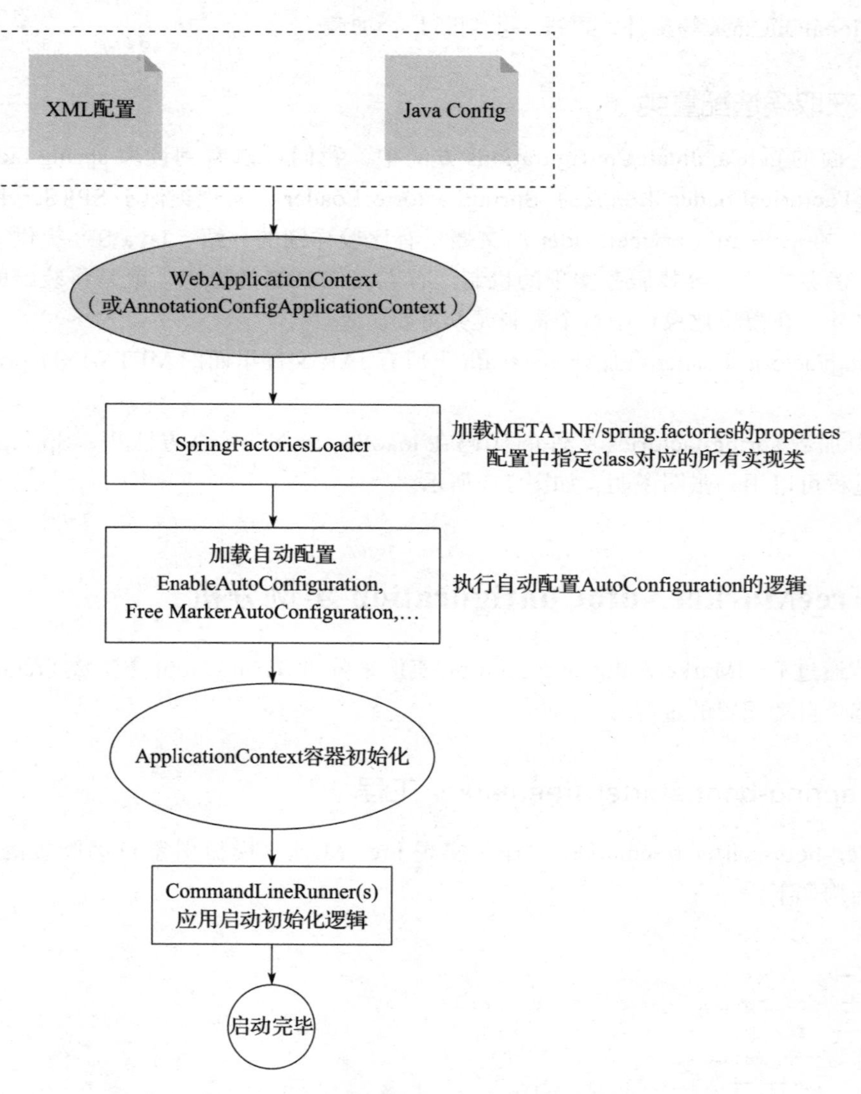

图 3-3　Spring Boot Autoconfigure 工作原理图

Spring Boot 通过 starter 对项目的依赖进行统一管理 .starter 利用了 Maven 的传递依赖解析机制，把常用库聚合在一起，组成了针对特定功能而定制的依赖 starter。

我们可以使用 IDEA 提供的 Maven 依赖图分析的功能，得到 spring-boot-starter-freemarker 依赖的 module，如图 3-4 所示。

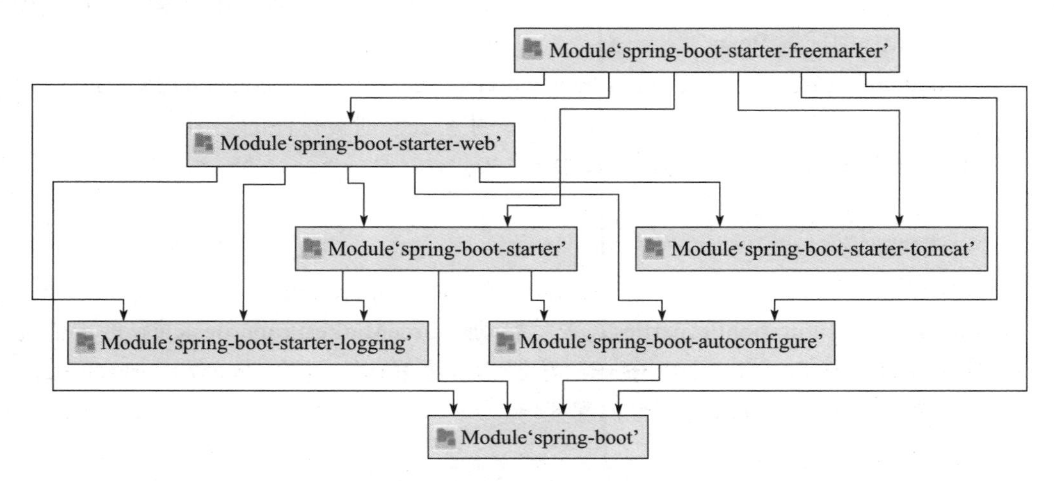

图 3-4　spring-boot-starter-freemarker 依赖的 module

从图中我们可以看出其中的依赖关系。

当 Spring Boot Application 中自动配置 EnableAutoConfiguration 的相关类执行完毕之后，Spring Boot 会进一步解析对应类的配置信息。如果我们配置了 spring-boot-starter-freemarker，Maven 就会通过这个 starter 所依赖的 spring-boot-autoconfigure，自动传递到 spring-boot-autoconfigure 工程中。

3.4.2　spring-boot-autoconfigure 工程

我们来简单分析一下 spring-boot-autoconfigure 工程中关于 FreeMarker 自动配置的逻辑实现。FreeMarker 自动配置的入口类是：

```
org.springframework.boot.autoconfigure.freemarker.FreeMarkerAutoConfiguration
```

这个配置类中导入了 FreeMarkerServletWebConfiguration、FreeMarkerReactiveWebConfiguration、FreeMarkerNonWebConfiguration 配置类：

```
@Configuration
@ConditionalOnClass({ freemarker.template.Configuration.class,
        FreeMarkerConfigurationFactory.class })
@EnableConfigurationProperties(FreeMarkerProperties.class)
@Import({ FreeMarkerServletWebConfiguration.class,
        FreeMarkerReactiveWebConfiguration.class, FreeMarkerNonWebConfigura
```

```
        tion.class })
public class FreeMarkerAutoConfiguration
```

其中：

- @Configuration 是 org.springframework.context.annotation 包里面的注解。我们已经知道用 @Configuration 注解该类，等价于 XML 中配置 Bean；用 @Bean 标注方法等价于 XML 中配置 Bean。
- @ConditionalOnClass 是 org.springframework.boot.autoconfigure.condition 包里面的注解。意思是当类路径下有指定的类的条件下，才会去注册被标注的类为一个 Bean。在上面的代码中的意思就是，当类路径中有：

  ```
  freemarker.template.Configuration.class
  FreeMarkerConfigurationFactory.class
  ```

  这两个类的时候，才会去配置 FreeMarkerAutoConfiguration。
- @EnableConfigurationProperties，表示启动对 FreeMarkerProperties.class 的内嵌配置支持，自动将 FreeMarkerProperties 注册为一个 Bean。这个 FreeMarkerProperties 类里面就是关于 FreeMarker 属性的配置：

```
@ConfigurationProperties(prefix = "spring.freemarker")
public class FreeMarkerProperties extends AbstractTemplateViewResolverProperties {

    public static final String DEFAULT_TEMPLATE_LOADER_PATH = "classpath:/
        templates/";

    public static final String DEFAULT_PREFIX = "";

    public static final String DEFAULT_SUFFIX = ".ftl";
    …

}
```

接下来，我们来看一下 FreeMarkerServletWebConfiguration 这个类。该类主要是用于配置基于 servlet web context 的 Freemarker 的配置。这个类的代码如下：

```
@Configuration
@ConditionalOnWebApplication(type = ConditionalOnWebApplication.Type.SERVLET)
@ConditionalOnClass({ Servlet.class, FreeMarkerConfigurer.class })
@AutoConfigureAfter(WebMvcAutoConfiguration.class)
class FreeMarkerServletWebConfiguration extends AbstractFreeMarkerConfiguration {

    protected FreeMarkerServletWebConfiguration(FreeMarkerProperties pro
        perties) {
        super(properties);
    }

    @Bean
    @ConditionalOnMissingBean(FreeMarkerConfig.class)
    public FreeMarkerConfigurer freeMarkerConfigurer() {
```

```
        FreeMarkerConfigurer configurer = new FreeMarkerConfigurer();
        applyProperties(configurer);
        return configurer;
    }

    @Bean
    public freemarker.template.Configuration freeMarkerConfiguration(
            FreeMarkerConfig configurer) {
        return configurer.getConfiguration();
    }

    @Bean
    @ConditionalOnMissingBean(name = "freeMarkerViewResolver")
    @ConditionalOnProperty(name = "spring.freemarker.enabled", matchIfMis
        sing = true)
    public FreeMarkerViewResolver freeMarkerViewResolver() {
        FreeMarkerViewResolver resolver = new FreeMarkerViewResolver();
        getProperties().applyToMvcViewResolver(resolver);
        return resolver;
    }

    @Bean
    @ConditionalOnMissingBean
    @ConditionalOnEnabledResourceChain
    public ResourceUrlEncodingFilter resourceUrlEncodingFilter() {
        return new ResourceUrlEncodingFilter();
    }
}
```

其中：

1）当该应用是基于Servlet的Web应用时，Spring容器内有Servlet.class、FreeMarkerConfigurer.class类实例存在。

2）Spring容器中不存在freeMarkerViewResolver的Bean。

3）应用程序的属性配置文件中没有匹配到"spring.freemarker.enabled"。

当1）2）3）三个条件都满足，则初始化freeMarkerViewResolver这个Bean。

我们也可以自定义自己的starter，以及实现对应的@MyEnableAutoConfiguration。Spring Boot有很多第三方starter，其自动配置的原理基本都是这样，比如mybatis-spring-boot-starter的MybatisAutoConfiguration，源码在如下地址：https://github.com/mybatis/spring-boot-starter，阅读源码可加深理解。

3.5　本章小结

为了更加鲜明地体会到Spring Boot带来的改变，我们首先介绍了一个传统的基于SSM的Java Web项目的完整开发过程，然后使用Spring Boot来完成同样的工作。通过对比学习，加深理解Spring Boot是怎样通过自动配置实现“极简化配置”的。

第Ⅱ部分 Part 2

Spring Boot 项目综合实战

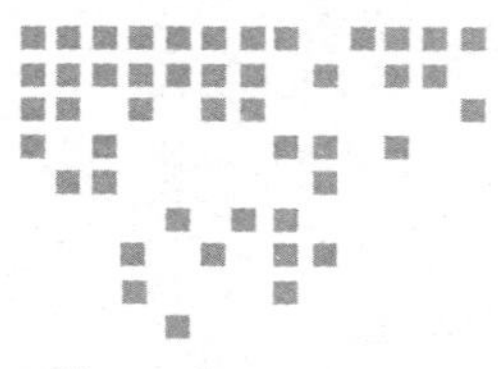

Chapter 4 第4章

Spring Boot 集成 MyBatis 数据库层开发

MyBatis 是一款优秀的持久层框架，它支持定制化 SQL、存储过程以及高级映射。MyBatis 避免了几乎所有的 JDBC 代码和手动设置参数以及获取结果集的操作。MyBatis 可以使用简单的 XML 或注解来配置和映射原生信息，将接口和 Java 的 POJO（Plain Old Java Objects，普通的 Java 对象）映射成数据库中的记录。

本章介绍 Spring Boot 集成 MyBatis 数据库层开发的内容。

4.1 Java EE 分层架构

无论是经典的 Java EE 架构，还是 Spring 轻量级的 Java EE 架构，系统基本都是分层的，如图 4-1 所示。

我们从下往上依次讲解 Java EE 分层架构。

- **模型**（Model）层也叫领域对象（Domain Object）层。领域驱动建模也是专门的一个方向。很多时候，业务领域模型清晰地建立了，后面的业务逻辑实现起来就会水到渠成。这一层主要由一系列的 POJO（Plain Old Java Object）组成。
- **数据访问对象**（Data Access Object，Dao）层主要提供对应 Model 层中的领域对象映射到数据库表的 CRUD 操作。在经典 Java EE 应用中，Dao 层叫 Eao 层。这里的 E 是 Entity。作用是一样的，Eao 层完成对实体 Entity 的 CRUD 操作。
- **业务逻辑**（Service）层主要是综合使用 Model 对象和 Dao 提供的 CRUD 接口，同时结合具体的业务流程来实现具体的业务逻辑。

- **控制器**（Controller）层提供一系列控制器，用以拦截并调用 Service 层的接口处理用户请求。最后，把处理结果传送到视图 View 层。
- **视图**（View）层主要是由一系列视图模板页面组成。例如，传统的 JSP 页面使用 Velocity 视图模板引擎的 vm 页面，使用 Freemarker 的 ftl 页面等。在该层完成用户的交互（例如，表单输入、按钮点击等操作），以及向用户展现输出结果的界面。

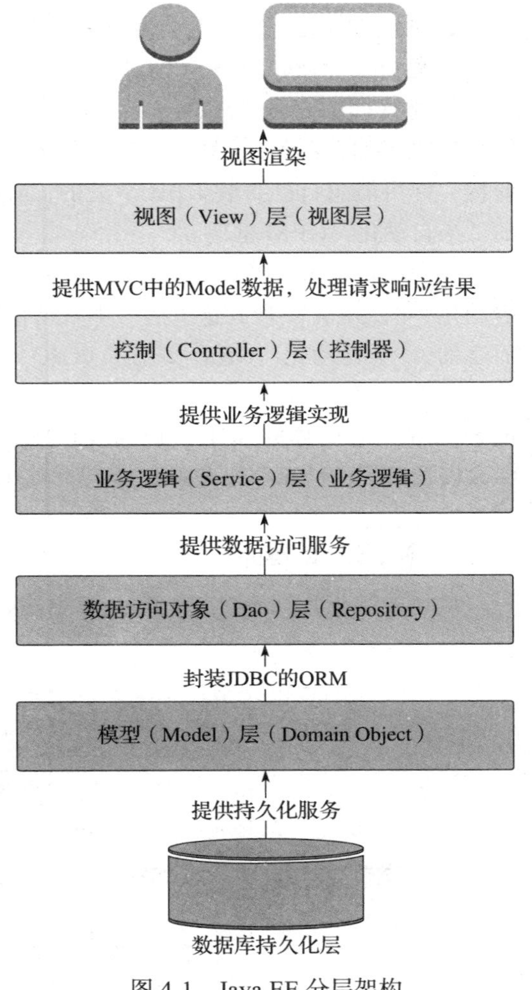

图 4-1　Java EE 分层架构

有了对 Java EE 系统架构的整体了解，让我们来看一下“半自动化”ORM 框架——MyBatis。

4.2　MyBatis 简介

本节带领大家对 MyBatis 进行整体概览。

4.2.1　概述

MyBatis 源自于 Apache 开源项目 iBatis，2010 年改名为 MyBatis。iBatis 是一个基于 Java 的持久层框架，提供的持久层框架包括 SQL Maps 和 Data Access Objects（DAO）。

MyBatis 采用面向对象编程的方式对数据库进行 CRUD 的操作，这是的应用程序中对关系数据库的操作更加方便简单。MyBatis 支持使用 XML 描述符配置文件和注解两种方式执行 SQL 语句。“简单灵活”是 MyBatis 在对象关系映射工具上的最大优势。

最早的主流 ORM（对象关系映射）框架 Hibernate，对数据库结构提供了完整的封装，提供了从 POJO 到数据库表的全套映射机制。程序员往往只需定义 POJO 到数据库表的映射关系，即可通过 Hibernate 提供的方法完成持久层操作，不需要对 SQL 熟练掌握，Hibernate 会根据指定的存储逻辑，自动生成对应的 SQL 并调用 JDBC 接口加以执行。

这样的机制无往不利，但是，在一些特定的环境下，这种一站式的解决方案却捉襟见肘。比如下面这些的应用场景：

- 出于安全考虑，只提供特定的 SQL（或存储过程）以获取所需数据，数据库具体的表结构不予公开。

- 开发规范中要求，所有牵涉到业务逻辑部分的数据库操作，必须在数据库层由存储过程实现（例如，金融行业的软件有在开发规范中严格指定）。
- 系统数据处理量巨大，对性能要求极为苛刻，需要高度优化 SQL 语句才能达到系统性能设计指标。

面对这样的需求，Hibernate 的自动化 SQL 的方案不再适用。当然，你可以直接使用 JDBC 进行数据库操作，只是拖沓的数据库访问代码，乏味的字段读取操作令人厌烦。这个时候，你可能想自己封装 JDBC。是的，MyBatis 就是满足上面这些特定场景的已经封装好的“半自动化”ORM 框架。

相对 Hibernate 和 ApacheOJB 等“一站式”ORM 解决方案而言，MyBatis 是一种“半自动化”的 ORM 实现。这里的“半自动化”是相对于 Hibernate 等“全自动化”ORM 实现而言的。“全自动化”ORM 实现了 POJO 和数据库表之间的映射，以及 SQL 的自动生成和执行。而 MyBatis 只解决 POJO 与 SQL 之间的映射关系，并不会为程序员在运行期自动生成 SQL 执行。具体的 SQL 代码需要程序员编写，然后通过映射配置文件，将 SQL 所需的参数以及返回的结果字段映射到指定的 POJO。

4.2.2 MyBatis 框架组成

从功能的业务流程维度来描述 MyBatis 的整体架构如图 4-2 所示。

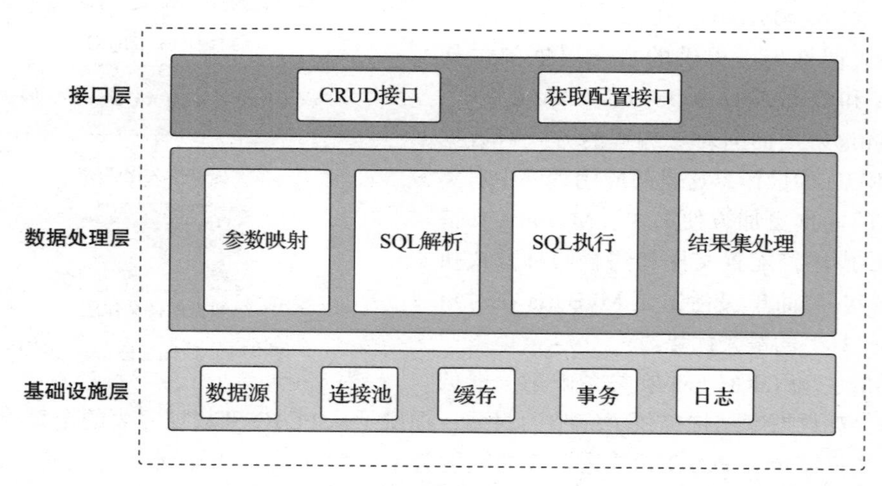

图 4-2 MyBatis 的整体架构图

下面对各个层分别作简要介绍。

1. 接口层

MyBatis 首先封装了对数据库的访问，把对数据库的会话和事务控制放到 SqlSession 对象中。直接使用 MyBatis 执行数据库操作的示例代码如下：

```
//MyBatis 配置文件
String resource = "mybatis-config.xml";
InputStream inputStream = Resources.getResourceAsStream(resource);
//创建 sqlSessionFactory 工厂对象
SqlSessionFactory sqlSessionFactory = new SqlSessionFactoryBuilder().build
    (inputStream);
//打开会话
SqlSession session = sqlSessionFactory.openSession();
try {
//执行 select 操作
Blog blog = session.selectOne(" com.easy.springboot.mybatis.example.Blog
    Mapper.selectBlog", 101);
} finally {
    //关闭会话
    session.close();
}
```

其中，SqlSessionFactory、SqlSession 这是 MyBatis 接口层的核心类，尤其是 SqlSession 它实现了常用的数据库操作的 API，这几个类都在 org.apache.ibatis.session 包下。SqlSession 接口的方法定义如下所示：

```
▼ SqlSession
    clearCache():void
    close():void
    commit():void
    commit(boolean):void
    delete(String):int
    delete(String, Object):int
    flushStatements():List<BatchResult>
    getConfiguration():Configuration
    getConnection():Connection
    getMapper(Class<T>):T
    insert(String):int
    insert(String, Object):int
    rollback():void
    rollback(boolean):void
    select(String, Object, ResultHandler):void
    select(String, Object, RowBounds, ResultHandler):void
    select(String, ResultHandler):void
    selectCursor(String):Cursor<T>
    selectCursor(String, Object):Cursor<T>
    selectCursor(String, Object, RowBounds):Cursor<T>
    selectList(String):List<E>
    selectList(String, Object):List<E>
    selectList(String, Object, RowBounds):List<E>
    selectMap(String, Object, String):Map<K, V>
    selectMap(String, Object, String, RowBounds):Map<K, V>
    selectMap(String, String):Map<K, V>
    selectOne(String):T
    selectOne(String, Object):T
    update(String):int
    update(String, Object):int
```

从上面的列表，我们可以看到在 SqlSession 接口中定义了对数据库的基本 CRUD 操作：select、update、insert、delete 等，还有数据库事务的回滚 rollback 和事务提交 commit 的操作，以及关于数据库连接、Configuration 配置、缓存、Mapper 等操作的 API。

Configuration 是 MyBatis 中的核心配置类。这个 Configuration 类中定义了需要配置的所有属性。了解这些参数的意义，对正确使用 MyBatis 非常必要。详细的内容可参考官方文档：http://www.mybatis.org/mybatis-3/configuration.html。Configuration 对象与 DefaultSqlSessionFactory 是 1-1 的关联关系，在一个 DefaultSqlSessionFactory 衍生出来的所有 SqlSession 作用域里，Configuration 对象是全局唯一的。同时我们看到，在 SqlSession 接口中定义了 getConfiguration() 接口来获得 Configuration 对象，因此除了配置文件之外，我们还可以在程序里动态配置 Configuration 的属性，这也正是 Spring Boot 集成 MyBatis 的启动器 mybatis-spring-boot-starter 背后的实现原理。

2. 数据处理层

应用启动的时候，MyBatis 会去解析两种配置文件：

- SqlMapConfig.xml（另外一种方式是使用基于 Java Config 注解的方式）。
- Mapper.xml（使用 MyBatis-Generator 自动生成的一系列的 Mapper.xml 文件，也可以使用注解方式，在 Java 方法上直接写 SQL 以及字段和返回结果的映射关系）。

其中，SqlMapConfig.xml 是在 XMLConfigBuilder 类中完成解析。Mapper.xml 是在 XMLMapperBuilder 中解析完成的。XMLMapperBuilder 对 Statement 的解析（即 Mapper.xml 中的 SELECT、INSERT、UPDATE、DELETE 标签定义的 SQL）则是委托给 XMLStatementBuilder 来完成。XMLStatementBuilder 中的核心解析代码在 parseStatementNode() 方法中。在 Mapper.xml 的解析是比较复杂的，涉及 PreparedMapping、ResultMapping、LanguageDriver、缓存（flushCache/useCache）、自动映射等一系列对象的构造和处理。感兴趣的朋友可以去阅读 MyBatis 的源码（https://github.com/mybatis/mybatis-3）。

MyBatis 的执行流程如图 4-3 所示。

3. 基础设施层

MyBatis 的基础设施层包含了日志、IO、反射、异常、缓存、数据源 & 连接池、事务管理、类型映射等模块，下节我们分析介绍。

4.2.3 MyBatis 基础设施

1. 日志

MyBatis 使用了自己定义的一套 logging 接口，根据开发者常使用的日志框架—— Log4j、log4j2、Apache Commons Log、java.util.logging、slf4j、stdout（控制台）等分别提供了适配器。这些适配器的代码我们可以在 org.apache.ibatis.logging 包下面看到。

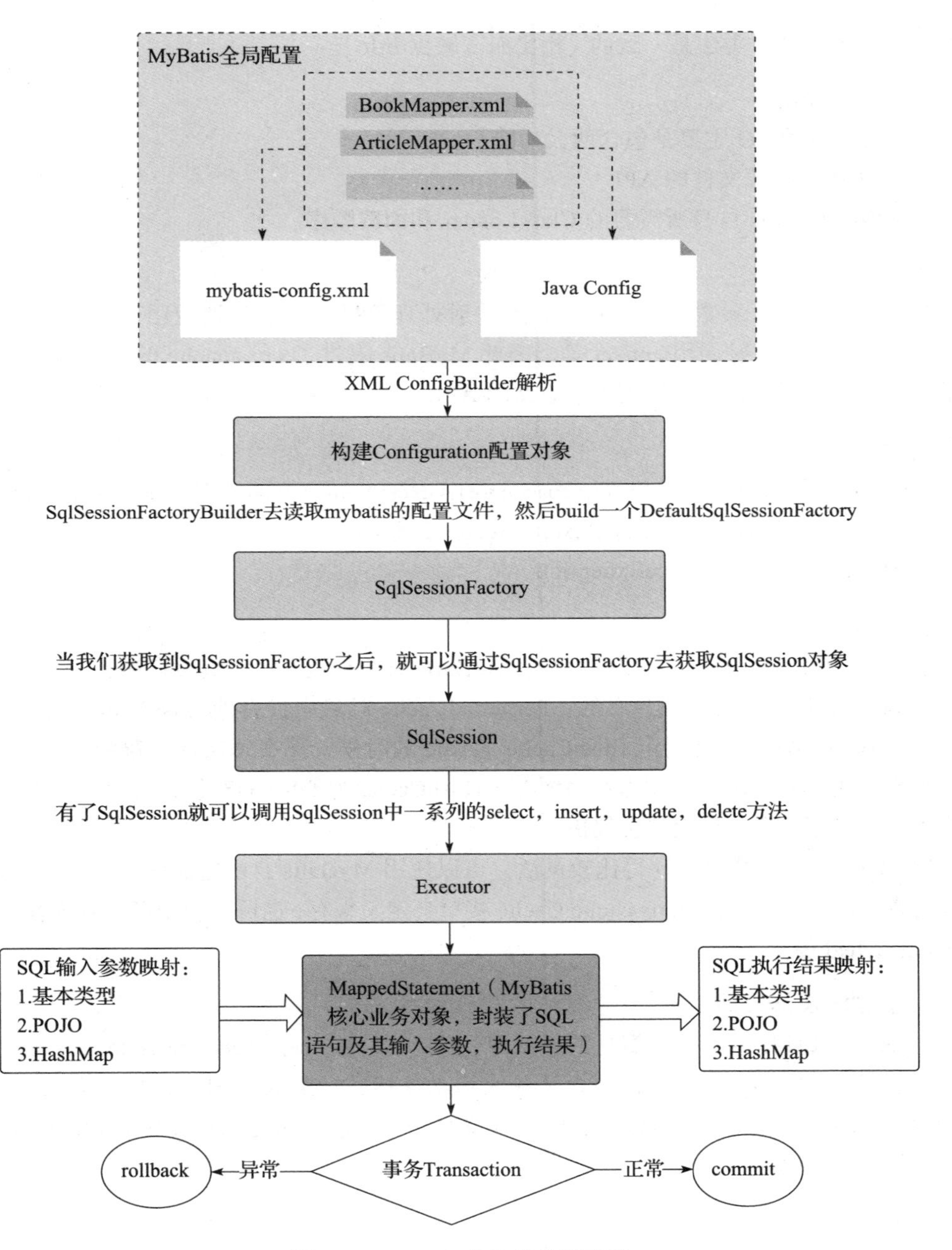

图 4-3　MyBatis 整体的执行流程

由于各日志框架的 Log 级别分类法有所不同（比如 java.util.logging.Level 提供的是 All、FINEST、FINER、FINE、CONFIG、INFO、WARNING、SEVERE、OFF 这九个级别，与通常的日志框架分类法不太一样），MyBatis 统一提供 trace、debug、warn、error 四个级别，

这基本与主流框架分类法是一致的（相比而言缺少 Info）。

2. 输入 / 输出

MyBatis 里的 I/O 主要是包含两大功能：

- 提供读取资源文件的 API。
- 封装 MyBatis 自身所需要的 ClassLoader 和加载顺序。

3. 反射

在 MyBatis 中，参数映射处理、结果映射处理等操作中大量地使用了反射技术，去读取 Class 元数据、反射调用 get/set 等，因此 MyBatis 提供了 org.apache.ibatis.reflection 对常见的反射操作的封装，以提供更简洁方便的 API。

4. 异常

MyBatis 的异常体系不复杂，org.apache.ibatis.exceptions 下就几个类，如 Exception Factory、Persistence Exception、Too Many Result Exception 等。

主要使用的是 PersistenceException。

5. 缓存

缓存是 MyBatis 里比较重要的部分，MyBatis 有两种缓存：一级缓存和二级缓存。

默认情况下一级缓存是开启的。PerpetualCache 对象中持有的 HashMap 即为一级缓存内容。当我们在方法级别指定 flushCache 为 true 的时候，那么每次调用都清理缓存，这样等于每次调用都要重新读数据库和写缓存。flushCache 为 false 的时候，每次调用不清缓存，除了第一次调用较慢，后面都会很快。

MyBatis 的二级缓存设计得比较灵活，可以使用 MyBatis 自己定义的二级缓存实现；也可以通过实现 org.apache.ibatis.cache.Cache 接口自定义缓存；也可以使用第三方内存缓存库，如 Memcached 等。

6. 数据源 & 连接池

MyBatis 自身提供了一个简易的数据源 / 连接池，在 org.apache.ibatis.datasource 下，主要实现类是 PooledDataSource，包含了最大活动连接数、最大空闲连接数、最长取出时间（避免某个线程过度占用）、连接不够时的等待时间等。这些属性都定义在 PooledDataSource 类中，相关的代码如下：

```
public class PooledDataSource implements DataSource {
    …
    //最大活跃连接数
    protected int poolMaximumActiveConnections = 10;
    //最大闲置连接数
    protected int poolMaximumIdleConnections = 5;
    //最大检查时间
    protected int poolMaximumCheckoutTime = 20000;
```

```
    //等待时间
    protected int poolTimeToWait = 20000;
    //连接容错数
    protected int poolMaximumLocalBadConnectionTolerance = 3;
    protected String poolPingQuery = "NO PING QUERY SET";
    protected boolean poolPingEnabled;
    protected int poolPingConnectionsNotUsedFor;

    private int expectedConnectionTypeCode;
    …
}
```

7. 事务

MyBatis 对事务的处理相对简单，TransactionIsolationLevel 中定义了几种隔离级别：

```
public enum TransactionIsolationLevel {
    //无事务
    NONE(Connection.TRANSACTION_NONE),
    //读提交
    READ_COMMITTED(Connection.TRANSACTION_READ_COMMITTED),
    //读未提交
    READ_UNCOMMITTED(Connection.TRANSACTION_READ_UNCOMMITTED),
    //可重复读
    REPEATABLE_READ(Connection.TRANSACTION_REPEATABLE_READ),
    //顺序读
    SERIALIZABLE(Connection.TRANSACTION_SERIALIZABLE);

    private final int level;

    private TransactionIsolationLevel(int level) {
        this.level = level;
  }

    public int getLevel() {
        return level;
    }
}
```

不支持内嵌事务这样较复杂的场景，同时由于其是持久层的缘故，所以真正在应用开发中会委托 Spring 来处理事务来实现真正的隔离。

8. 类型映射

MyBatis 的类型映射在包 org.apache.ibatis.type 下面。其中通过 TypeAliasRegistry 类来注册并维护与 Java 基本类型的映射关系。这个映射关系存储在一个 Map<String，Class<?>> 对象 TYPE_ALIASES 中，相关代码如下：

```
public class TypeAliasRegistry {
    // 存储类型映射关系的 HashMap
    private final Map<String, Class<?>> TYPE_ALIASES = new HashMap<String,
        Class<?>>();
```

```
    public TypeAliasRegistry() {
// String 类型
        registerAlias("string", String.class);
// 基本类型的包装类型
        registerAlias("byte", Byte.class);
        registerAlias("long", Long.class);
        registerAlias("short", Short.class);
        registerAlias("int", Integer.class);
        registerAlias("integer", Integer.class);
        registerAlias("double", Double.class);
        registerAlias("float", Float.class);
        registerAlias("boolean", Boolean.class);
// 基本类型的包装数组类型
        registerAlias("byte[]", Byte[].class);
        registerAlias("long[]", Long[].class);
        registerAlias("short[]", Short[].class);
        registerAlias("int[]", Integer[].class);
        registerAlias("integer[]", Integer[].class);
        registerAlias("double[]", Double[].class);
        registerAlias("float[]", Float[].class);
        registerAlias("boolean[]", Boolean[].class);
// 基本类型
        registerAlias("_byte", byte.class);
        registerAlias("_long", long.class);
        registerAlias("_short", short.class);
        registerAlias("_int", int.class);
        registerAlias("_integer", int.class);
        registerAlias("_double", double.class);
        registerAlias("_float", float.class);
        registerAlias("_boolean", boolean.class);
// 基本数组类型
        registerAlias("_byte[]", byte[].class);
        registerAlias("_long[]", long[].class);
        registerAlias("_short[]", short[].class);
        registerAlias("_int[]", int[].class);
        registerAlias("_integer[]", int[].class);
        registerAlias("_double[]", double[].class);
        registerAlias("_float[]", float[].class);
        registerAlias("_boolean[]", boolean[].class);
// 常用的日期、十进制数、大整数、对象类等内置类型
        registerAlias("date", Date.class);
        registerAlias("decimal", BigDecimal.class);
        registerAlias("bigdecimal", BigDecimal.class);
        registerAlias("biginteger", BigInteger.class);
        registerAlias("object", Object.class);
// 常用的日期、十进制数、大整数、对象类等内置类型的数组
        registerAlias("date[]", Date[].class);
        registerAlias("decimal[]", BigDecimal[].class);
        registerAlias("bigdecimal[]", BigDecimal[].class);
        registerAlias("biginteger[]", BigInteger[].class);
        registerAlias("object[]", Object[].class);
// 基本集合类类型
```

```
        registerAlias("map", Map.class);
        registerAlias("hashmap", HashMap.class);
        registerAlias("list", List.class);
        registerAlias("arraylist", ArrayList.class);
        registerAlias("collection", Collection.class);
        registerAlias("iterator", Iterator.class);
// 结果集类型
        registerAlias("ResultSet", ResultSet.class);
    }
    …
}
```

针对 Java8，MyBatis 单独添加了一个映射类 Java8TypeHandlersRegistrar：

```
@UsesJava8
public abstract class Java8TypeHandlersRegistrar {

    public static void registerDateAndTimeHandlers(TypeHandlerRegistry
        registry) {
        registry.register(Instant.class, InstantTypeHandler.class);
        registry.register(LocalDateTime.class, LocalDateTimeTypeHandler.
            class);
        registry.register(LocalDate.class, LocalDateTypeHandler.class);
        registry.register(LocalTime.class, LocalTimeTypeHandler.class);
        registry.register(OffsetDateTime.class, OffsetDateTimeTypeHan
            dler.class);
        registry.register(OffsetTime.class, OffsetTimeTypeHandler.class);
        registry.register(ZonedDateTime.class, ZonedDateTimeTypeHandler.
            class);
        registry.register(Month.class, MonthTypeHandler.class);
        registry.register(Year.class, YearTypeHandler.class);
        registry.register(YearMonth.class, YearMonthTypeHandler.class);
        registry.register(JapaneseDate.class, JapaneseDateTypeHandler.class);
    }

}
```

在 JdbcType 枚举中维护了一份与 JDBC 类型的映射关系：

```
public enum JdbcType {
ARRAY(Types.ARRAY),
    BIT(Types.BIT),
    TINYINT(Types.TINYINT),
    SMALLINT(Types.SMALLINT),
    INTEGER(Types.INTEGER),
    BIGINT(Types.BIGINT),
    FLOAT(Types.FLOAT),
    REAL(Types.REAL),
    DOUBLE(Types.DOUBLE),
    NUMERIC(Types.NUMERIC),
    DECIMAL(Types.DECIMAL),
    CHAR(Types.CHAR),
    VARCHAR(Types.VARCHAR),
```

```
    LONGVARCHAR(Types.LONGVARCHAR),
    DATE(Types.DATE),
    TIME(Types.TIME),
    TIMESTAMP(Types.TIMESTAMP),
    BINARY(Types.BINARY),
    VARBINARY(Types.VARBINARY),
    LONGVARBINARY(Types.LONGVARBINARY),
    NULL(Types.NULL),
    OTHER(Types.OTHER),
    BLOB(Types.BLOB),
    CLOB(Types.CLOB),
    BOOLEAN(Types.BOOLEAN),
    CURSOR(-10),                          // Oracle
    UNDEFINED(Integer.MIN_VALUE + 1000),
    NVARCHAR(Types.NVARCHAR),           // JDK6
    NCHAR(Types.NCHAR),                 // JDK6
    NCLOB(Types.NCLOB),                 // JDK6
    STRUCT(Types.STRUCT),
    JAVA_OBJECT(Types.JAVA_OBJECT),
    DISTINCT(Types.DISTINCT),
    REF(Types.REF),
    DATALINK(Types.DATALINK),
    ROWID(Types.ROWID),                 // JDK6
    LONGNVARCHAR(Types.LONGNVARCHAR), // JDK6
    SQLXML(Types.SQLXML),               // JDK6
    DATETIMEOFFSET(-155);               // SQL Server 2008

    public final int TYPE_CODE;
    private static Map<Integer,JdbcType> codeLookup = new HashMap<Integer,
        JdbcType>();

    static {
        for (JdbcType type : JdbcType.values()) {
            codeLookup.put(type.TYPE_CODE, type);
        }
    }

    JdbcType(int code) {
        this.TYPE_CODE = code;
    }

    public static JdbcType forCode(int code)  {
        return codeLookup.get(code);
    }

}
```

每个基本类型都有一个对应的 TypeHandler，这些 Handler 的列表如下：

```
ArrayTypeHandler.java
BaseTypeHandler.java
BigDecimalTypeHandler.java
BigIntegerTypeHandler.java
```

```
BlobByteObjectArrayTypeHandler.java
BlobInputStreamTypeHandler.java
BlobTypeHandler.java
BooleanTypeHandler.java
ByteArrayTypeHandler.java
ByteObjectArrayTypeHandler.java
ByteTypeHandler.java
CharacterTypeHandler.java
ClobReaderTypeHandler.java
ClobTypeHandler.java
DateOnlyTypeHandler.java
DateTypeHandler.java
DoubleTypeHandler.java
EnumOrdinalTypeHandler.java
EnumTypeHandler.java
FloatTypeHandler.java
InstantTypeHandler.java
IntegerTypeHandler.java
JapaneseDateTypeHandler.java
LocalDateTimeTypeHandler.java
LocalDateTypeHandler.java
LocalTimeTypeHandler.java
LongTypeHandler.java
MonthTypeHandler.java
NClobTypeHandler.java
NStringTypeHandler.java
ObjectTypeHandler.java
OffsetDateTimeTypeHandler.java
OffsetTimeTypeHandler.java
ShortTypeHandler.java
SqlDateTypeHandler.java
SqlTimeTypeHandler.java
SqlTimestampTypeHandler.java
StringTypeHandler.java
TimeOnlyTypeHandler.java
TypeHandler.java
UnknownTypeHandler.java
YearMonthTypeHandler.java
YearTypeHandler.java
ZonedDateTimeTypeHandler.java
```

MyBatis 通过 parameterType 指定输入参数的类型，类型可以是简单类型、hashmap、pojo 的包装类型。使用 resultType 进行输出映射，只有查询出来的列名和 pojo 中的属性名一致，该列才可以映射成功：

- 如果查询出来的列名和 pojo 中的属性名全部不一致，不创建 pojo 对象。
- 只要查询出来的列名和 pojo 中的属性有一个一致，就会创建 pojo 对象。

如果查询出来的列名和 pojo 的属性名不一致，可以通过定义一个 resultMap 对列名和 pojo 属性名之间作一个映射关系。MyBatis 中使用 resultMap 完成高级输出结果映射。

上面讲了这么多的理论基础知识，下面我们开始项目实战。

4.3 项目实战

本节讲解如何使用 Spring Boot MyBatis Starter 来极简化集成 Spring Boot 和 MyBatis 进行数据库 dao 层代码的开发。我们将介绍使用 Mapper.xml 和注解的方式编写 SQL。

4.3.1 使用 Spring Boot CLI 创建工程

可以方便地使用 IDEA 集成的 Spring Initializr 界面和 Pivotal Web Services 提供的网页版 Spring Initializr 工具界面 http://start.spring.io/ 来创建 Spring Boot 项目。这个操作过程只要打开界面按照提示操作即可快速上手。我们在这里就不多介绍。

这里介绍一种更加极简地使用 Spring Boot CLI 命令行方式来创建 Spring Boot 项目。Spring Boot CLI 是一个命令行工具，如果你想快速创建 Spring Boot 原型工程，可以使用它。它还可以直接运行 Groovy 脚本。

4.3.2 Spring Boot 命令行 CLI 简介

首先我们来安装 CLI 运行环境。安装的版本是：Spring CLI v2.0.0.BUILD-SNAPSHOT。直接去 Spring 官方仓库：http://repo.spring.io/snapshot/org/springframework/boot/spring-boot-cli/，下载 2.0.0.BUILD-SNAPSHOT 目录中的 Spring Boot CLI zip 包。解压后的目录如下：

```
spring-2.0.0.BUILD-SNAPSHOT$ tree
.
├── INSTALL.txt
├── LICENCE.txt
├── bin
│   ├── spring
│   └── spring.bat
├── legal
│   └── open_source_licenses.txt
├── lib
│   └── spring-boot-cli-2.0.0.BUILD-SNAPSHOT.jar
└── shell-completion
    ├── bash
    │   └── spring
    └── zsh
        └── _spring

6 directories, 8 files
```

在 HOME 目录下执行 vim .bashrc 把 spring 命令行环境配置到 PATH 变量中：

```
export SPRING_BOOT_HOME=/Users/jack/soft/spring-2.0.0.BUILD-SNAPSHOT
export PATH=$PATH:$SPRING_BOOT_HOME/bin
```

执行 $source .bashrc，然后去命令行中验证 spring 命令的版本：

```
$ spring --version
```

```
Spring CLI v2.0.0.BUILD-SNAPSHOT
```

至此，Spring Boot CLI 命令行环境安装完成。

8

Spring CLI 常用命令如下：

```
$ spring
usage: spring [--help] [--version]
        <command> [<args>]

Available commands are:

    run [options] <files> [--] [args]
        Run a spring groovy script

    grab
        Download a spring groovy script's dependencies to ./repository

      jar [options] <jar-name> <files>
        Create a self-contained executable jar file from a Spring Groovy
            script

      war [options] <war-name> <files>
        Create a self-contained executable war file from a Spring Groovy
            script

      install [options] <coordinates>
        Install dependencies to the lib/ext directory

      uninstall [options] <coordinates>
        Uninstall dependencies from the lib/ext directory

      init [options] [location]
        Initialize a new project using Spring Initializr (start.spring.io)

      shell
        Start a nested shell

Common options:

      -d, --debug Verbose mode
        Print additional status information for the command you are running


See 'spring help <command>' for more information on a specific command.
```

可以通过 spring help run 命令来查看 spring run 命令的说明。

spring init 命令可以用来创建 Spring Boot 项目。具体的使用说明可以通过如下命令来查看

```
$ spring help init
```

例如，使用 $spring init demo 命令创建一个最简单的 Spring Boot 项目：

```
$ spring init demo
Using service at https://start.spring.io
Project extracted to '/Users/jack/KotlinSpringBoot/gs/demo'
```

我们将得到一个标准 Spring Boot 工程。另外，我们可以通过下面的命令来获取目前 Spring Boot CLI 所支持的初始化项目的所有选项：

```
$ spring init -list
```

通过命令 $spring init -list 查看命令的使用细节，参见表 4-1 和表 4-2。

表 4-1 项目构建类型

Id	说明	标签
gradle-build	Generate a Gradle build file	build: gradle, format: build
gradle-project	Generate a Gradle based project archive	build: gradle, format: project
maven-build	Generate a Maven pom.xml	build: maven, format: build
maven-project *	Generate a Maven based project archive	build: maven, format: project

* 表示是默认值。

表 4-2 创建项目参数选项表

Id	说明	默认值
artifactId	project coordinates (infer archive name)	demo
bootVersion	spring boot version	1.5.9.RELEASE
description	project description	Demo project for Spring Boot
groupId	project coordinates	com.example
javaVersion	language level	1.8
language	programming language	java
name	project name (infer application name)	demo
packageName	root package	com.example.demo
packaging	project packaging	jar
type	project type	maven-project
version	project version	0.0.1-SNAPSHOT

集成 Spring Boot 和 MyBatis 进行开发的命令如下：

```
$ mkdir demo_springboot_with_mybatis
$ cd demo_springboot_with_mybatis
$ spring init \
-d=web,mybatis,mysql \
-b=2.0.0.M7 \
-a=demo_springboot_with_mybatis \
-g=com.easy.springboot \
-j=1.8 \
```

```
-l=kotlin \
--build=gradle \
-n=demo_springboot_with_mybatis \
-p=jar \
-x
```

参数说明见表 4-3。

表 4-3　Starter 的参数

参数	说明
-d=web, mybatis, mysql	项目起步依赖 dependencies: web, mybatis, mysql（逗号分开）
-b=2.0.0.M7	Spring Boot 的 version 使用 2.0.0.M7
-a=demo_springboot_with_mybatis	设置 artifactId 的值为 demo_springboot_with_mybatis
-g=com.easy.springboot	设置 groupId 的值为 com.easy.springboot
-j=1.8	设置 JDK 版本 1.8
-l=kotlin	设置编程语言 Kotlin
-n=demo_springboot_with_mybatis	项目名称
-p=jar	打包方式
-x	项目文件包解压（Extract）

命令执行的结果如下：

```
Using service at https://start.spring.io
Project extracted to '/Users/jack/KotlinSpringBoot/demo_springboot_with_mybatis'
```

执行完命令，我们即可得到一个使用 MyBatis 的标准 Spring Boot 工程。把项目导入 IDEA 中，等待 Gradle 构建完毕。可以看到项目的依赖树，如下所示。

```
▼ demo_springboot_with_mybatis (auto-import enabled)
  ▼ Source Sets
    ▼ main
      ▼ Dependencies
        ▼ org.mybatis.spring.boot:mybatis-spring-boot-starter:1.3.1 (Compile)
          ▼ org.springframework.boot:spring-boot-starter-jdbc:2.0.0.M7 (Compile)
              org.springframework.boot:spring-boot-starter:2.0.0.M7 (Compile)
            ► org.springframework:spring-jdbc:5.0.2.RELEASE (Compile)
            ► com.zaxxer:HikariCP:2.7.4 (Compile)
          ► org.springframework.boot:spring-boot-starter:2.0.0.M7 (Compile)
          ▼ org.mybatis.spring.boot:mybatis-spring-boot-autoconfigure:1.3.1 (Compile)
              org.springframework.boot:spring-boot-autoconfigure:2.0.0.M7 (Compile)
            org.mybatis:mybatis:3.4.5 (Compile)
            org.mybatis:mybatis-spring:1.3.1 (Compile)
        ► org.springframework.boot:spring-boot-starter-web:2.0.0.M7 (Compile)
        ► org.jetbrains.kotlin:kotlin-stdlib-jre8:1.1.61 (Compile)
        ► org.jetbrains.kotlin:kotlin-reflect:1.1.61 (Compile)
          mysql:mysql-connector-java:5.1.44 (Runtime)
    ▼ test
      ▼ Dependencies
          demo_springboot_with_mybatis:main (Compile)
```

```
▶ org.mybatis.spring.boot:mybatis-spring-boot-starter:1.3.1 (Compile)
▶ org.springframework.boot:spring-boot-starter-web:2.0.0.M7 (Compile)
▶ org.jetbrains.kotlin:kotlin-stdlib-jre8:1.1.61 (Compile)
▶ org.jetbrains.kotlin:kotlin-reflect:1.1.61 (Compile)
▶ org.springframework.boot:spring-boot-starter-test:2.0.0.M7 (Compile)
  mysql:mysql-connector-java:5.1.44 (Runtime)
```

我们可以看到在 org.mybatis.spring.boot: mybatis-spring-boot-starter: 1.3.1 起步器依赖中引入了下面的这些核心依赖：

- org.springframework.boot: spring-boot-starter: 2.0.0.M7
- org.springframework.boot: spring-boot-starter-jdbc: 2.0.0.M7
- org.mybatis: mybatis: 3.4.5
- org.mybatis: mybatis-spring: 1.3.1

这个时候，我们直接启动应用会看到控制台日志报错：

```
Error starting ApplicationContext. To display the conditions report re-run
    your application with 'debug' enabled.
2018-01-10 00:00:58.319 ERROR 22964 --- [           main] o.s.b.d.Logging
    FailureAnalysisReporter    :

***************************
APPLICATION FAILED TO START
***************************

Description:

Cannot determine embedded database driver class for database type NONE

Action:

If you want an embedded database please put a supported one on the classpath.
    If you have database settings to be loaded from a particular profile
    you may need to active it (no profiles are currently active).
```

因为我们还没有配置数据库信息。Spring Boot 的自动配置无法完成。下面我们来配置数据库连接信息。

4.3.3 配置 application.properties

在 Spring Boot 工程的全局配置文件配置 application.properties 中添加 datasource 的配置如下：

```
spring.datasource.driver-class-name=com.mysql.jdbc.Driver
spring.datasource.url=jdbc:mysql://127.0.0.1:3306/reakt?characterEncoding=
    utf8&charaterSetResults=utf8&useSSL=false
spring.datasource.username=root
spring.datasource.password=root
```

再次重新启动应用，我们可以看到控制台关于‘dataSource’这个 Bean 的创建成功的日志如下：

```
2018-01-10 00:05:58.925  INFO 23014 --- [           main] o.s.j.e.a.Annot
    ationMBeanExporter        : Bean with name 'dataSource' has been autode
        tected for JMX exposure
2018-01-10 00:05:58.936  INFO 23014 --- [           main] o.s.j.e.a.Anno
    tationMBeanExporter        : Located MBean 'dataSource': registering
        with JMX server as MBean [com.zaxxer.hikari:name=dataSource,type=
            HikariDataSource]
2018-01-10 00:05:59.102  INFO 23014 --- [           main] o.s.b.w.embedded.
    tomcat.TomcatWebServer  : Tomcat started on port(s): 8080 (http) with
        context path ''
2018-01-10 00:05:59.112  INFO 23014 --- [           main] d.DemoSpringboo
    tWithMybatisApplicationKt : Started DemoSpringbootWithMybatisApplication
        Kt in 9.049 seconds (JVM running for 10.61)
```

从上面的 AnnotationMBeanExporter 那行日志，我们可以看出 Spring Boot 中默认使用的数据源连接池是 HikariDataSource。

4.3.4　使用 IDEA 中自带的连接数据库客户端

为了方便后面讲解和演示数据库的表结构和测试数据，我们使用 IDEA 中自带的连接数据库客户端工具在 GUI 界面中操作数据库。配置示意图如图 4-4 所示。

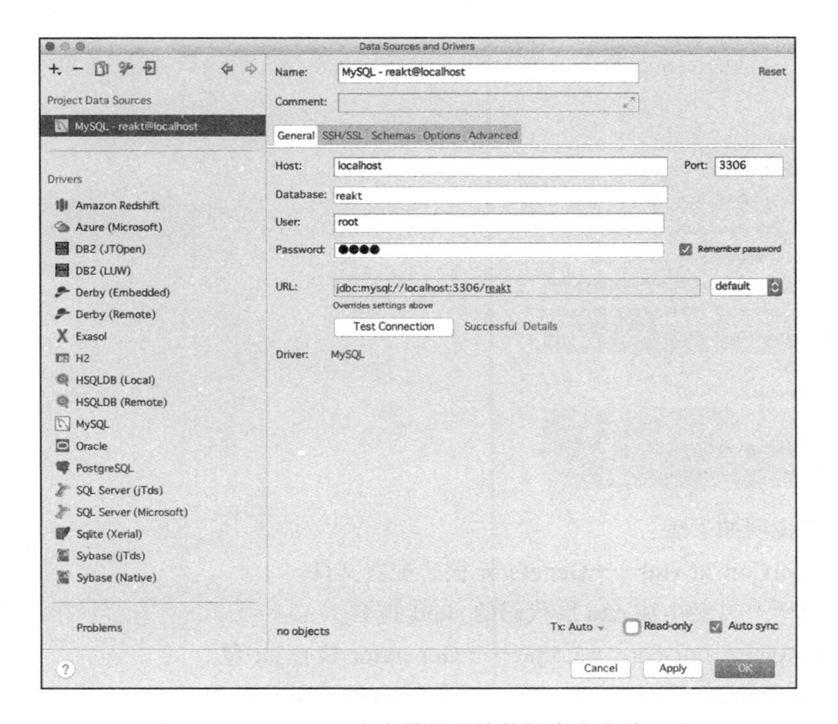

图 4-4　配置 IDEA 中自带的连接数据库客户端工具

IDEA 中自带的连接数据库客户端工具 GUI 界面使用如图 4-5 所示。

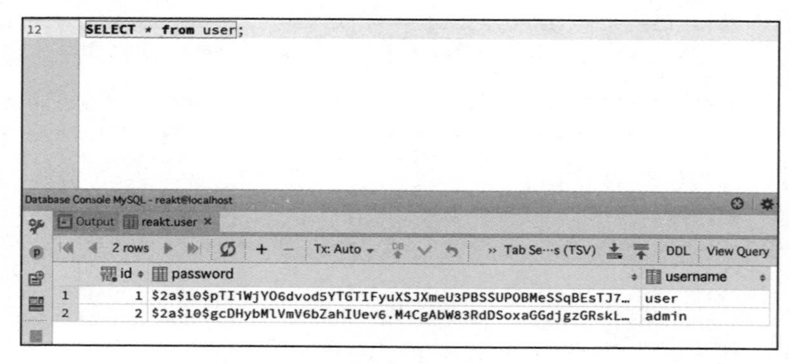

图 4-5 GUI 界面

4.3.5 使用 MyBatis Generator 生成 dao 层代码

使用 MyBatis 通常我们使用 MyBatis Generator 来自动生成 dao 层代码。使用 MyBatis Generator 依赖的 jar 包和配置文件如下所示。

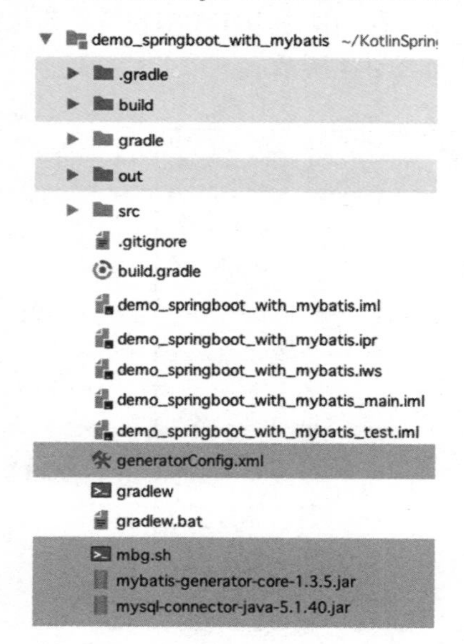

对应文件说明如下表：

- generatorConfig.xml——Generator 核心配置文件。
- mbg.sh——自动生成 dao 层代码的 shell 脚本。
- mybatis-generator-core-1.3.5.jar——Generator 核心 jar 包。
- mysql-connector-java-5.1.40.jar——MySQL 的 JDBC 驱动。

其中，generatorConfig.xml 完整代码参考本章的工程源代码。

提示　更多配置项可参考 mybatis generator 官方文档：http://www.mybatis.org/generator/configreference/xmlconfig.html

生成代码的执行脚本 mbg.sh 如下：

```
#!/usr/bin/env bash
java -jar mybatis-generator-core-1.3.5.jar -configfile generatorConfig.
    xml -overwrite
```

在项目根目录执行上面的脚本：

```
demo_springboot_with_mybatis$ ./mbg.sh
...
MyBatis Generator finished successfully, there were warnings.
```

生成的代码如下所示：

```
▼ com
  ▼ easy
    ▼ springboot
      ▼ demo_springboot_with_mybatis
        ▼ dao
            ArticleMapper.java
            BookMapper.java
            CategoryMapper.java
            RoleMapper.java
            TagMapper.java
            UserMapper.java
            UserRolesMapper.java
        ▼ mapper
            ArticleMapper.xml
            BookMapper.xml
            CategoryMapper.xml
            RoleMapper.xml
            TagMapper.xml
            UserMapper.xml
            UserRolesMapper.xml
        ▼ model
            Article.java
            Book.java
            Category.java
            Role.java
            Tag.java
            User.java
            UserRolesKey.java
► gradle
► out
▼ src
```

把生成的代码放到项目对应目录下面，如下所示：

```
▼ src
  ▼ main
      java
    ▼ kotlin
      ▼ com
        ▼ easy
```

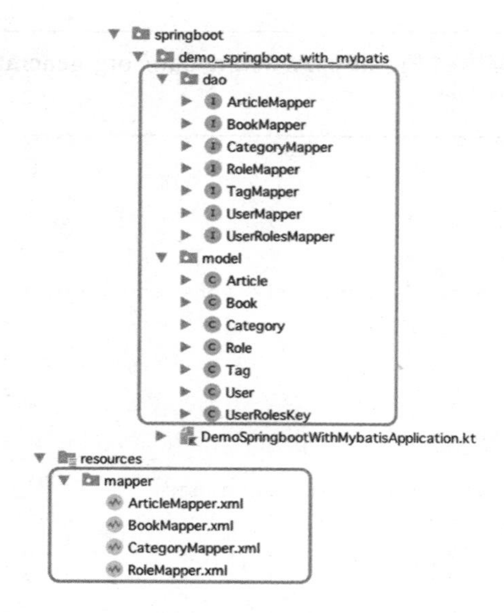

4.3.6 设置 MyBatis 同时使用 Mapper.xml 和注解

在 application.properties 配置中添加 mybatis.mapper-locations 配置如下：

```
mybatis.mapper-locations=classpath:/mapper/*.xml
```

这样，我们就可以充分使用 MBG 自动生成的 Mapper.xml 文件了。例如，自动生成的 UserMapper.java、UserMapper.xml、User.java 等源代码参考示例工程。

4.3.7 使用 @Select 注解

使用 @Select 注解直接在方法上写 SQL，UserMapper.java 中的代码如下：

```
public interface UserMapper {
    …

    @Select("SELECT * FROM user")
    List<User> findAll();
}
```

UserService.kt 代码如下：

```
interface UserService {
    fun findAll(): List<User>
}
UserServiceImpl.kt 代码如下：@Service
class UserServiceImpl : UserService {
    @Autowired lateinit var userMapper: UserMapper
    override fun findAll(): List<User> {
        return userMapper.findAll()
    }
}
```

UserController.kt 代码如下：

```
@RestController
@RequestMapping("/user")
class UserController {
    @Autowired lateinit var userService: UserService

    @GetMapping("/findAll")
    fun findAll(): List<User> {
        return userService.findAll()
    }

}
```

我们可以在每个 Mapper.java 接口上标注 @Mapper 注解，这样 Spring Boot 会去自动扫描 com.easy.springboot.demo_springboot_with_mybatis 包下面的这些 Mapper 接口，并由 Spring 自动创建对应的 Mapper 接口的实现类 Bean。

还有一种方式，就是不需要单独在每个 Mapper.java 接口上面使用 @Mapper 注解，而是统一在 Spring Boot 入口类上使用 @MapperScan 注解告诉 Spring Boot 所有 Mapper 接口在具体哪个包下面，代码如下：

```
@SpringBootApplication
@MapperScan("com.easy.springboot.demo_springboot_with_mybatis.dao")
class DemoSpringbootWithMybatisApplication

fun main(args: Array<String>) {
    runApplication<DemoSpringbootWithMybatisApplication>(*args)
}
```

整体的代码组织结构参考实例工程。重启应用，在浏览器中输入 http://127.0.0.1:8080/user/findAll，可以看到响应输出。

4.3.8　使用 MyBatis 分页插件 pagehelper

在业务接口中通常会有分页获取数据的场景。MyBatis 的标准分页插件是 pagehelper。

在 build.gradle 配置文件中添加 pagehelper-spring-boot-starter 依赖：

```
compile group: 'com.github.pagehelper', name: 'pagehelper-spring-boot-starter',
    version: '1.2.3'
```

下面我们来实现这样的一个把 user 表中所有用户全部查找出来的 HTTP 接口：http://127.0.0.1:8080/user/findAll，它的输出如下：

```
[
    {
        "id": 1,
        "password": "$2a$10$pTIiWjYO6dvod5YTGTIFyuXSJXmeU3PBSSUPOBMeSSqBEsTJ7PmW2",
        "username": "user"
```

```
    },
    {
        "id": 2,
        "password": "$2a$10$gcDHybMlVmV6bZahIUev6.M4CgAbW83RdDSoxaGGdjgzGRskLYVy6",
        "username": "admin"
    }
]
```

首先，我们来配置一下数据库表到 Java Bean 的自动驼峰命名转换。在使用 @Mapper 注解方式代替 Mapper.xml 配置文件，那么在使用 @Select 等注解配置 SQL 语句的情况下，如何配置下划线风格的数据库字段名到驼峰命名风格的 JavaBean 实体类属性命名转换？

我们知道设置 MyBatis 的自动驼峰命名转换，在 XML 中可以直接配置 mapUnderscore-ToCamelCase 属性。同样的配置，在 Spring Boot 的配置文件 application.properties 中，加入下面的配置项即可：

```
mybatis.configuration.map-underscore-to-camel-case=true
```

使用 Spring Boot 后，你会越来越喜欢用注解方式进行配置来代替 XML 配置文件方式。MyBatis 中也可以完全使用注解，避免使用 XML 方式配置。但是为了更加便利，我们通常还会使用 MBG 自动生成的 CRUD 基本方法。这个时候最佳方案就是同时使用 MyBatis 注解和 Mapper.xml 配置文件。

4.3.9 MyBatis 插件机制

在讲到下面的 PageHelper 分页接口的实现之前，我们先来熟悉一下 MyBatis 的插件机制。MyBatis 的插件机制是通过拦截器动态代理实现的。动态代理的大致执行过程通常是：

拦截器代理类对象→拦截器→目标方法

当我们调用 ParameterHandler、ResultSetHandler、StatementHandler、Executor 的对象的时候，实际上使用的是 Plugin 这个代理类的对象，这个类实现了 InvocationHandler 接口。在调用上述被代理类的方法时，就会执行 Plugin 的 invoke 方法。Plugin 在 invoke 方法中根据 @Intercepts 的配置信息（方法名，参数等）动态判断是否需要拦截该方法，最后使用需要拦截的方法 Method 封装成 Invocation，并调用 Invocation 的 proceed 方法。

例如 Executor 的执行大概是这样的流程：

Executor → Plugin → Interceptor → Invocation

Executor.Method → Plugin.invoke → Interceptor.intercept → Invocation.proceed → method.invoke

4.3.10 实现分页接口

本节介绍如何使用 MyBatis 分页插件 PageHelper 实现分页查询接口。

我们已经在 build.gradle 配置文件中添加了 pagehelper-spring-boot-starter 依赖。下面我

们直接通过代码实例来讲解。

CategoryMapper.java 代码如下：

```
public interface CategoryMapper {
    ...
    @Select("SELECT * FROM category")
    List<Category> findAll();
}
```

就是这里的 SQL 语句：

```
SELECT * FROM category
```

后面会被 PageHelper 拦截，通过 Plugin 的动态代理机制在后面添加上 limit 等子句。CategoryService.kt 代码如下：

```
interface CategoryService {
    fun page(pageNo: Int, pageSize: Int): PageInfo<Category>
}
```

其中，PageInfo 对象中定义的属性字段如下：

```
public class PageInfo<T> implements Serializable {
    private static final long serialVersionUID = 1L;
    //当前页
    private int pageNum;
    //每页的数量
    private int pageSize;
    //当前页的数量
    private int size;

    //由于 startRow 和 endRow 不常用，这里说个具体的用法
    //可以在页面中 "显示 startRow 到 endRow 共 size 条数据 "

    //当前页面第一个元素在数据库中的行号
    private int startRow;
    //当前页面最后一个元素在数据库中的行号
    private int endRow;
    //总记录数
    private long total;
    //总页数
    private int pages;
    //结果集
    private List<T> list;

    //前一页
    private int prePage;
    //下一页
    private int nextPage;

    //是否为第一页
    private boolean isFirstPage = false;
```

```
    //是否为最后一页
    private boolean isLastPage = false;
    //是否有前一页
    private boolean hasPreviousPage = false;
    //是否有下一页
    private boolean hasNextPage = false;
    //导航页码数
    private int navigatePages;
    //所有导航页号
    private int[] navigatepageNums;
    //导航条上的第一页
    private int navigateFirstPage;
    //导航条上的最后一页
    private int navigateLastPage;

    public PageInfo() {
    }

    /**
     * 包装Page对象
     *
     * @param list
     */
    public PageInfo(List<T> list) {
        this(list, 8);
    }
    …
}
```

CategoryServiceImpl.kt 代码如下：

```
@Service
class CategoryServiceImpl : CategoryService {
    @Autowired lateinit var categoryMapper: CategoryMapper

    override fun page(pageNo: Int, pageSize: Int): PageInfo<Category> {
        // 设置分页参数
        PageHelper.startPage<Category>(pageNo, pageSize)
        val list = categoryMapper.findAll()
        // 使用PageInfo对象包装分页返回结果
        return PageInfo(list)
    }

}
```

CategoryController.kt 代码如下：

```
@RestController
@RequestMapping("/category")
class CategoryController {
    @Autowired lateinit var categoryService: CategoryService

    @GetMapping("/page")
```

```
    fun page(@RequestParam(value = "pageNo", defaultValue = "0") pageNo: Int,
             @RequestParam(value = "pageSize", defaultValue = "10") pageSize:
                 Int): PageInfo<Category> {
        return categoryService.page(pageNo, pageSize)
    }

}
```

重启应用，浏览器打开 http://127.0.0.1:8080/category/page?pageNo=0&pageSize=10，响应输出如下：

```
{
    "pageNum": 0,
    "pageSize": 10,
    "size": 10,
    "startRow": 1,
    "endRow": 10,
    "total": 3229,
    "pages": 323,
    "list": [
        {
            "id": 1,
            "code": "110",
            "detail": "",
            "gmtCreate": "2017-12-29T14:24:58.000+0000",
            "gmtModify": "2017-12-29T14:24:58.000+0000",
            "isDeleted": 0,
            "name": "数学",
            "type": 1
            },……
    ],
    "prePage": 0,
    "nextPage": 1,
    "isFirstPage": false,
    "isLastPage": false,
    "hasPreviousPage": false,
    "hasNextPage": true,
      ……
    "firstPage": 1,
    "lastPage": 8
}
```

4.3.11　PageHelper 工作原理

PageHelper 是通过 MyBatis 的 plugin 机制实现了 Interceptor 接口，这个过程如图 4-6 所示。

PageHelper 通过拦截器获取到同一线程中的预编译好的 SQL 语句后，将 SQL 语句重新包装成具有分页功能的 SQL 语句，例如 com.github.pagehelper.dialect.helper.MySqlDialect 这个类，关于这个逻辑的代码如下：

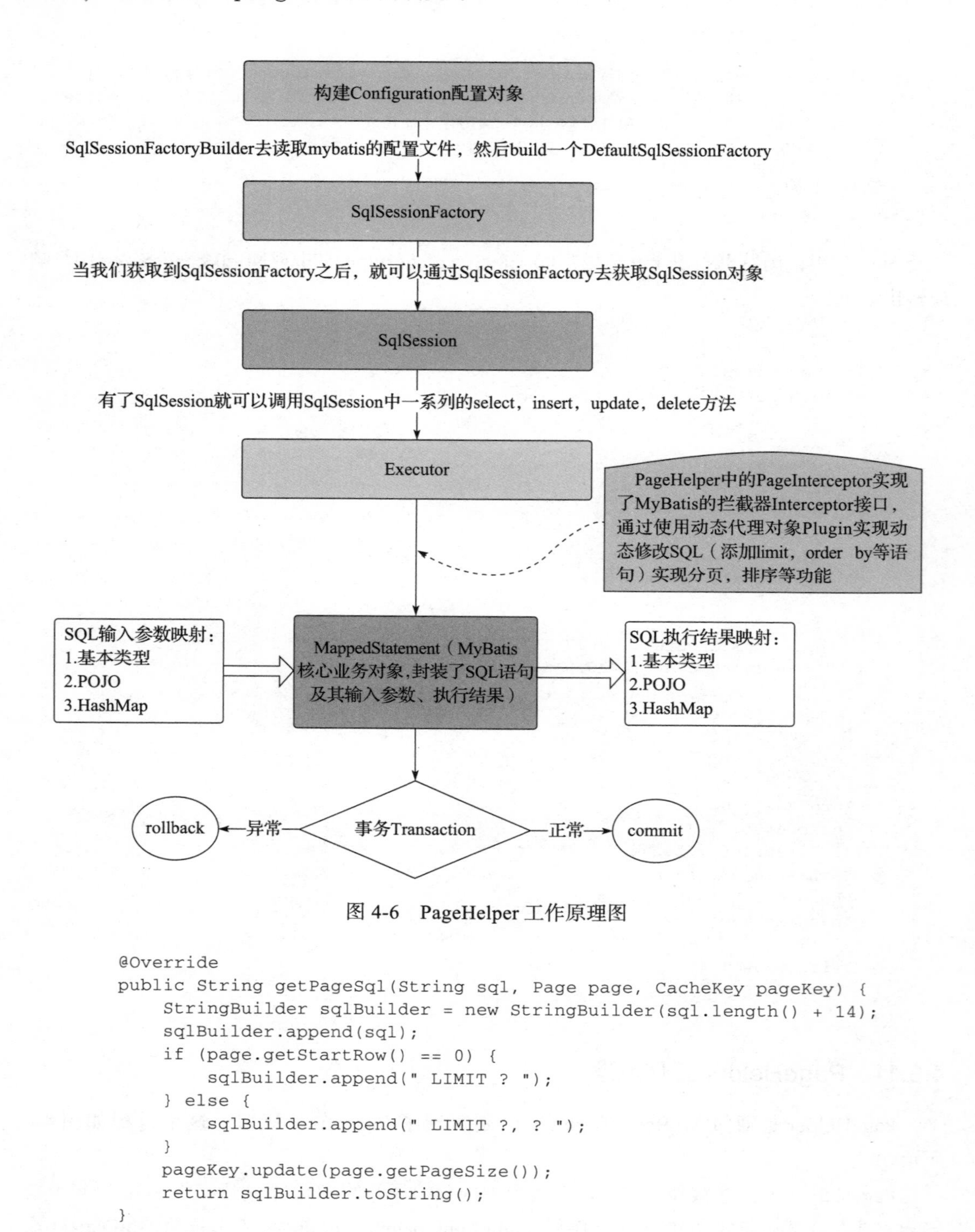

图 4-6 PageHelper 工作原理图

```
@Override
public String getPageSql(String sql, Page page, CacheKey pageKey) {
    StringBuilder sqlBuilder = new StringBuilder(sql.length() + 14);
    sqlBuilder.append(sql);
    if (page.getStartRow() == 0) {
        sqlBuilder.append(" LIMIT ? ");
    } else {
        sqlBuilder.append(" LIMIT ?, ? ");
    }
    pageKey.update(page.getPageSize());
    return sqlBuilder.toString();
}
```

接着将返回的新 SQL 再次赋值给下一步操作，所以实际执行的 SQL 语句就是有了分页功能的 SQL 语句。

例如，这段代码：

```
override fun page(pageNo: Int, pageSize: Int): PageInfo<Category> {
    PageHelper.startPage<Category>(pageNo, pageSize)
    val list = categoryMapper.findAll()
    return PageInfo(list)
}
```

在真正执行 findAll 之前，会首先进入 org.apache.ibatis.binding.MapperProxy 的如下方法：

```
invoke(Object proxy, Method method, Object[] args)
```

Debug 现场如下面的系列图所示：

```
getPageSql:80, MySqlDialect (com.github.pagehelper.dialect.helper)
getPageSql:181, AbstractHelperDialect (com.github.pagehelper.dialect)
getPageSql:97, PageHelper (com.github.pagehelper)
intercept:129, PageInterceptor (com.github.pagehelper)
invoke:61, Plugin (org.apache.ibatis.plugin)
query:-1, $Proxy87 (com.sun.proxy)
selectList:148, DefaultSqlSession (org.apache.ibatis.session.defaults)
selectList:141, DefaultSqlSession (org.apache.ibatis.session.defaults)
invoke0:-1, NativeMethodAccessorImpl (sun.reflect)
invoke:62, NativeMethodAccessorImpl (sun.reflect)
invoke:43, DelegatingMethodAccessorImpl (sun.reflect)
invoke:497, Method (java.lang.reflect)
invoke:433, SqlSessionTemplate$SqlSessionInterceptor (org.mybatis.spring)
selectList:-1, $Proxy73 (com.sun.proxy)
selectList:230, SqlSessionTemplate (org.mybatis.spring)
executeForMany:137, MapperMethod (org.apache.ibatis.binding)
execute:75, MapperMethod (org.apache.ibatis.binding)
invoke:59, MapperProxy (org.apache.ibatis.binding)
findAll:-1, $Proxy74 (com.sun.proxy)
page:16, CategoryServiceImpl (com.easy.springboot.demo_springboot_with_mybatis.service
page:20, CategoryController (com.easy.springboot.demo_springboot_with_mybatis.controlle
invoke0:-1, NativeMethodAccessorImpl (sun.reflect)
invoke:62, NativeMethodAccessorImpl (sun.reflect)
invoke:43, DelegatingMethodAccessorImpl (sun.reflect)
```

Debug 断点 1 进入 MapperProxy 类中的 invoke 方法，如图 4-7 所示。

Debug 断点 2 执行 sqlSessionProxy.<E> selectList（statement,parameter）方法，如图 4-8 所示。

Debug 断点 3 进入到 com.github.pagehelper.dialect.helper. MySqlDialect 类的 getPageSql 方法，就是在这个方法里动态添加了 LIMIT，如图 4-9 所示。

Debug 断点 4 进入到 MyBatis 通用分页拦截器 PageInterceptor 中的 intercept 方法。如图 4-10 所示。

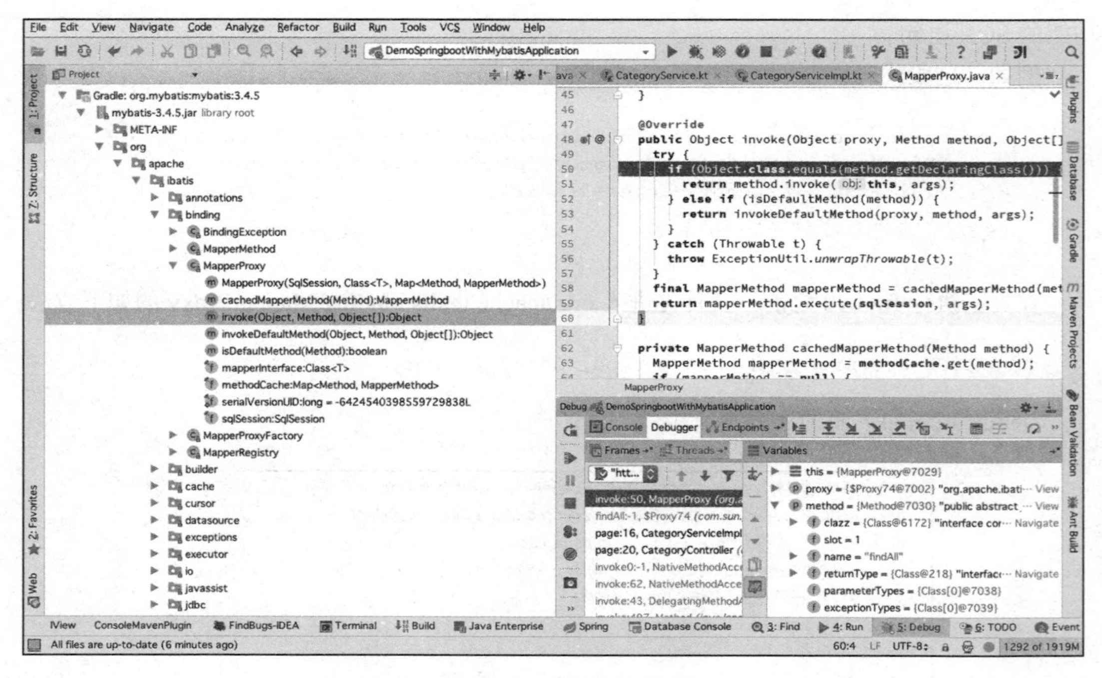

图 4-7 进入 MapperProxy 类中的 invoke 方法

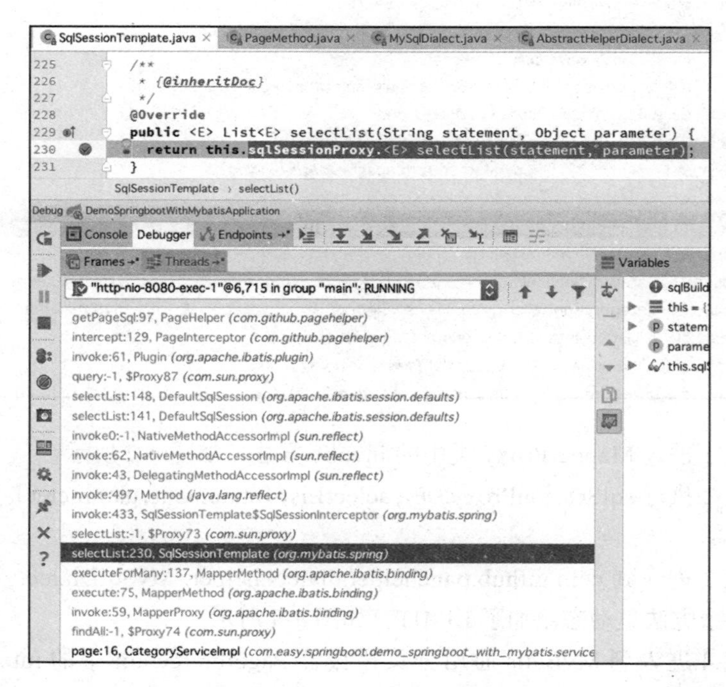

图 4-8 执行 sqlSessionProxy.<E> selectList（statement，parameter）方法

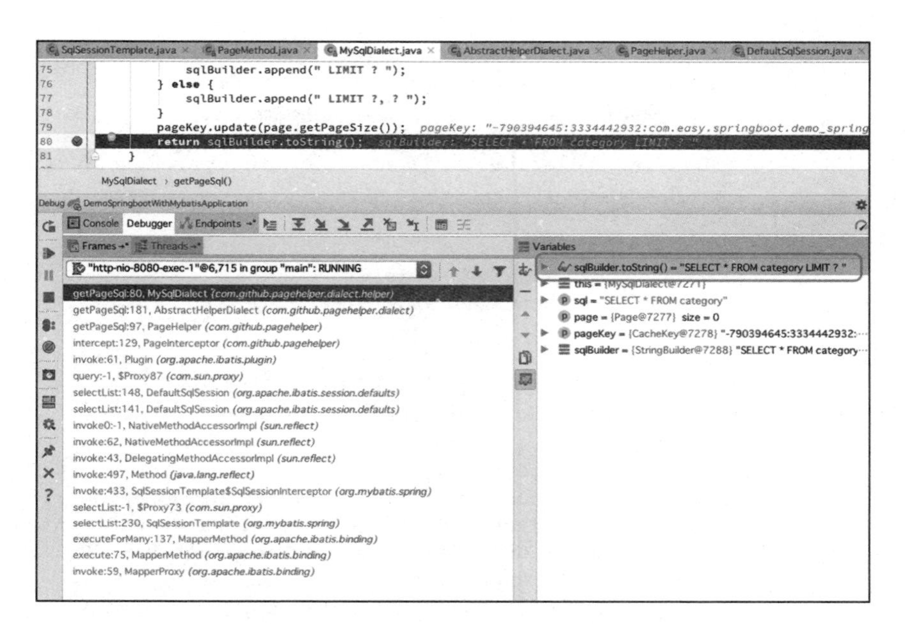

图 4-9　MySqlDialect 类的 getPageSql 方法

```
intercept:136, PageInterceptor (com.github.pagehelper)
invoke:61, Plugin (org.apache.ibatis.plugin)
query:-1, $Proxy87 (com.sun.proxy)
selectList:148, DefaultSqlSession (org.apache.ibatis.session.defaults)
selectList:141, DefaultSqlSession (org.apache.ibatis.session.defaults)
invoke0:-1, NativeMethodAccessorImpl (sun.reflect)
invoke:62, NativeMethodAccessorImpl (sun.reflect)
invoke:43, DelegatingMethodAccessorImpl (sun.reflect)
invoke:497, Method (java.lang.reflect)
invoke:433, SqlSessionTemplate$SqlSessionInterceptor (org.mybatis.spring)
selectList:-1, $Proxy73 (com.sun.proxy)
selectList:230, SqlSessionTemplate (org.mybatis.spring)
executeForMany:137, MapperMethod (org.apache.ibatis.binding)
```

图 4-10　PageInterceptor 中的 intercept 方法

这个核心的 intercept 方法的代码如下：

```
public Object intercept(Invocation invocation) throws Throwable {
    try {
        Object[] args = invocation.getArgs();
        MappedStatement ms = (MappedStatement) args[0];
        Object parameter = args[1];
        RowBounds rowBounds = (RowBounds) args[2];
        ResultHandler resultHandler = (ResultHandler) args[3];
        Executor executor = (Executor) invocation.getTarget();
        CacheKey cacheKey;
        BoundSql boundSql;
```

```
        //由于逻辑关系，只会进入一次
        if(args.length == 4){
            //4 个参数时
            boundSql = ms.getBoundSql(parameter);
            cacheKey = executor.createCacheKey(ms, parameter, rowBounds,
                boundSql);
        } else {
            //6 个参数时
            cacheKey = (CacheKey) args[4];
            boundSql = (BoundSql) args[5];
        }
        List resultList;
        //调用方法判断是否需要进行分页，如果不需要，直接返回结果
        if (!dialect.skip(ms, parameter, rowBounds)) {
            //反射获取动态参数
            String msId = ms.getId();
            Configuration configuration = ms.getConfiguration();
            Map<String, Object> additionalParameters = (Map<String,
                Object>) additionalParametersField.get(boundSql);
            //判断是否需要进行 count 查询
            if (dialect.beforeCount(ms, parameter, rowBounds)) {
                String countMsId = msId + countSuffix;
                Long count;
                //先判断是否存在手写的 count 查询
                MappedStatement countMs = getExistedMappedStatement
                    (configuration, countMsId);
                if(countMs != null){
                    count = executeManualCount(executor, countMs, para
                        meter, boundSql, resultHandler);
                } else {
                    countMs = msCountMap.get(countMsId);
                    //自动创建
                    if (countMs == null) {
                        //根据当前的 ms 创建一个返回值为 Long 类型的 ms
                        countMs = MSUtils.newCountMappedStatement(ms,
                            countMsId);
                        msCountMap.put(countMsId, countMs);
                    }
                    count = executeAutoCount(executor, countMs,
                        meter, boundSql, rowBounds, resultHandler);
                }
                //处理查询总数
                //返回 true 时继续分页查询，false 时直接返回
                if (!dialect.afterCount(count, parameter, rowBounds)) {
                    //当查询总数为 0 时，直接返回空的结果
                    return dialect.afterPage(new ArrayList(), rowBounds);
                }
            }
            //判断是否需要进行分页查询
            if (dialect.beforePage(ms, parameter, rowBounds)) {
                //生成分页的缓存 key
```

```
                    CacheKey pageKey = cacheKey;
                    //处理参数对象
                    parameter = dialect.processParameterObject(ms, parameter,
                        boundSql, pageKey);
                    //调用方言获取分页 sql
                    String pageSql = dialect.getPageSql(ms, boundSql,parameter,
                        rowBounds, pageKey);
                    BoundSql pageBoundSql = new BoundSql(configuration,
                        pageSql, boundSql.getParameterMappings(), parameter);
                    //设置动态参数
                    for (String key : additionalParameters.keySet()) {
                        pageBoundSql.setAdditionalParameter(key, additional
                            Parameters.get(key));
                    }
                    //执行分页查询
                    resultList = executor.query(ms, parameter, RowBounds.
                        DEFAULT, resultHandler, pageKey, pageBoundSql);
                } else {
                    //不执行分页的情况下，也不执行内存分页
                    resultList = executor.query(ms, parameter, RowBounds.
                        DEFAULT, resultHandler, cacheKey, boundSql);
                }
            } else {
                //rowBounds 用参数值，不使用分页插件处理时，仍然支持默认的内存分页
                resultList = executor.query(ms, parameter, rowBounds,
                    resultHandler, cacheKey, boundSql);
            }
            return dialect.afterPage(resultList, parameter, rowBounds);
        } finally {
            dialect.afterAll();
        }
    }
```

Debug 断点 5 最后进入 org.mybatis.spring.SqlSessionTemplate$SqlSessionInterceptor 类的 invoke 方法代码中。在这个方法中返回了真正分页之后返回的 result。如图 4-11 所示。

上面的代码执行完毕后，会再次回到 MapperProxy.invoke 方法代码中。最终，回到 CategoryServiceImpl. page 方法中：

```
@Service
class CategoryServiceImpl : CategoryService {
    @Autowired lateinit var categoryMapper: CategoryMapper

    override fun page(pageNo: Int, pageSize: Int): PageInfo<Category> {
        PageHelper.startPage<Category>(pageNo, pageSize)
        val list = categoryMapper.findAll()
        return PageInfo(list)
    }

}
```

最终得到的 list 的值。

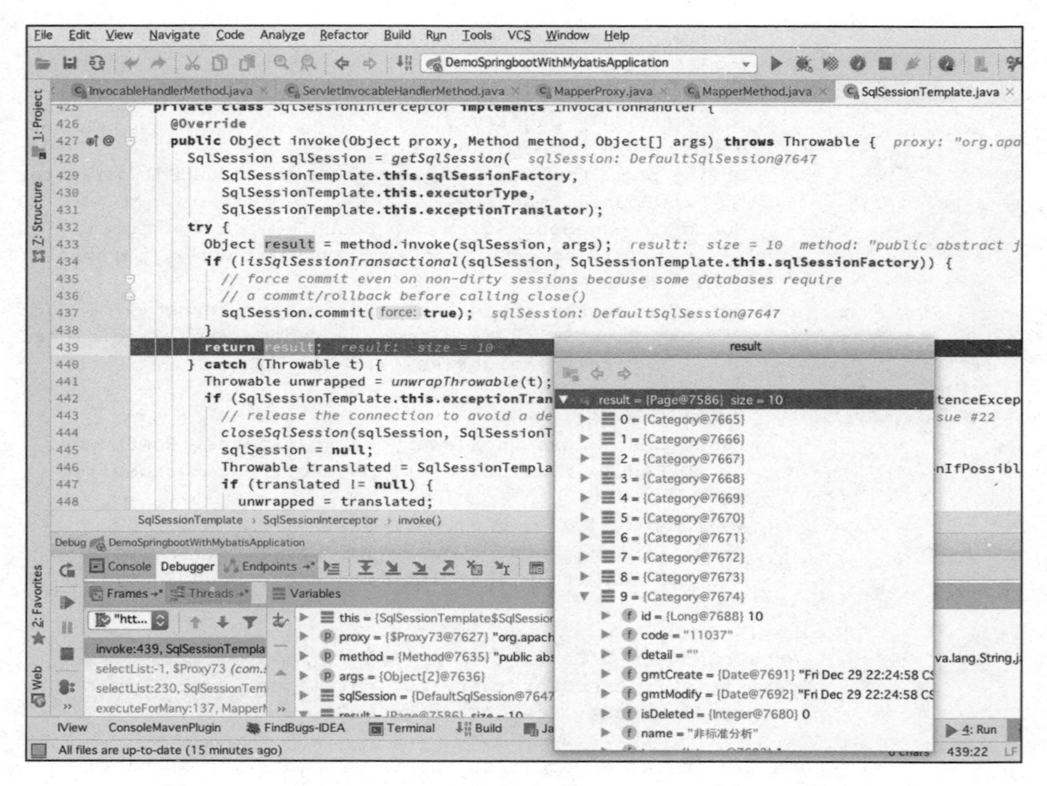

图 4-11 SqlSessionTemplate$SqlSessionInterceptor 类的 invoke 方法

4.3.12 多表关联查询级联

通常情况下，我们还会关联多个表进行查询。本节我们介绍如何使用 MyBatis 的 @Results、@Result 和 @Many 注解进行多表级联查询。

我们的数据库表结构如下。

user 表如下：

Field	Type	Null	Key	Default	Extra
id	bigint(20)	NO	PRI	<null>	auto_increment
password	varchar(255)	YES		<null>	
username	varchar(50)	YES	UNI	<null>	

role 表如下：

Field	Type	Null	Key	Default	Extra
id	bigint(20)	NO	PRI	<null>	auto_increment
role	varchar(50)	YES	UNI	<null>	

user_roles 表如下：

Field	Type	Null	Key	Default	Extra
user_id	bigint(20)	NO	PRI	<null>	
roles_id	bigint(20)	NO	PRI	<null>	

user 表如下：

id	password	username
1	$2a$10$pTIiWjYO6dvod…	user
2	$2a$10$gcDHybMlVmV6b…	admin

role 表如下：

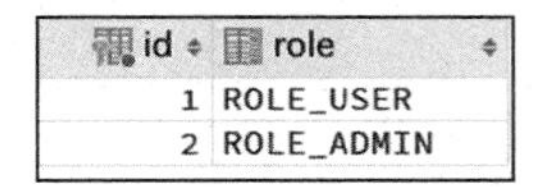

id	role
1	ROLE_USER
2	ROLE_ADMIN

user_roles 表如下：

user_id	roles_id
1	1
2	1
2	2

UserDto.kt 代码如下：

```
class UserDto: User() {
    lateinit var roles: List<Role>
}
```

UserMapper.java 代码如下：

```
public interface UserMapper {
    …
    @Select("SELECT * FROM user")
    @Results({
            @Result(id = true, column = "id", property = "id"),
            @Result(column = "username", property = "username"),
            @Result(column = "password", property = "password"),
            @Result(column = "id", property = "roles",
                    many = @Many(
                            select = "com.easy.springboot.demo_springboot_with_my
                               batis.dao.RoleMapper.listByUserId",
                            fetchType = FetchType.EAGER)),
    })
    List<UserDto> listUserDto();

}
```

其中，@Many 注解中的 RoleMapper.listByUserId（Long userId）方法的入参由 column = "id" 指定的来自 User 表对应的 User 对象的 id。关联的 SQL 是：

```
@Select("SELECT * FROM user")
```

RoleMapper.listByUserId（Long userId）代码如下：

```
    @Select("SELECT r.* FROM role r \n" +
            "JOIN user_roles ur ON r.id = ur.roles_id\n" +
            "JOIN user u ON u.id = ur.user_id\n" +
            "WHERE u.id=#{userId}")
    List<Role> listByUserId(Long userId);
```

UserService.kt 和 UserServiceImpl.kt 代码如下：

```
interface UserService {
    fun findAll(): List<User>
    fun listUserDto(): List<UserDto>
}

@Service
class UserServiceImpl : UserService {

    @Autowired lateinit var userMapper: UserMapper
    override fun findAll(): List<User> {
        return userMapper.findAll()
    }

    override fun listUserDto(): List<UserDto> {
        return userMapper.listUserDto()
    }

}
```

UserController.kt 代码如下：

```
@RestController
@RequestMapping("/user")
class UserController {
    @Autowired lateinit var userService: UserService

    @GetMapping("/findAll")
    fun findAll(): List<User> {
        return userService.findAll()
    }

    @GetMapping("/listUserDto")
    fun listUserDto(): List<UserDto> {
        return userService.listUserDto()
    }

}
```

浏览器打开 http://127.0.0.1:8080/user/listUserDto，可以看到响应输出如下：

```
[
    {
        "id": 1,
        "password": "$2a$10$pTIiWjYO6dvod5YTGTIFyuXSJXmeU3PBSSUPOBMeSSqBEsTJ7PmW2",
        "username": "user",
        "roles": [
            {
                "id": 1,
                "role": "ROLE_USER"
            }
        ]
    },
    {
        "id": 2,
        "password": "$2a$10$gcDHybMlVmV6bZahIUev6.M4CgAbW83RdDSoxaGGdjgzGRskLYVy6",
        "username": "admin",
        "roles": [
            {
                "id": 1,
                "role": "ROLE_USER"
            },
            {
                "id": 2,
                "role": "ROLE_ADMIN"
            }
        ]
    }
]
```

@Many 注解

MyBatis 中的 @Many 注解表示一对多的关系。这个注解的定义如下：

```
@Documented
@Retention(RetentionPolicy.RUNTIME)
@Target({})
public @interface Many {
    String select() default "";
    FetchType fetchType() default FetchType.DEFAULT;}
```

在上面的多表级联查询的代码中：

```
    @Select("SELECT * FROM user")
    @Results({
            @Result(id = true, column = "id", property = "id"),
            @Result(column = "username", property = "username"),
            @Result(column = "password", property = "password"),
            @Result(column = "id", property = "roles",
            many = @Many(select =    "com.easy.springboot.demo_spring
                boot_with_mybatis.dao.RoleMapper.
listByUserId",fetchType = FetchType.EAGER)),
```

```
    })
    List<UserDto> listUserDto();
```

我们使用了 @Many 注解的 select 属性来指向一个完全限定名方法：

```
com.easy.springboot.demo_springboot_with_mybatis.dao.RoleMapper.listByUserId
```

该方法将返回一个 List 对象，赋值给 property = "roles" 这个属性。也就是对应 List<UserDto> listUserDto() 方法返回值类型 UserDto 中的 roles 属性。使用 column="id"，把 user 数据表中的 id 列值作为输入参数传递给 RoleMapper.listByUserId 方法。

4.4 本章小结

使用 MyBatis 可以非常灵活地实现 SQL，同时有非常实用的 MyBatis Generator 代码生成工具和简单方便的分页插件 PageHelper 帮助开发者完成工作。综合使用基于 XML 的 Mapper 文件和注解的方式，可使 dao 层代码的开发更加高效。

但是，由于 MyBatis 毕竟还是“半自动化”的 ORM 框架，在一些简单通用的场景下（例如，没有复杂条件查询），我们更加愿意去使用“自动化”的 ORM 框架。

在下一章中我们将介绍如何集成 Spring Boot 和 Spring Data JPA 进行数据库层的开发。

本章示例项目源代码 https://github.com/KotlinSpringBoot/demo_springboot_with_mybatis

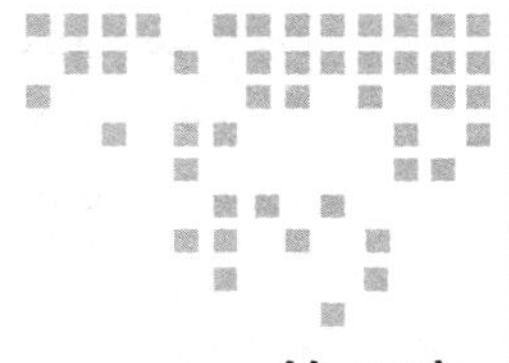

第 5 章　Chapter 5

Spring Boot 集成 JPA 数据库层开发

在上一章中，我们使用 MyBatis 实现了高效开发。

本章将介绍一个更加自动化、封装“更高级”的框架——Spring Data JPA。本章先简单介绍了 PA 的基本概念及架构，然后通过实战介绍集成 Spring Data JPA 进行数据方层的开发。

5.1　JPA 简介

JPA(Java Persistence API）是 Sun 官方提出的 Java 持久化规范（JSR 338: JavaTM Persistence2.2，这些接口的命名包空间是 javax.persistence.*，详细内容参考：https://github.com/javaee/jpa-spec）。

JPA 的出现主要是为了简化持久化开发工作和整合 ORM 技术，结束 Hibernate、TopLink、JDO 等 ORM 框架各自为营的局面。JPA 是在充分吸收了现有 ORM 框架的基础上发展而来的，易于使用，伸缩性强。

JPA 支持面向对象的高级特性，如类之间的继承、多态和类之间的复杂关系，这使得开发者能最大限度地使用面向对象的模型设计企业应用，而不需要自行处理这些特性在关系数据库中的持久化。

JPA 整体架构如图 5-1 所示。

JPA 架构中的核心组成参见表 5-1。

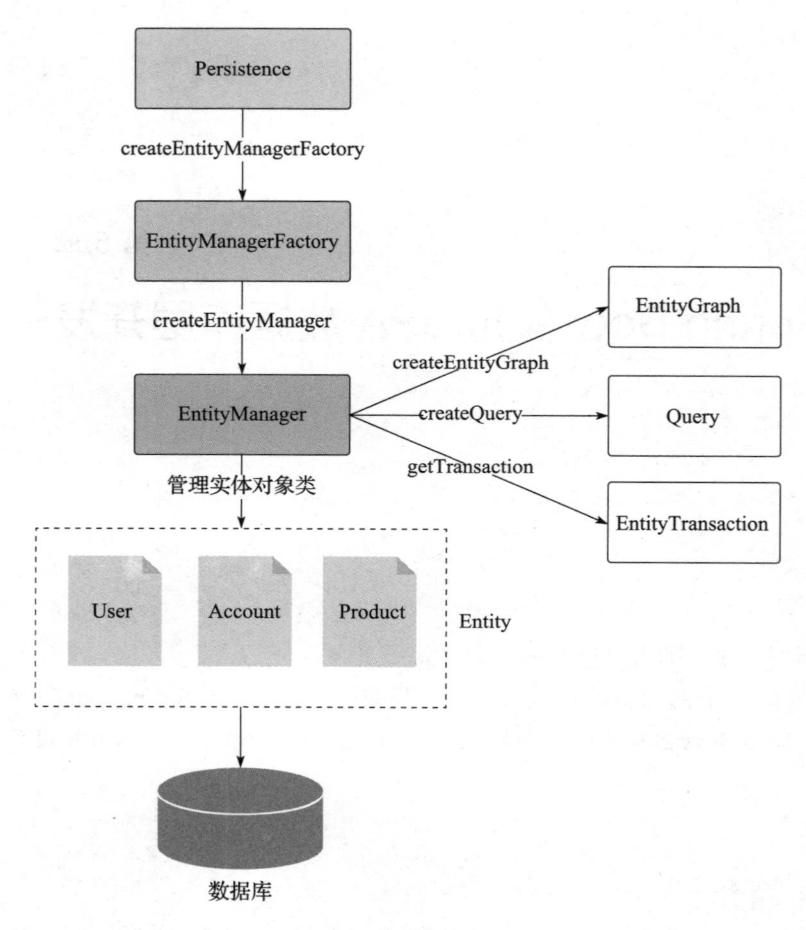

图 5-1 JPA 架构

表 5-1 JPA 的核心组成

类	描述
Persistence	这个类包含静态方法来获取 EntityManagerFactory 实例
EntityManagerFactory	EntityManager 的工厂类，创建并管理多个 EntityManager 实例
EntityManager	这是一个接口，它管理持久化操作的对象，它的工作原理类似工厂的查询实例
Entity	Entity 是持久性对象，是存储在数据库中的记录
EntityGraph	是指定查找操作或查询路径和边界的更好的解决方案（JPA2.1 中引入）
Query	连接每个 JPA 供应商，能够获得符合标准的关系对象
EntityTransaction	它与 EntityManager 是一对一的关系，对于每一个 EntityManager，操作是由 EntityTransaction 类维护的

其中：

- EntityManagerFactory 和 EntityManager 的关系是一对多。

- EntityManager 和 EntityTransaction 之间的关系是一对一。对于每个 EntityManager 操作，只有一个 EntityTransaction 实例。
- EntityManager 和 Query 之间的关系是一对多。使用一个 EntityManager 实例可以执行多个查询。
- EntityManager 和 Entity 之间的关系是一对多。一个 EntityManager 实例可以管理多个实体。

实例对象的生命周期有以下 4 种状态：

- New：新创建的实例对象，没有 identity 值。
- Managed：在持久化上下文中受管理的对象。
- Detached：游离于持久化上下文之外的实例对象。
- Removed：被删除的实例对象。

JPA 中详细定义了随着 EntityManager 的各种操作时，实例对象生命周期状态间的转换关系。JPA 中定义了 4 种实体之间的关联关系 OneToOne、OneToMany、ManyToOne、ManyToMany，提供了对应的注解。后面的项目实战将详细介绍。

5.1.1 JPA 生态

JPA 的主要设计者是 Hibernate 的设计者。JPA 是一种规范，不是产品，而 Hibernate 是一种 ORM 技术的产品。JPA 有点像 JDBC，为各种不同的 ORM 技术提供一个统一的接口，方便把应用移植到不同的 ORM 技术上。

低耦合一直是开发人员在软件设计上追求的目标，使用 JPA 可以把应用完全从 Hibernate 中解脱出来了。JPA 的宗旨是为 POJO 提供持久化标准规范，能够脱离容器独立运行，方便开发和测试。Hibernate3.2+、TopLink10.1.3 以及 OpenJPA 都提供了 JPA 的实现。

Spring Data JPA 在 JPA 规范的基础下提供了 Repository 层的实现。虽然 ORM 框架都实现了 JPA 规范，但是在不同 ORM 框架之间切换要编写的代码会有差异，而 Spring Data JPA 封装了这些差异性。并且 Spring Data JPA 对 Repository 层封装得很好，代码写起来更加简洁。

另外 Spring Data JPA 并不是一个标准意义上的框架，Spring Data JPA 只是简化了对 JPA 的使用，并没有实现 JPA 规范。JPA 本身就是一个规范，是一组接口，要用的话还是需要结合 JPA 的实现一起用，比如 Hibernate，或者其他的 JPA 实现（Apache OpenJPA、TopLink、JDO 等）。JPA 不是一种新的 ORM 框架，JPA 的出现是为了规范现有的 ORM 技术，而不是取代现有的 Hibernate、TopLink 等 ORM 框架。JPA 就和 JDBC 一样，提供一种通用的访问各个 ORM 实现产品的桥梁工具。

Spring Data JPA、JPA 以及 ORM 框架之间的关系如图 5-2 所示。

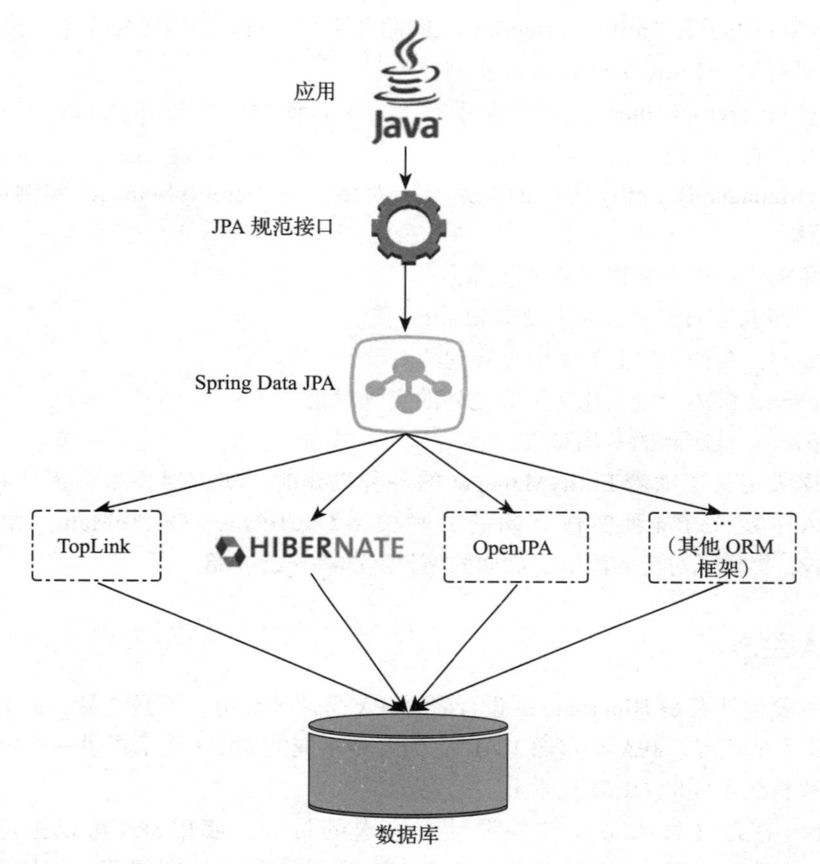

图 5-2 Spring Data JPA、JPA 以及 ORM 框架之间的关系

5.1.2 JPA 技术栈

JPA 的总体思想和现有 Hibernate、TopLink、JDO 等 ORM 框架大体一致。总的来说，JPA 包括以下 3 方面的技术：

- ORM 映射元数据——JPA 支持 XML 和注解两种元数据的形式。元数据描述对象和表之间的映射关系，框架据此将实体对象持久化到数据库表中。在 Spring Boot 集成 Spring Data JPA 的过程中，我们使用 Hibernate 的实现，同时采用基于注解的方式编写 JPQL（或原生 SQL）查询语句。
- Java 持久化 API——JPA 提供的持久化 API，用来操作实体对象，执行 CRUD 操作，框架在后台替我们完成所有的事情，让我们从烦琐的 JDBC 和 SQL 代码中解脱出来。
- JPA 查询语言——JPQL（Java Persistence Query Language）是 JPA 的结构化查询语言，类似于 HQL，都是通过面向对象而非面向数据库的查询语言去查询数据，从而实现数据库表名称跟实体名称 EntityName 的解耦。

5.2 ORM 框架概述

目前 ORM 框架的产品非常多，除了几个大公司、组织的产品外，其他一些小团队也在推出自己的 ORM 框架。目前流行的 ORM 框架有如下这些产品：

- Entity EJB　这是一直备受争议的组件技术。事实上，EJB 为 Java EE 的蓬勃发展赢得了极高的声誉，EJB 作为一种重量级、高花费的 ORM 技术具有不可比拟的优势。EJB3.1 采取了低侵入式的设计，增加了 Annotation，具有极大的吸引力。
- Hibernate　是 JBoss 的持久层解决方案，整个 HIbernate 项目投入了 Jboss 的怀抱，而 JBoss 又加入了 RedHat 组织，所以现在 Hibernate 属于 RedHat 的一部分。Hibernate 灵巧的设计、优秀的性能，还有其丰富的文档都是其风靡全球的重要因素。
- MyBatis　Apache 软件基金组织的子项目。与其说它是一种 ORM 框架，不如说它是一种“ SQL Mapping”框架。曾经在 J2EE 的开发中扮演非常重要的角色，但因为不支持存粹的面向对象操作，因此现在逐渐被取代。但是在一些公司中，它依然占有一席之地，特别是一些对数据访问特别灵活的地方。iBatis 更加灵活，它允许开发人员直接编写 SQL 语句。
- TopLink　Oracle 公司的产品，作为一个遵循 OTN 协议的商业产品，TopLink 在开发过程中可以自由地下载和使用，但是一旦用于商业产品，则需要收取费用。由于这一点，TopLink 的市场占有率不高。
- OBJ　Apache 软件基金组织的子项目。另一个开源的 ORM 框架，可以说是 Apache 作为 iBatis 之后的取代产品，也是非常优秀的 O/R Mapping 框架，但是由于 Hibernate 的光芒太盛，所以并未广泛使用，而且由于 OJB 的开发文档不是很多，这也影响了 OJB 的流行。

5.3 Hibernate 简介

JPA 需要用 Provider 来实现功能，Hibernate 就是 JPA Provider 中很强的一个。可以简单的理解为 JPA 是标准接口，Hibernate 是实现。

Hibernate 框架主要是通过三个组件来实现：hibernate-annotation、hibernate-entitymanager 和 hibernate-core，如下所示：

- hibernate-annotation　是 Hibernate 支持 annotation 方式配置的基础，包括标准的 JPA annotation 以及 Hibernate 自身特殊功能的 annotation。
- hibernate-core　是 Hibernate 的核心实现，提供了 Hibernate 所有的核心功能。
- hibernate-entitymanager　实现了标准的 JPA，可以把它看成 hibernate-core 和 JPA 之间的适配器，它并不直接提供 ORM 的功能，而是对 hibernate-core 进行封装，使得 Hibernate 符合 JPA 的规范。

总的来说，JPA 是持久化规范，而 Hibernate 框架实现了 JPA。

Hibernate 整体架构如图 5-3 所示。

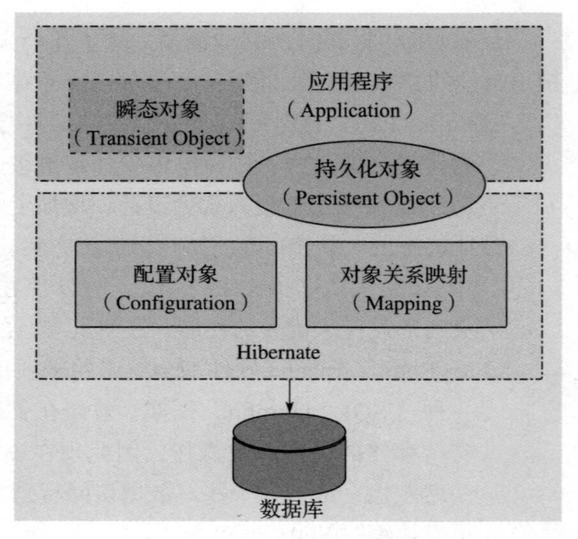

图 5-3 Hibernate 整体架构

Hibernate 核心模块介绍如下。

- 瞬态对象（Transient Object） 表示与当前会话不相关的持久类的实例。持久化对象与会话关联，一旦会话关闭，它们将被分离，并可以在任何应用层中自由使用。
- 配置对象（Configuration） Configuration 表示使用 Hibernate（可以是属性文件 hibernate.properties 或 XML 文件 hibernate.cfg.xml）所需的配置。Configuration 通常在应用程序初始化期间创建一次，用来连接到数据库，并创建 SessionFactory。代码示例如下：

```
Configuration config = new Configuration().configure();
                                        // 默认读取 resource 目录下的 hibernate.cfg.xml
Configuration config = new Configuration().configure("hibernate.cfg.xml");
                                        // 指定配置文件
```

  现在这个 Configuration 对象已经包括所有 Hibernate 运行期的参数，通过 Configuration 实例的 buildSessionFactory() 方法可以构建一个唯一的 SessionFactory：

```
SessionFactory sessionFactory = config.buildSessionFactory();
```

  另外，Configuration 实例是一个启动期间的对象，一旦 sessionFactory 创建完成就被丢弃了。

- 对象关系映射（Mapping） Java POJO 类和数据库表之间的映射是使用 XML 配置文件或基于注解来提供的。

Hibernate 详细架构如图 5-4 所示。下面简要介绍详细架构图中的核心组件。

1. 会话工厂（SessionFactory）

SessionFactory 负责初始化 Hibernate，它充当数据库源的代理，并负责创建 Session 对象。从名字就能推断出，这里用到了工厂模式。因为一般情况下，一个项目通常只需要一个 SessionFactory 就够，当需要操作多个数据库时，可以为每个数据库指定一个 SessionFactory。它是线程安全的。

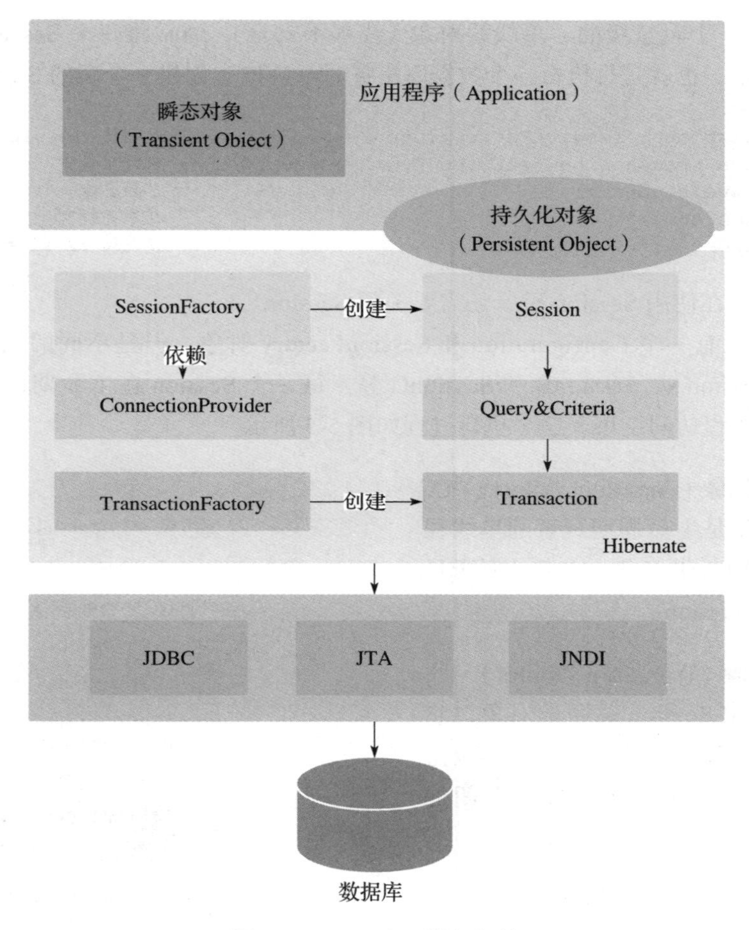

图 5-4　Hibernate 详细架构

创建一个 SessionFactory 一般有两种方式：

- 从 XML 文件读取配置信息构建 SessionFactory。
- 从 Java 属性文件读取配置信息构建 SessionFactory。

在 Spring Boot 集成 Spring Data JPA（默认使用 Hibernate ORM 实现）的时候，自动配置采用基于 JavaConfig 注解的方式来创建 SessionFactory。

2. 会话（Session）

Hibernate 中的 Session 并不是 Java Web 应用程序中所说的 Session，一般把 HttpSession 对象称为用户会话，用来保存用户会话过程中的状态信息。而 Hibernate 中的 Session 用来表示应用程序和数据库的一次交互（会话），在这个 Session 中，包含了一般的持久化方法（CRUD），封装了 JDBC 连接、事务等。同时，Session 还保存了持久化对象的缓存，用于遍历实体对象图，或者表示查找对象。

Session 是一个轻量级的、单线程对象（线程不安全），通常将每个 Session 实例和一个数据库事务绑定，也就是每执行一个数据库事务，都应该先创建一个新的 Session 实例：

```
Session session = sessionFactory.openSession();      //创建 Session
Transaction transaction= session.beginTransaction(); //开启事务
session.save(student);                               //只要传一个 Student 的对象实例
transaction.commit();                                //事务提交
session.close();                                     //关闭 session
```

我们看到，在使用 Session 后，还需要关闭 Session。

整个系统只有一个 Configuration 和 SessionFactory 对象，但是不同的用户访问不同的 Session，如 SessionA、SessionB 和 SessionC 等。每一个 Session 在生命期内都可以进行多个操作，直到线程访问结束。这个执行过程如图 5-5 所示。

3. 持久化对象（Persistent Object，PO）

持久化对象是生命周期短暂的单线程对象，包含了持久化状态，它们从属于且仅从属于一个 Session。

4. 瞬态对象（Transient Object）

没有与特定 Session 关联的对象，它们可能是刚被程序实例化，还没来得及持久化的对象，或者是一个已经被关闭的 Session 所有实例化。

5. 事务工厂（TransationFactory）

事务实例的工厂，用于创建事务。对应用程序不可见，支持扩展实现。

6. 事务（Transaction）

提供事务操作的支持，是一个单线程对象，代表一次原子操作（表示一批不可分割的事务操作，成功则提交，失败则回滚），Hibernate 事务是对底层具体的 JDBC、JTA 以及 CORBA 事务的抽象。一个 Session 在某些情况下可能跨越多个事务。

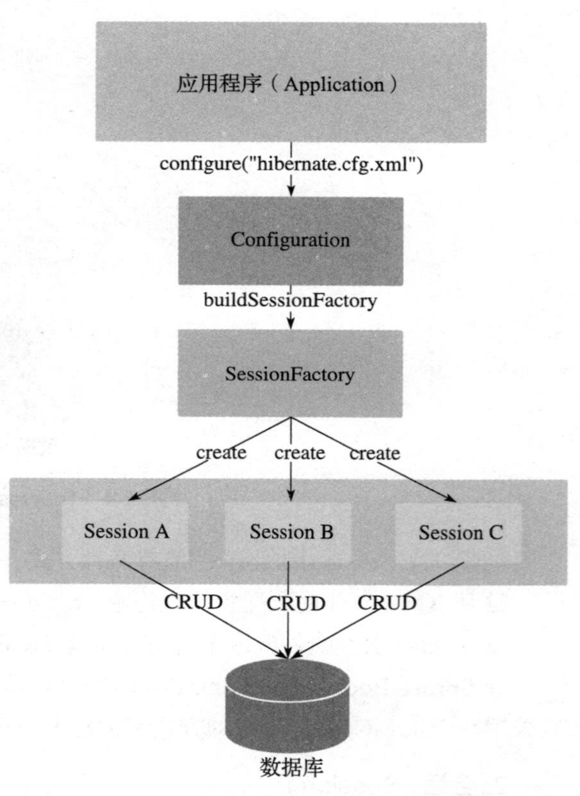

图 5-5 Hibernate Session 工作流程

7. 连接提供者（ConnectionProvider）

JDBC 连接池，从底层的 Datasource 或者 DriverManager 抽象而来。对应用程序不可见，但是支持扩展实现。

8. Query 和 Criteria 接口

负责执行各种数据库查询操作。Session 对象生成 Query 对象或者 Transaction 对象；可通过 Session 对象的 get()、load()、save()、update()、delete() 和 saveOrUpdate() 等方法对 PO 进行加载、保存、更新、删除等操作。

在查询的情况下，通过 Session 对象生成 Query 对象，然后利用 Query 对象执行查询操作；如果没有异常，Transaction 对象将提交这些操作结果到数据库中。如果有异常，Transaction 对象将执行回滚操作。

Hibernate 的主要优点如下。

- 面向对象的 ORM 编程。Hibernate 提供完整的对象 / 关系映射（Object/Relational Mapping）功能，以自然的面向对象方法思想（包括继承、多态、关联、组合和 Java 集合框架）来操作数据库表。它使用时只需要操纵对象，使开发更对象化，抛弃了数据库中心的思想，完全采用面向对象思想。
- 轻量级。Hibernate 的优点是 JDBC 的轻量级的对象封装（没有侵入性），它是一个独立的对象持久层框架。Hibernate 能用在任何 JDBC 能使用的场合，例如 dao 层代码。
- 移植性好。Hibernate 支持连接不同数据库的 SQL 方言。
- 缓存机制。提供一级缓存和二级缓存。
- 简洁的 HQL 编程。HQL 语法非常类似于 SQL 语法。SQL 的语法因为 SQL 很简单被广泛使用。SQL 直接针对关系数据库表、记录和字段，而 HQL 使用 Java 类和属性实例来编写 SQL 逻辑代码。
- 高性能。Hibernate 支持延迟初始化，提供多样化的数据获取策略、自动版本和时间戳的乐观锁定。
- 可靠。Hibernate 运行稳定性好且高质量。
- 可扩展性。Hibernate 是高度可配置和可扩展的。

事物都有两面性。Hibernate 的缺点如下：

- Hibernate 在大批量数据处理时有弱势。
- 针对单一对象简单的增删查改，适合用 Hibernate；而对于批量的修改、删除，不适合用 Hibernate；要使用数据库的特定优化机制的时候，不适合用 Hibernate。这个时候，我们通常就会使用 MyBatis 这样的“半自动化”ORM 框架。

从抽象层次去看，对于数据库的操作：Hibernate 是面向对象的，而 MyBatis 是面向关系的。两者要面对的领域和要解决的问题根本不同：面向对象致力于解决计算机逻辑问题，而关系模型致力于解决数据的高效存取问题。

提示　更多关于 Hibernate 的介绍可以参考：

- https://github.com/hibernate/hibernate-orm

- http://hibernate.org/orm/
- http://docs.jboss.org/hibernate/orm/current/quickstart/html_single/

5.4 Spring Data JPA 简介

Spring Data 项目是 Spring 的一个子项目，旨在统一和简化各类型持久化存储。Spring Data 使得访问数据变得简单，包括关系型和非关系型、并行计算框架、基于云的数据服务等。Spring Data JPA 是其中之一，本节主要介绍 Spring Data JPA。

随着 NoSQL 和 BigData 的兴起，出现了越来越多的新技术，比如非关系型数据库、Map-Reduce 框架等。Spring Data 项目的产生正是为了让 Spring 开发者能更方便地使用这些新技术，Spring Data 项目的生态如图 5-6 所示，这些项目分别为不同的技术提供支持，其主要功能参见表 5-2。

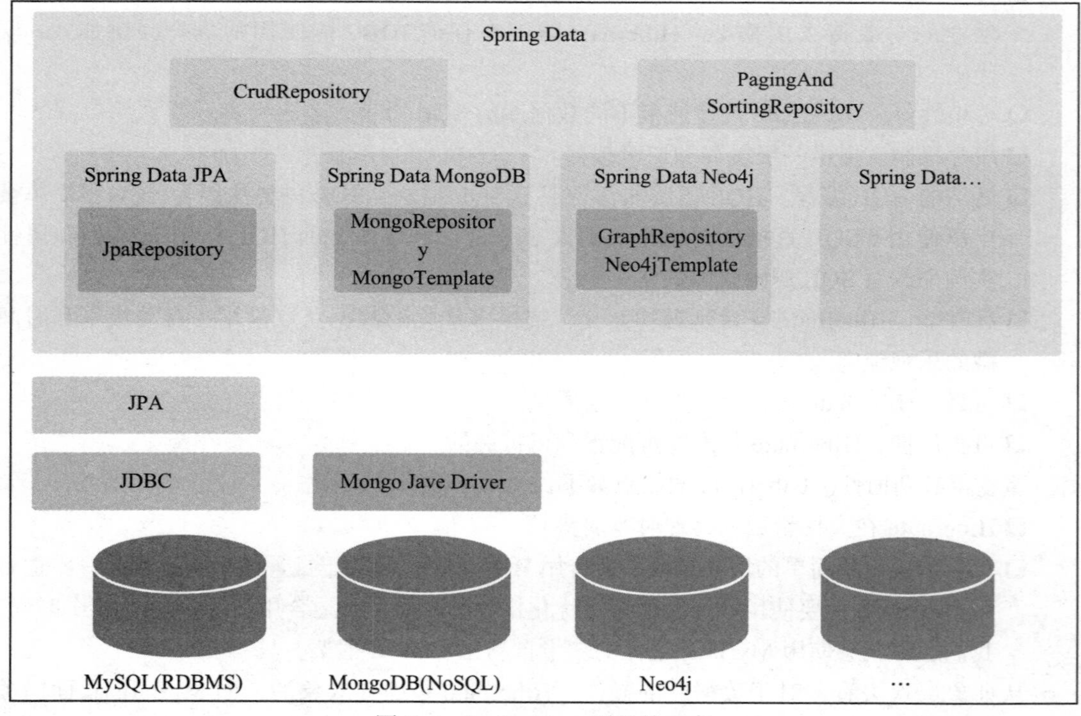

图 5-6 Spring Data 项目的生态

表 5-2 Spring Data 项目功能

项目	主要功能
Commons	提供共享基础框架，适合各个子项目使用，支持跨数据库持久化
GemFire	简化配置和访问 GemFire

（续）

项目	主要功能
JPA	简化基于 JPA 的 ORM 框架的使用，可以极简地创建数据访问 Repositories
Hadoop	提供基于 Spring 的 Hadoop 作业配置和一个 POJO 编程模型的 MapReduce 作业
KeyValue	支持 Map Repositories 和 SPI，简化键值存储操作。集成 Redis 和 Riak，提供多个常用场景下的简单封装
Document	集成文档数据库：CouchDB 和 MongoDB，并提供基本的配置映射和资料库支持
JDBC	提供 JDBC-based Repositories 的支持
JDBC Extensions	支持 Oracle RAD、高级队列和高级数据类型
LDAP	提供 Spring LDAP 的 Spring Data Repository 支持
MongoDB	提供基于 Spring 的 MongoDB 数据库访问的对象文档（object-document）存储的支持
REST	将 Spring Data Repositories 导出为超媒体驱动（hypermedia-driven）的 REST 资源
Redis	提供简化配置和使用 Redis 的支持
Cassandra	集成支持 Apache Cassandra 的使用
Solr	集成支持 Apache Solr 的使用
Aerospike	集成支持 Aerospike 的使用
ArangoDB	集成支持 ArangoDB 的使用
Couchbase	集成支持 Couchbase 的使用
DynamoDB	集成支持 DynamoDB 的使用
Elasticsearch	集成支持 Elasticsearch 的使用
Hazelcast	集成支持 Apache Hazelcast 的使用
Jest	集成支持基于 Elasticsearch 的 Jest REST client 的使用
Neo4j	集成支持 Neo4j 的 object-graph 存储操作
Vault	集成支持 Spring Data KeyValue 的使用
Spring Content	将内容与 Spring Data 实体关联，并将其存储在多个不同的存储区中，包括文件系统、S3、数据库或 Mongo 的 GridFS

通过 Spring Data，开发者可以用 Spring 提供的相对一致的方式来访问位于不同类型的数据存储中的数据。

无论是哪种持久化存储，数据访问对象（Data Access Objects，DAO）通常都会提供实体对象的 CRUD 操作、查询、排序和分页方法等。而 Spring Data 则提供了基于这些层面的统一接口（CrudRepository 和 PagingAndSortingRepository）以及对持久化存储的实现。这些接口定义在 spring-data-commons 中。

其中 CrudRepository 的接口如下：

```
CrudRepository
    count():long
    delete(T):void
    deleteAll():void
    deleteAll(Iterable<? extends T>):void
    deleteById(ID):void
    existsById(ID):boolean
```

```
findAll():Iterable<T>
findAllById(Iterable<ID>):Iterable<T>
findById(ID):Optional<T>
save(S):S
saveAll(Iterable<S>):Iterable<S>
```

PagingAndSortingRepository 的接口如下：

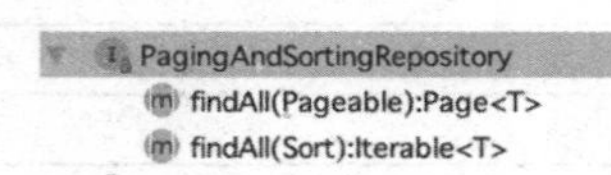

Spring Data JPA 是 Spring Data 的子模块。Spring Data JPA 的目标是实现数据访问层代码的极简化，它极大的简化了基于 JPA 规范的 ORM 框架的使用。可以在几乎不用编写任何代码的情况下，实现对数据库的访问和基本 CRUD 操作，还包括分页、排序等常用的功能。

使用 Spring Data JPA，我们只需要通过实现 JPA 提供的 Repository 接口，就立即拥有了通用的 CRUD 以及分页、排序等 API 接口（对应的 JPA 接口实现框架，如 Hibernate，将会自动实现查询方法）。

如果还需要自定义个性化的查询方法，在接口中直接实现。除了支持面向对象的 JPQL 语句，还支持原生 SQL 的查询。不需要 XML 配置文件，完全使用注解。

5.5 项目实战

本节介绍如何使用 Spring Boot 集成 Spring Data JPA 进行数据库层的开发。

5.5.1 Spring Data JPA 提供的接口

Spring Data JPA 提供的接口如表 5-3 所示。

表 5-3 Spring Data JPA 提供的接口

接口	说明
Repository	最顶层的接口，是一个空的接口，目的是统一所有 Repository 的类型，且能让组件扫描的时候自动识别
CrudRepository	是 Repository 的子接口，提供 CRUD 的功能
PagingAndSortingRepository	是 CrudRepository 的子接口，添加分页和排序的功能
JpaRepository	是 PagingAndSortingRepository 的子接口，增加了一些实用的功能，比如：saveAll、findAll、deleteInBatch 等批量操作
JpaSpecificationExecutor	负责查询的接口
Specification	Spring Data JPA 提供的一个查询规范，要做复杂的查询，只需围绕这个规范来设置查询条件即可，其中提供了 where、and、not、or、toPredicate 等条件

Spring Data JPA 是 Spring 基于 ORM 框架、JPA 规范封装的一套 JPA 应用框架，可使开

发者用极简的代码实现对数据的访问和操作。它提供了包括增删改查等在内的常用功能，且易于扩展！学习并使用 Spring Data JPA 可以极大地提高开发效率。

关于 Spring Data JPA 的更多内容可参考：

spring-data-jpa 项目空间：https://github.com/spring-projects/spring-data-jpa

参考文档：https://docs.spring.io/spring-data/jpa/docs/2.0.1.RELEASE/reference/html/

5.5.2　创建项目

spring-boot-starter-data-jpa 就是 Spring Boot 集成 Spring Data JPA 的起步依赖。创建使用 Spring Boot JPA Starter 集成 Spring Boot 和 JPA 开发的项目的命令行如下：

```
$ mkdir demo_springboot_with_jpa
$ cd demo_springboot_with_jpa
$ spring init \
-d=web,data-jpa,mysql \
-b=2.0.0.M7 \
-a=demo_springboot_with_jpa \
-g=com.easy.springboot \
-j=1.8 \
-l=kotlin \
--build=gradle \
-n=demo_springboot_with_jpa \
-p=jar \
-x
```

执行上面的命令行，将得到一个 Spring Boot 工程。

5.5.3　配置数据库连接

在工程根目录下新建一个 app.sql 文件。此时 IDEA 会提示配置数据库 DataSource，选择 MySQL，如图 5-7 所示。

配置数据库连接如图 5-8 所示。

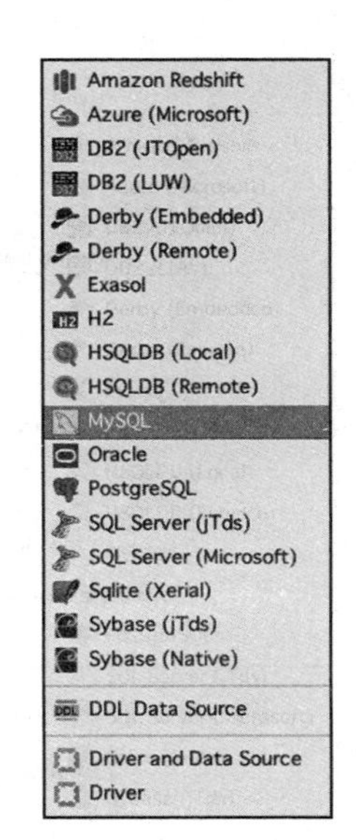

图 5-7　新建 MySQL 数据库连接

5.5.4　自动生成 Entity 实体类代码

配置完数据库连接后，可以在右侧的工具栏看到“Database”选项，如图 5-9 所示。

选中要自动生成实体类代码的表，依次选择 Scripted Extensions → Generate POJOs.groovy，如图 5-10 所示。

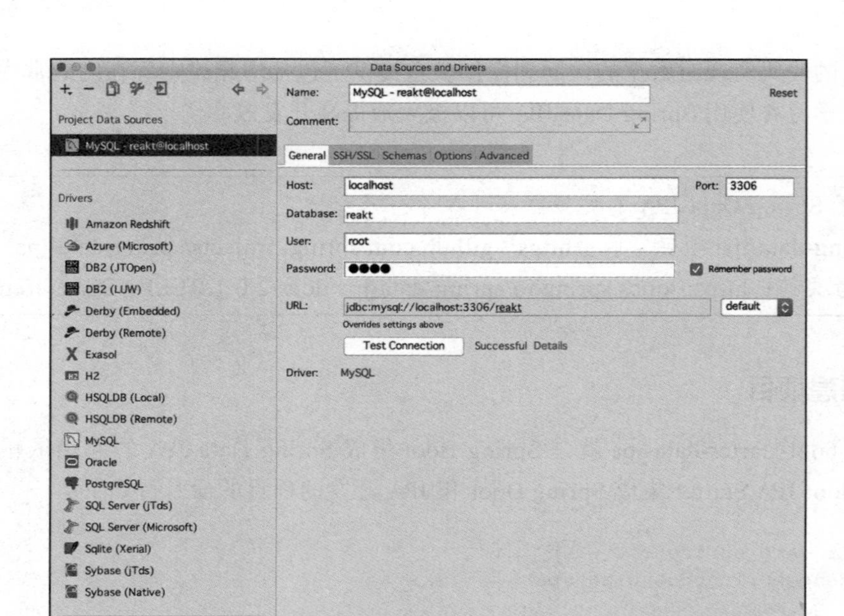

图 5-8 配置数据库连接

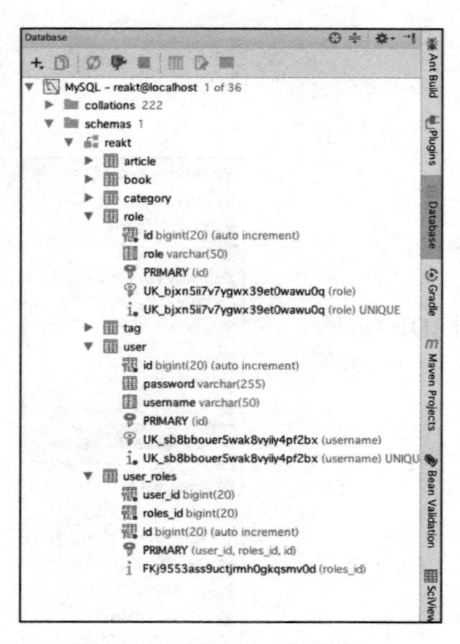

图 5-9 右侧的工具栏的“Database”选项

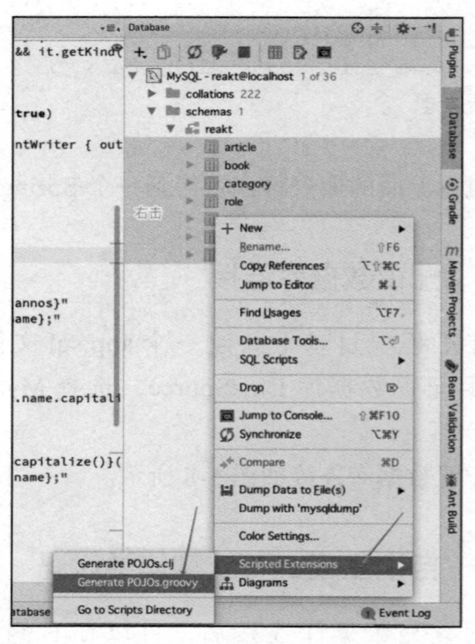

图 5-10 依次选择 Scripted Extensions > Generate POJOs.groovy

指定放置实体类代码目标的目录：/Users/jack/KotlinSpringBoot/demo_springboot_with_ jpa/src/main/kotlin/com/easy/springboot/demo_springboot_with_jpa/entity

生成的实体类代码如下所示。

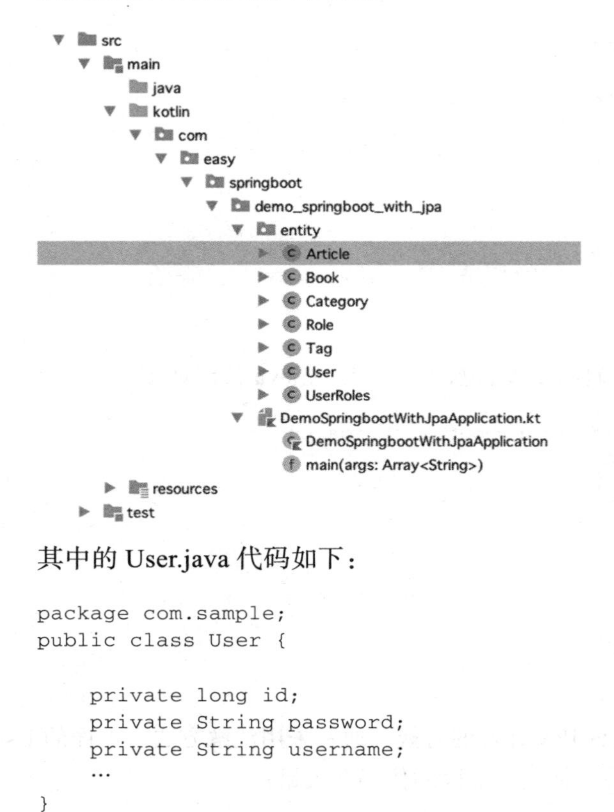

其中的 User.java 代码如下：

```
package com.sample;
public class User {

    private long id;
    private String password;
    private String username;
    ...
}
```

我们发现自动生成的 POJO 部分不满足我们的要求。例如：

- package com.sample；包中没有导入 import javax.persistence.* 包。
- 没有在对应的实体类上添加 @Entity 注解。

我们可以定制自动生成 POJO 代码的 Generate POJOs.groovy 脚本。首先进入脚本目录，如图 5-11 所示。

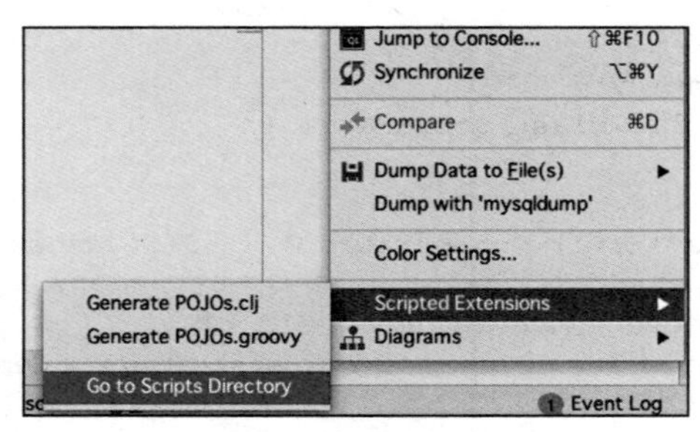

图 5-11　进入脚本目录

其中的 Generate POJOs.groovy 就是用来自动生成代码的。我们自定义的 POJOs.groovy 脚本代码如下：

```
packageName = "com.easy.springboot.demo_springboot_with_jpa.entity;"

def generate(out, className, fields) {
    out.println "package $packageName"
    out.println "import javax.persistence.*;"
    out.println ""
    out.println ""
    out.println "@Entity"
    out.println "public class $className {"
    ...
}
```

然后，重新执行上面生成 POJO 代码的步骤，新生成的 User.java 代码如下：

```
package com.easy.springboot.demo_springboot_with_jpa.entity;
import javax.persistence.*;
@Entity
public class User {

    private long id;
    private String password;
    private String username;
    …
}
```

在使用 Spring Boot 集成 Spring Data JPA 开发的时候，通常采用注解方法。上面的 User 代码还缺少 @Id 来指定表的主键。如果不指定，上面的代码会报错：

```
Caused by: org.hibernate.AnnotationException: No identifier specified for
    entity: com.easy.springboot.demo_springboot_with_jpa.entity.User
```

在自动生成 POJO 的 Groovy 脚本中添加 @Id 注解的生成代码：

```
def generate(out, className, fields) {
    out.println "package $packageName"
    out.println "import javax.persistence.*;"
    out.println ""
    out.println ""
    out.println "@Entity"
    out.println "public class $className {"
    out.println ""
    fields.each() {
        if (it.annos != "") out.println "    ${it.annos}"
        if(it.name =="id"){
            out.println "    @Id" //添加@Id注解
            out.println "    @GeneratedValue(strategy = GenerationType.IDENTITY)"
        }
        out.println "    private ${it.type} ${it.name};"
```

```
    }
    ..
}
```

然后，重新生成 POJO，这个时候所得到的 User.java 代码如下：

```
package com.easy.springboot.demo_springboot_with_jpa.entity;
import javax.persistence.*;
@Entity
public class User {

    @Id
    @GeneratedValue(strategy = GenerationType.IDENTITY)
    private long id;
    private String password;
    private String username;
    …
}
```

其实，还可以通过上面的自动生成实体类 POJO 代码的方式，去编写生成 dao 层的 Repository 接口代码、service 层代码和 controller 层代码。这里就不详细展开。

在后面的章节中，我们将展示怎样开发一个 Gradle 插件去自动生成这些代码。

5.5.5　配置项目数据源信息

在 application.properties 中配置项目数据源信息的代码如下：

```
# 数据库连接信息
spring.datasource.url=jdbc:mysql://localhost:3306/reakt?useUnicode=true&
characterEncoding=UTF8&useSSL=false
spring.datasource.username=root
spring.datasource.password=root
spring.datasource.driverClassName=com.mysql.jdbc.Driver
# 指定 DBMS
spring.jpa.database=MYSQL
# 是否记录 sql 执行日志
spring.jpa.show-sql=true
# Hibernate ddl auto (create, create-drop, update)
spring.jpa.hibernate.ddl-auto=none
#spring.jpa.hibernate.ddl-auto=update
# stripped before adding them to the entity manager
spring.jpa.properties.hibernate.dialect=org.hibernate.dialect.MySQL5Dialect
```

这个时候，我们可以成功启动应用。因为我们是从数据库表反向生成实体类 Entity 代码的，所以 spring.jpa.hibernate.ddl-auto 的属性配置为 none：

```
spring.jpa.hibernate.ddl-auto=none
```

如果是在开发测试环境中，可以设置为 create、update 等属性值，这样支持从实体类代码自动去数据库中创建表结构。

5.5.6 实现查询接口

下面来实现一个查询接口：http://127.0.0.1:8080/user/findAll，响应输出如下：

```
[
    {
        "id": 1,
        "password": "$2a$10$pTIiWjYO6dvod5YTGTIFyuXSJXmeU3PBSSUPOBMeSSqBEsTJ7PmW2",
        "username": "user"
    },
    {
        "id": 2,
        "password": "$2a$10$gcDHybMlVmV6bZahIUev6.M4CgAbW83RdDSoxaGGdjgzGRskLYVy6",
        "username": "admin"
    }
]
```

1. 实现代码

其中，UserDao.kt 代码如下：

```
interface UserDao : JpaRepository<User, Long>
```

UserController.kt 代码如下：

```
@RestController
@RequestMapping("/user")
class UserController {
    @Autowired lateinit var userDao: UserDao

    @GetMapping("/findAll")
    fun findAll(): List<User> {
        return userDao.findAll()
    }
}
```

2. 代码结构

代码组织结构如下所示。

```
▼ src
  ▼ main
      java
    ▼ kotlin
      ▼ com
        ▼ easy
          ▼ springboot
            ▼ demo_springboot_with_jpa
              ▼ controller
                ▼ UserController
                    findAll(): List<User>
                    userDao: UserDao
              ▼ dao
                  UserDao
              ▼ entity
                ▶ Article
```

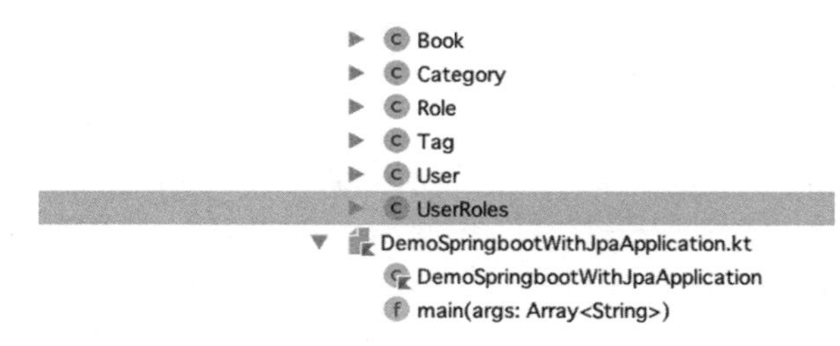

3. findAll 执行原理

我们在 interface UserDao: JpaRepository<User, Long> 接口代码中，并没有编写一行实现代码，而就拥有了常用的 CRUD 方法（由 JPA 框架在运行时自动创建）。这就是使用 Spring Data JPA 的极简之处。例如，这个 findAll 方法由 JpaRepository 声明，JpaRepository 声明的方法 API 如下所示：

```
JpaRepository
  deleteAllInBatch():void
  deleteInBatch(Iterable<T>):void
  findAll():List<T>
  findAll(Example<S>):List<S>
  findAll(Example<S>, Sort):List<S>
  findAll(Sort):List<T>
  findAllById(Iterable<ID>):List<T>
  flush():void
  getOne(ID):T
  saveAll(Iterable<S>):List<S>
  saveAndFlush(S):S
```

这些方法由框架在运行时实现。通过打断点，可以发现 findAll 方法首先经过 JdkDynamicAopProxy，然后进入了 SimpleJpaRepository 实现类中。SimpleJpaRepository 类中提供了常用的 CRUD 方法的具体实现。

4. 运行测试

在浏览器输入 http://127.0.0.1:8080/user/findAll，可以看到如上文所述的响应输出。

5.5.7　分页查询

实现 category 表数据的分页查询，数据库中的测试数据如图 5-12 所示。

上面自动生成的 Category.java 代码如下：

```
package com.easy.springboot.demo_springboot_with_jpa.entity;
import javax.persistence.*;

@Entity
public class Category {

    @Id
    @GeneratedValue(strategy = GenerationType.IDENTITY)
    private long id;
    private String code;
```

```
    private String detail;
    private java.sql.Timestamp gmtCreate;
    private java.sql.Timestamp gmtModify;
    private long isDeleted;
    private String name;
    private long type;
    …
}
```

CategoryDao.kt 代码如下：

```
interface CategoryDao: JpaRepository<Category,Long>
```

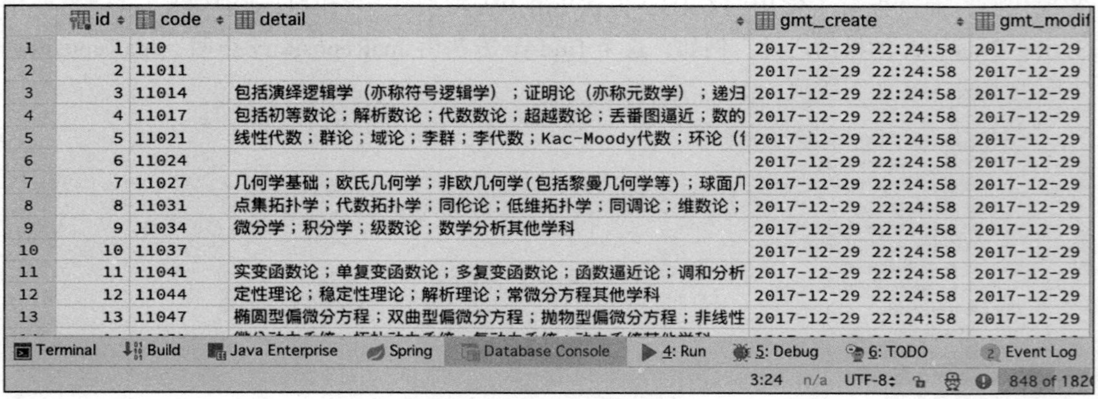

	id	code	detail	gmt_create	gmt_modif
1	1	110		2017-12-29 22:24:58	2017-12-29
2	2	11011		2017-12-29 22:24:58	2017-12-29
3	3	11014	包括演绎逻辑学（亦称符号逻辑学）；证明论（亦称元数学）；递归	2017-12-29 22:24:58	2017-12-29
4	4	11017	包括初等数论；解析数论；代数数论；超越数论；丢番图逼近；数的	2017-12-29 22:24:58	2017-12-29
5	5	11021	线性代数；群论；域论；李群；李代数；Kac-Moody代数；环论（	2017-12-29 22:24:58	2017-12-29
6	6	11024		2017-12-29 22:24:58	2017-12-29
7	7	11027	几何学基础；欧氏几何学；非欧几何学(包括黎曼几何学等)；球面几	2017-12-29 22:24:58	2017-12-29
8	8	11031	点集拓扑学；代数拓扑学；同伦论；低维拓扑学；同调论；维数论；	2017-12-29 22:24:58	2017-12-29
9	9	11034	微分学；积分学；级数论；数学分析其他学科	2017-12-29 22:24:58	2017-12-29
10	10	11037		2017-12-29 22:24:58	2017-12-29
11	11	11041	实变函数论；单复变函数论；多复变函数论；函数逼近论；调和分析	2017-12-29 22:24:58	2017-12-29
12	12	11044	定性理论；稳定性理论；解析理论；常微分方程其他学科	2017-12-29 22:24:58	2017-12-29
13	13	11047	椭圆型偏微分方程；双曲型偏微分方程；抛物型偏微分方程；非线性	2017-12-29 22:24:58	2017-12-29

Terminal　Build　Java Enterprise　Spring　Database Console　4: Run　5: Debug　6: TODO　Event Log

3:24　n/a　UTF-8　848 of 182

图 5-12　数据库中的测试数据

CategoryController.kt 代码如下：

```
@RestController
@RequestMapping("/category")
class CategoryController {
    @Autowired lateinit var categoryDao: CategoryDao

    @GetMapping("/page")
    fun findAll(
            @RequestParam(value = "pageNo", defaultValue = "0") pageNo: Int,
            @RequestParam(value = "pageSize", defaultValue = "10") pageSize: Int
    ): Page<Category> {
        return categoryDao.findAll(PageRequest.of(pageNo, pageSize))
    }
}
```

完整代码参考示例工程。重新启用应用，在浏览器输入 http://127.0.0.1:8080/category/page?pageNo = 0&pageSize = 3，响应输出如下：

```
{
    "content": [
        {
            "id": 1,
```

```
            "code": "110",
            "detail": "",
            "gmtCreate": "2017-12-29T14:24:58.000+0000",
            "gmtModify": "2017-12-29T14:24:58.000+0000",
            "isDeleted": 0,
            "name": " 数学 ",
            "type": 1
        },
        …
    ],
    "pageable": {
        "sort": {
            "sorted": false,
            "unsorted": true
        },
        "offset": 0,
        "pageSize": 3,
        "pageNumber": 0,
        "unpaged": false,
        "paged": true
    },
    "last": false,
    "totalElements": 3229,
    "totalPages": 1077,
    "size": 3,
    "number": 0,
    "numberOfElements": 3,
    "sort": {
        "sorted": false,
        "unsorted": true
    },
    "first": true
}
```

5.5.8　多表级联查询

我们要查询的两张表分别是 user、role，它们之间的关联表是 user_roles，表结构如下所示。

```
▼ MySQL - reakt@localhost 1 of 36
  ▶ collations 222
  ▼ schemas 1
    ▼ reakt
      ▶ article
      ▶ book
      ▶ category
      ▼ role
          id bigint(20) (auto increment)
          role varchar(50)
          PRIMARY (id)
          UK_bjxn5ii7v7ygwx39et0wawu0q (role)
          UK_bjxn5ii7v7ygwx39et0wawu0q (role) UNIQUE
      ▶ tag
      ▼ user
          id bigint(20) (auto increment)
          password varchar(255)
          username varchar(50)
          PRIMARY (id)
```

```
UK_sb8bbouer5wak8vyiiy4pf2bx (username)
UK_sb8bbouer5wak8vyiiy4pf2bx (username) UNIQUE
▼ user_roles
    user_id bigint(20)
    roles_id bigint(20)
    id bigint(20) (auto increment)
    PRIMARY (user_id, roles_id, id)
    FKj9553ass9uctjrmh0gkqsmv0d (roles_id)
```

在 user 表里添加 Set<Role> roles 成员：

```
@Entity
public class User {
    …
    @ManyToMany(
            targetEntity = Role.class,
            cascade = {CascadeType.PERSIST, CascadeType.MERGE},
            fetch = FetchType.LAZY)
    @JoinTable(name = "user_roles", joinColumns = { @JoinColumn(name ="user_id" )},
        inverseJoinColumns = { @JoinColumn(name = "roles_id") })
    @OrderBy("id")
    private Set<Role> roles;

    public Set<Role> getRoles() {
        return roles;
    }

    public void setRoles(Set<Role> roles) {
        this.roles = roles;
    }

    …

}
```

这个时候，重启应用，会发现控制台报错：

```
Caused by: org.hibernate.DuplicateMappingException: Table [user_roles] con
    tains physical column name [user_id] referred to by multiple physical
    column names: [user_id], [userId] at …
```

DuplicateMappingException，即复映射异常。意思是说，user_roles 物理表中的 user_id 列引用指向多个列名：user_id、userId。这个时候，可以通过 @Column 注解来显式指定列名：

```
@Entity
public class UserRoles {

    @Column(name="user_id")
    private long userId;
    @Column(name="roles_id")
    private long rolesId;
    @Id
    @GeneratedValue(strategy = GenerationType.IDENTITY)
    private long id;
```

```
    ...

}
```

此时，重启应用，可以发现启动成功。在浏览器输入 http://127.0.0.1:8080/user/findAll，响应输出如下：

```
[
    {
        "id": 1,
        "password": "$2a$10$pTIiWjYO6dvod5YTGTIFyuXSJXmeU3PBSSUPOBMeSSqBEsTJ7PmW2",
        "username": "user",
        "roles": [
            {
                "id": 1,
                "role": "ROLE_USER"
            }
        ]
    },
    {
        "id": 2,
        "password": "$2a$10$gcDHybMlVmV6bZahIUev6.M4CgAbW83RdDSoxaGGdjgzGRskLYVy6",
        "username": "admin",
        "roles": [
            {
                "id": 1,
                "role": "ROLE_USER"
            },
            {
                "id": 2,
                "role": "ROLE_ADMIN"
            }
        ]
    }
]
```

5.5.9　级联类型

JPA 允许从父实体传播到子级的状态转换。为此，JPA 接口规范中的 javax.persistence.CascadeType 枚举类定义了各种级联类型：

```
public enum CascadeType {
    ALL,
    PERSIST,
    MERGE,
    REMOVE,
    REFRESH,
    DETACH
}
```

我们用表 5-4 简单说明。

表 5-4 级联类型 CascadeType

CascadeType	说明
ALL	级联所有实体状态转换
PERSIST	级联实体持久化操作。对象会进入持久化状态，对该对象的操作会自动同步到数据库。指定 cascade = CascadeType.PERSIST 在实体类关联的实体字段上，那么保存该实体类时会级联保存该实体类关联的实体
MERGE	级联实体合并操作。对象会进入被管理状态，对该对象（可以是任何状态的对象）的操作会自动同步到数据库
REMOVE	级联实体删除操作。该 CascadeType.REMOVE 使我们能够沿父实体逐个删除级联的子实体
REFRESH	级联实体刷新操作。当对父实体级联的对象进行操作并保存时，会关联保存子实体到数据库
DETACH	级联实体分离操作。分离所有相关联的实体，该实体已在数据库中，对象将处于分离状态，对该对象的操作不会同步到数据库

5.5.10 模糊搜索接口

本节通过使用 @Query 注解来实现一个模糊搜索接口，搜索结果以分页的方式返回。CategoryDao.kt 的代码如下：

```
interface CategoryDao : JpaRepository<Category, Long> {
    @Query("select c from #{#entityName} c where c.name like %:searchText%")
    fun search(searchText: String, pageable: Pageable): Page<Category>
}
```

其中的 search() 方法说明如下：

- @Query：注解是定义在 org.springframework.data.jpa.repository 包下面的注解。该注解里面可以写 JPQL 或者写原生 SQL（设置 nativeQuery = true）。
- #{#entityName}：代替实体 Category 的名称，而 Spring Data JPA 会自动根据 Category 实体上对应的 @Entity 注解，自动将实体名称填入 JPQL 语句中。这帮助我们解决了项目中很多 Dao 接口的方法除了实体类名称不同，其他操作都相同的问题。
- searchText：JPQL 中的查询参数占位符。

CategoryController.kt 代码如下：

```
@RestController
@RequestMapping("/category")
class CategoryController {
    @Autowired lateinit var categoryDao: CategoryDao

    ...

    @GetMapping("/search")
    fun search(
            @RequestParam(value = "searchText", defaultValue = "") search
                Text: String,
            @RequestParam(value = "pageNo", defaultValue = "0") pageNo: Int,
            @RequestParam(value = "pageSize", defaultValue = "10") pageSize: Int
```

```
    ): Page<Category> {
        return categoryDao.search(searchText, PageRequest.of(pageNo, pageSize))
    }
}
```

重启应用，在浏览器输入 http://127.0.0.1:8080/category/search?searchText = 计算机 &pageNo = 0&pageSize = 3，可以得到如下分页数据：

```
{
    "content": [
        {
            "id": 25,
            "code": "11085",
            "detail": "",
            "gmtCreate": "2017-12-29T14:24:58.000+0000",
            "gmtModify": "2017-12-29T14:24:58.000+0000",
            "isDeleted": 0,
            "name": " 计算机数学 ",
            "type": 1
        },
        …
    ],
    "pageable": {
        "sort": {
            "sorted": false,
            "unsorted": true
        },
        "offset": 0,
        "pageSize": 3,
        "pageNumber": 0,
        "unpaged": false,
        "paged": true
    },
    "last": false,
    "totalPages": 9,
    "totalElements": 27,
    "size": 3,
    "number": 0,
    "numberOfElements": 3,
    "sort": {
        "sorted": false,
        "unsorted": true
    },
    "first": true
}
```

5.5.11　JPQL 语法基础

JPQL（Java Persistence Query Language）与 SQL 的语法很相似。它有一个类似 SQL 的 JPA 查询的语法，这是一个重要的优势，因为 SQL 是一种非常强大的查询语言，我们对它很熟悉。

SQL 和 JPQL 之间的主要区别是，SQL 直接与关系数据库中的表、记录和字段名称关联；而 JPQL 与 Java 实体类对象及其属性名称关联。JPQL 是完全面向对象的，具备继承、多态和关联等特性，和 Hibernate HQL 很相似。

使用 JPQL 需要注意以下几个问题。

- 实体名和属性区分大小写：JPQL 语句除了 Java 类和属性名称外，查询都是大小写不敏感的。
- JPQL 中的保留关键字不区分大小写。
- 标识实体的别名，也不区分大小写。
- 实体的别名可以使用 AS 关键字来标识，AS 关键字也可以省略。

查询时如果使用了实体的别名，在 SELECT 查询时便可以引用该实体的属性（而不是表的字段名），如下所示：

```
SELECT c.name FROM Customer c
```

如果查询的实体属性是另一个实体（一对一关系），也可以通过属性后“.”来获得实体的属性，如下所示：

```
SELECT c.address FROM Customer c
SELECT c.address.street FROM Customer c（多级属性访问）
```

1. JPQL 函数

JPQL 中也定义了一些常用的函数，这些函数可以针对字符型、数值型和日期型数值使用。

字符串函数通常可在查询时使用，例如查询顾客姓名的长度大于 10 的 JPQL 语句如下：

```
SELECT c FROM Customer c WHERE LENGTH(c.name)>10
```

JPQL 中提供的字符串函数主要有：

- CONCAT（str1，str2）：返回连接两个字符串的值。
- SUBSTRING（str，start，len）：返回字符串的一段，start 为字符串的开始索引位（第一个字符位置为 1），len 为截取的长度。
- TRIM（str）：去掉字符串的首尾的空格。
- LENGTH（str）：返回字符串的长度。
- LOWER（String）：转为小写字母。
- UPPER（String）：转为大写字母。

数值函数通常也可以在查询时使用，如下：

```
SELECT c FROM Customer c WHERE ABS(c.asset)>20
```

JPQL 中提供的主要数值函数有：

- ABS（num）：返回数的绝对值。
- SQRT（num）：返回数的平方。

- MOD（int，int）：取模。
- SIZE：返回集合类的总数。

JPA 提供了三种获取系统当前时间格式的方法：

- CURRENT_DATE
- CURRENT_TIME
- CURRENT_TIMESTAMP

2. 子查询

当一个查询条件依赖于另一个查询结果时，就需要使用子查询（嵌套查询），如下：

```
SELECT c FROM Customer c WHERE c.age > (SELECT AVG(c.age)FROM Customer c)
```

3. EXISTS 表达式

EXISTS 表达式用于判断子查询的结果，如果子查询的结果有一个或多个，则返回 true；如果子查询没有返回任何结果，则返回 false。语法如下：

```
[NOT] EXISTS （子查询表达式）
```

如查询系统当前日期之前的订单所属的客户：

```
SELECT c FROM Customer c WHERE EXISTS
(SELECT o FROM c.orders o WHERE o.createTime <CURRENT_DATE)
```

也可以在 EXISTS 关键字前加上“NOT”表示不存在，查询结果正好与 EXISTS 相反。

4. ALL 和 ANY 表达式

当子查询的返回结果有多个，顶层查询的条件使用 =、<、<=、>、>=、<> 这些比较来满足条件，就需要使用 ALL、ANY 和 SOME（ANY、SOME 与 NOT ALL 等价，只要存在一个即可）表达式，语法格式如下：

```
{ALL|ANY|SOME}(子查询)
```

例如，查询订单明细中存在 50 个以上订单的 JPQL，如下所示：

```
SELECT o FROM Order o WHERE 50 < ANY (SELECT l.quantityFROM o.lineItems l)
```

5. GROUP BY（分组）

分组查询是 JPQL 中很重要的查询，它可以按照指定的属性将数据分组，通常在统计数据时使用。分组查询通常使用“GROUP BY”表达式和“HAVING”表达式，分组查询的基本语法如下所示：

```
GROUP BY <分组子句>(HAVING<having 子句>)
```

“GROUPBY”关键字后指明分组的属性，“HAVING”可以对分组后的数据进行过滤，作用相当于 WHERE 子句，只能用在分组查询中，例如：

```
SELECT c FROM Customer c GROUP BY c.asset HAVING AVG(c.asset)>1000
```

6. ORDER BY（排序）

ORDER BY子句可以对查询结果进行排序，语法如下：

```
ORDER BY 排序属性 [ASC |DESC]{, 排序属性 [ASC |DESC]}*
```

其中“ORDERBY”关键字后指定排序的属性，多个属性排序用分号分割，例如：

```
SELECT c FROM Customer c ORDER BY c.id ASC,c.name DESC
```

7. MEMBER OF

MEMBER OF操作符用于判断一个实体是否包含在集合类对象中（MEMBER OF前可以添加NOT)，例如：

```
SELECT c FROM Customer c WHERE :order MEMBER OF c.orders
```

8. update（更新）

update语句用于执行数据更新操作，主要用于针对单个实体类的批量更新。以下语句将帐户余额不足万元的客户状态设置为“未偿付”：

```
update Customers c set c.status = '未偿付' where c.balance < 10000
```

9. delete（删除）

delete语句用于执行数据删除操作。以下语句删除不活跃的、没有订单的客户：

```
delete from Customers c where c.status = 'inactive' and c.orders is empty
```

10. JOIN（内连接）

内连接是最常用的连接方式，连接后关联数据不会出现null值，语法如下：

```
[INNER] JOIN
```

JPQL代码示例如下：

```
SELECT c,o FROM Customer c JOIN c.orders o
```

11. LEFT JOIN（左连接）

左连接也可以叫做左外连接，它是以左表为基础，关联右表。连接后右表中的数据可能为null值，语法如下：

```
LEFT [OUTER] JOIN
```

JPQL代码示例如下：

```
SELECT  c,o FROMCustomer c LEFT JOIN c.orders o
```

此时，查询结果是所有的客户，即使客户没有订单也在查询结果中。

12. JOIN FETCH（抓取连接）

实体属性的加载有两种方式，分为即时加载和懒加载。同样，对于实体关联的查询，也可以设置查询时的加载方式，这就是抓取连接，如下所示：

```
[LEFT|INNER] JOIN FETCH
```

在 JOIN 关键字后加上 FETCH 关键字，表示查询为抓取查询。

FETCH 连接主要针对实体属性为懒加载方式，使用内连接并没有加载关联的实体，例如：

```
SELECT c FROM Customer c JOIN c.orders o
```

此时在客户端调用 getOrders 方法将抛出异常。但若将关联查询设置为抓取方式：

```
SELECT c FROM Customer c JOIN FETCH
```

则查询结果的 customer 对性已加载所关联的 orders 属性。

13. DISTINCT（唯一性）

当进行连接查询时，通常会产生一些重复数据，若要去掉重复的数据，在关联查询时可以使用 DISTINCT 关键字，例如：

```
SELECT DISTINCT c FROM Customer c JOIN c.orders o
```

14. 比较操作符

WHERE 条件表达式可以包含比较操作符和逻辑操作符，如表 5-5 所示。

表 5-5 WHERE 表达式包含的操作符

操作符	说明
=	等于
>	大于
>=	大于等于
<	小于
<=	小于等于
<>	不等
[NOT] BETWEEN	在区间里，例如： `SELECT c FROM Customer WHRER c.asset BETWEEN 1000.0 and2000.0`
[NOT] LIKE	模糊匹配。LIKE 操作符用来查询匹配指定的字符串，匹配字符串的关键字符有以下两种： ● 下划线 “_”：表示匹配某一个字符 ● 百分号 “%”：表示匹配零个或多个字符 如果要查询的字符串存在 “_” 或 “%” 等特殊字符，则需要在匹配的字符串前加转义字符 “\”
[NOT] IN	IN 操作符可以查询在指定的多个值，语法如下： `[NOT] IN (值 {, 值 }*| 子查询)`

（续）

操作符	说 明

IS [NOT] NULL	NULL 操作符用来判断属性是否是 null 值。判断属性是 null，使用“IS NULL”关键字：
	`SELECT c FROM Customer c WHERE c.address IS NULL`
	判断属性不为 null，使用“IS NOT NULL”关键字：
	`SELECT c FROM Customer c WHERE c.address IS NOT NULL`
	NULL 操作符也可以用在参数查询中，用于判断输入的参数是否为 null 值：
	`SELECT c FROM Customer c WHERE :zip IS NOT NULLL ANDc.address.zip=:zip`

IS EMPTY	EMPTY 操作符用于判断实体的集合类属性是否为空。它与 NULL 操作符不太相同，主要针对属性是集合类的判断。
	判断集合类属性为空值，使用 IS EMPTY：
	`SELECT c FROM Customer c WHERE c.orders IS EMPTY`
	判断集合类属性不是空值，使用 IS NOT EMPTY：
	`SELECT c FROM Customer c WHERE c.orders IS NOT EMPTY`

AND	逻辑与
OR	逻辑或

JPQL 语法规范文档参考：https://docs.oracle.com/cd/E16764_01/apirefs.1111/e13046/ejb3_langref.html

5.5.12 JPA 常用注解

JPA 中常用的注解列表如表 5-6 所示。

表 5-6 JPA 中常用注解

注 解	描 述
@Entity	声明类为实体或表
@Table	声明表名
@Basic	指定非约束明确的各个字段
@Embedded	指定类或它的值是一个可嵌入的类的实例的实体的属性
@Id	指定的类的属性，用于识别一个表中的主键
@GeneratedValue	指定如何进行标识属性初始化。JPA 提供了四种主键生成器： • TABLE，由 JPA 提供者通过创建数据库表来记录生成的主键值 • SEQUENCE，由数据库 Sequence 对象提供主键值 • IDENTITY，由数据库的自增列提供主键值 • AUTO，由 JPA 提供者根据数据库自行决定生成算法
@Transient	指定的属性，它是不持久的，即该值永远不会存储在数据库中
@Column	指定持久属性列属性
@SequenceGenerator	定义使用 @GeneratedValue 注解的主键生成器。例如： `@GeneratedValue(strategy=GenerationType.SEQUENCE,generator="SEQ_Name")` @GeneratedValue 和 @SequenceGenerator 是 JPA 标准注解，GeneratedValue 用来定义主键生成策略，SequenceGenerator 用来定义一个生成主键的序列；它们要联合使用才有效

（续）

注　解	描　述
@TableGenerator	定义主键生成器。使用 @TableGenerator 标记定义表生成策略。该注解可以用在类名、方法名、和属性上。一旦在实体中标记，不仅可以在本实体中使用，在其他的实体中也可以引用，作用范围是整个 persistent unit 配置的实体类中
@AccessType	这种类型的注释用于设置访问类型。如果设置 @AccessType（FIELD），然后进入 FIELD 明智的。如果设置 @AccessType（PROPERTY），然后进入属性发生明智的
@JoinColumn	指定一个实体组织或实体的集合，用在多对一和一对多关联场景
@UniqueConstraint	指定唯一约束
@ColumnResult	SQL 查询 SELECT 子句中列的引用名称
@ManyToMany	定义了连接表之间的多对多、一对多的关系
@ManyToOne	定义了连接表之间的多对一的关系
@OneToMany	定义了连接表一对多的关系
@OneToOne	定义了连接表一对一的关系
@NamedQueries	指定命名查询的列表
@NamedQuery	指定命名查询

5.6　本章小结

使用 Spring Data JPA 可以用极简的代码快速实现功能丰富的 dao 层代码。同时，还可以通过 JPQL 灵活编写实现 SQL 查询代码，使用丰富的级联注解快速实现多表级联查询的逻辑。我们不再需要像 MyBatis 那样单独去配置 MyBatis Generator 代码生成工具和分页插件 PageHelper。这一切由 Spring Data JPA 和 IDEA 的 Database 数据库工具帮我们完成。

本章中我们是从数据库表来生成实体类代码，在下一章中我们演示如何使用代码自动创建数据表（这种方式在项目开发测试阶段非常有用），同时开发一个 Gradle 插件来自动生成 entity 层、dao 层、service 层的原型代码。

提示　本章实战项目源代码：https://github.com/KotlinSpringBoot/demo_springboot_with_jpa

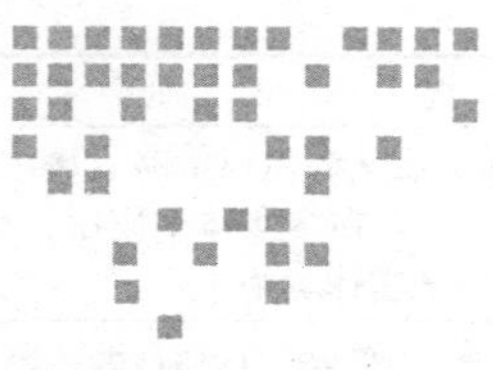

Chapter 6 第6章

Spring Boot Gradle 插件应用开发

在前面的章节中，我们都使用过 Gradle 构建 Spring Boot 应用，那么我们就有必要来深入了解一下 Spring Boot Gradle 插件背后的工作原理。

本章开发一个 Gradle 插件 com.easyspringboot.kor，主要功能是在 Spring Boot 工程编码过程中自动生成 model 层、dao 层、service 层、controller 层的样板代码文件，用来极简化 Spring Boot 应用开发过程中样板代码的编写。

6.1 Gradle 简介

Gradle 是一个项目自动化构建工具，主要负责依赖管理和任务执行。

Gradle 基于 Groovy DSL（Domain Specific Language，领域特定语言）来声明项目设置，抛弃了 Maven 基于 XML 的各种烦琐配置。XML 的结构是嵌套分层的，不利于项目配置程序的编写，同时当 XML 文件增长到十分庞大的时候，将变得难以管理。除了 XML 的繁冗、不够灵活等缺点之外，Maven 的学习曲线也相比更加陡峭。而 Gradle 使用基于 JVM 的 Groovy 动态类型编程语言来完成项目的自动化构建配置，能完成复杂系统的构建任务。毫无疑问，基于 Groovy 的 DSL 要比使用 XML 配置更加极简灵活。

类似于 Maven 的 pom.xml 文件，每个 Gradle 项目都需要有一个对应的 build.gradle 配置文件，我们通过在该文件中定义具体的构建任务来完成构建工作。其中的每个任务是可配置的，任务之间也可以依赖，还可以配置默认任务。Gradle 里有两个基本概念：

- 项目（Project）
- 任务（Task）

项目由多个任务组成，任务是一个构建中原子性的工作，例如编译、打包、执行等。一个典型的 Gradle 项目构建流程如图 6-1 所示。

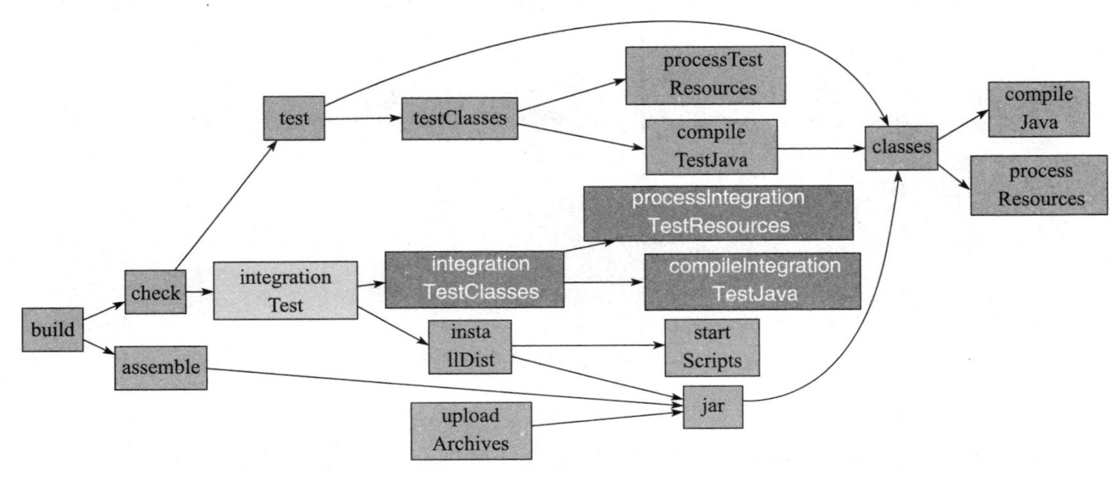

图 6-1 Gradle 项目构建流程

Gradle 支持的语言有 Java、Groovy、Kotlin 和 Scala 等。支持 Maven、Ivy 仓库，支持传递性依赖管理。Gradle 继承了 Apache Ant 和 Apache Maven 的优秀基因，例如：依赖管理、仓库、约定优于配置等概念。

提示 更多关于 Gradle 的内容，可参考文档：https://docs.gradle.org/current/userguide/userguide-_single.html。

Gradle 最显著的特性可以地概括为两点：极简、灵活。

基于 Groovy 的极简脚本，表达力丰富。之前在 Maven pom.xml 中复杂的事情，现在在 Gradle 中就是小菜一碟，比如修改现有的构建生命周期，只需要几行配置。同样的事情，在 Maven 中可能需要编写一个插件。

提示 Groovy 是一门基于 JVM 的动态类型编程语言。它的特性可以简单概括如下：

- Groovy 和 Java 一样，源码都会被编译成 class 字节码文件然后可以在 JVM 中运行。
- Groovy 兼容所有的 Java 语法，也就是说，你可以在 .groovy 中直接写 Java 代码，编译运行。
- Groovy 中自定义变量和方法都使用关键字 def 来定义。
- 闭包（这是在 Gradle 配置中经常使用的特性）。

更多关于 Groovy 的语法学习，可以参考：http://www.groovy-lang.org/documentation.html。

6.2 用 Gradle 构建生命周期

用 Gradle 构建系统的生命周期可分为：初始化、配置和运行三个阶段，如图 6-2 所示。

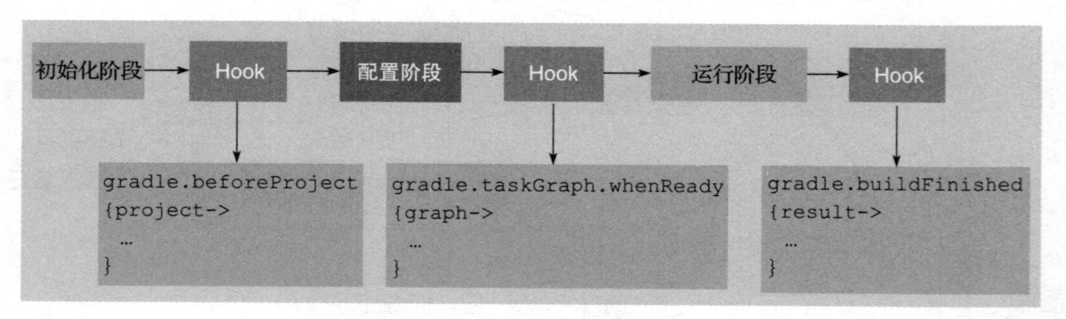

图 6-2 Gradle 构建生命周期

1. 初始化阶段

此阶段读取根工程中 setting.gradle 中的 include 信息，决定有哪几个工程加入构建。

2. 配置阶段

此阶段主要解析每个项目中的 build.gradle 配置文件，处理依赖关系和执行顺序等，脚本本身也需要依赖来完成自身的分析。

比如，在多项目构建的例子中，Gradle 会去解析每个子目录中的 build.gradle，每个项目都会被解析，其内部的任务也会被添加到一个有向无环图 (Directed Acyclic Graph，DAG) 里，用于解决执行过程中的依赖关系。图中的每一个任务称为一个节点，每一个节点通过边来连接，可以通过 dependsOn 或者隐式的依赖推导来创建依赖关系。DAG 图不会有环，就是说一个已经执行的任务不会再次执行，图 6-3 简要地展示了这个过程。

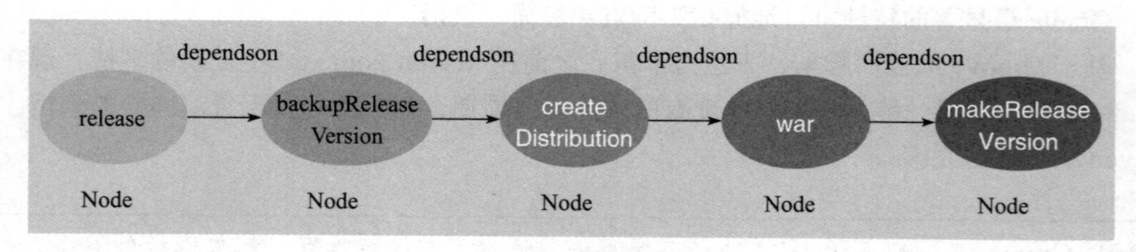

图 6-3 Gradle DAG 图

3. 运行阶段

根据 Gradle 命令传递过来的任务名称，执行相关依赖任务。要执行一个任务，只需要输入：

```
gradle [任务名]
```

Gradle 确保这个任务和它所依赖的任务都会执行，要执行多个任务只需要在后面添加多个任务名。此阶段真正构建项目并执行项目下的各个任务。Gradle 就会将这个任务链上的所有任务全部按依赖顺序执行一遍。任务链由 dependsOn 进行串联。

如同 Ant 一般，Gradle 给了用户足够的自由去定义自己的任务，不过同时 Gradle 也提供了类似 Maven 的约定优于配置方式，这是通过 Gradle 的 Java 插件实现的，从文档上看，Gradle 是推荐这种方式的。Java 插件定义了与 Maven 完全一致的项目布局：

```
src/main/java
src/main/resources
src/test/java
src/test/resources
```

而使用 Groovy 自定义项目布局更加方便简单：

```
sourceSets {
    main {
        java {
            srcDir 'src/java'
        }
        resources {
            srcDir 'src/resources'
        }
    }
}
```

Gradle Java 插件也定义了构建生命周期，包括编译主代码、处理资源、编译测试代码、执行测试、上传归档等等任务。

自动化依赖管理的基石是仓库，Maven 中央仓库已经成为了 Java 开发者不可或缺的资源，Gradle 的依赖管理直接使用 Maven 中央仓库，例如：

```
repositories {
    mavenLocal()
    mavenCentral()
    maven { url "https: //repo.spring.io/snapshot" }
    maven { url "https: //repo.spring.io/milestone" }
}
```

这段 Groovy 配置代码就是在 Gradle 中配置使用 Maven 本地仓库、中央仓库以及自定义的地址仓库。Gradle 下载后的文件被存储在 $USER_HOME/.gradle 目录下。这种实现的方法与 Maven 基本一样。可以说 Gradle 不仅最大限度地继承了 Maven 的很多理念，Maven 仓库资源也是直接拿来用。

Gradle 能够解析现有的 Maven POM，从而得到传递性依赖的信息，并且引入到当前项目中，在此基础上，它也支持排除传递性依赖或者干脆关闭传递性依赖，这一点是 Maven 所不具备的特性。

Gradle 项目使用 Maven 项目生成的资源已经不是个问题了，接着需要反过来考虑,Maven 用户是否能够使用 Gradle 生成的资源呢？或者更简单点，Gradle 项目生成的构件是否可以发布到 Maven 仓库中供人使用呢？这一点非常重要，因为如果做不到这一点，你可能就会丢失大量的用户。幸运的是 Gradle 再次给出了令人满意的答案。使用 Gradle 的 Maven 插件，用户就可以轻松地将项目构件上传到 Maven 仓库中：

```
apply plugin: 'maven'
...
uploadArchives {
    repositories{
        mavenDeployer {
            repository(url: "http:          //localhost:8888/nexus/content/repositories/
                                            //snapshots/") {
                authentication(userName: "admin", password: "admin")
                pom.groupId = "com.easy.springboot"
                pom.artifactId = "demo-gradle-plugin"
            }
        }
    }
}
```

上传本机仓库：

```
apply plugin: 'maven'
...
uploadArchives {
    repositories {
        mavenDeployer {
            repository(url: uri('/Users/jack/.m2/repository'))
        }
    }
}
```

6.3 Gradle 插件

Gradle 插件是 Gradle 的扩展，它对 project 做了一些配置，我们可以通过插件给 project 添加自己定制的 task 来实现自定义的项目构建逻辑。Gradle 自身带了一些插件，例如：java、application、maven、war、eclipse、idea 插件等。这些插件的详细介绍可以参考：https://docs.gradle.org/4.4.1/userguide/userguide.html。我们也可以自己定制开发 Plugin 插件，并发布到公共仓库供别人使用。在 Plugin 中，我们可以向 Project 中加入新的 Task，定义 configurations 和 property 等。例如 java 插件就定义了编译 Java 代码的任务 JavaCompile，指定了域对象 SourceSet，同时对项目目录结构 Java 源代码位于 src/main/java 的约定等。

常用的 Gradle 插件如表 6-1 所示。

表 6-1　常用的 Gradle 插件

插件 Id	项目应用	描述
Java	java-base	向项目添加 Java 编译、测试和绑定的功，它是许多其他 Gradle 插件的基础

（续）

插件 Id	项目应用	描述
Kotlin	java, kotlin-base	添加对构建 Kotlin 项目的支持
Org.springframework.boot	Spring Boot	添加对构建 Spring Boot 项目的支持
Groovy	java, groovy-base	添加对构建 Groovy 项目的支持
Scala	java, scala-base	添加对构建 Scala 项目的支持

例如，Java、Kotlin、Groovy、Scala 这些插件的添加，让 JVM 在编译和执行时对各种语言支持，org.springframework.boot 插件则是对构建 Spring Boot 项目提供支持。这些插件的使用如下：

```
apply plugin: 'java'
apply plugin: 'kotlin'
apply plugin: 'groovy'
apply plugin: 'scala'
apply plugin: 'kotlin-spring'
apply plugin: 'eclipse'
apply plugin: 'org.springframework.boot'
apply plugin: 'io.spring.dependency-management'
```

例如，新建一个 Gradle Java 项目，得到的目录结构如下：

```
demo_java_plugin$ tree
.
├── build.gradle
├── settings.gradle
└── src
    ├── main
    │   ├── java
    │   │   └── com
    │   │       └── easy
    │   │           └── springboot
    │   │               └── demo_java_plugin
    │   │                   └── HelloWorld.java
    │   └── resources
    └── test
        ├── java
        │   └── com
        │       └── easy
        │           └── springboot
        │               └── demo_java_plugin
        │                   └── HelloWorldTest.java
        └── resources
```

我们可以看到在 build.gradle 配置文件中有如下一行：

```
apply plugin: 'java'
```

即应用 Gradle Java 插件。我们可以在 IDEA 侧边栏看到 Tasks 列表，如图 6-4 所示。

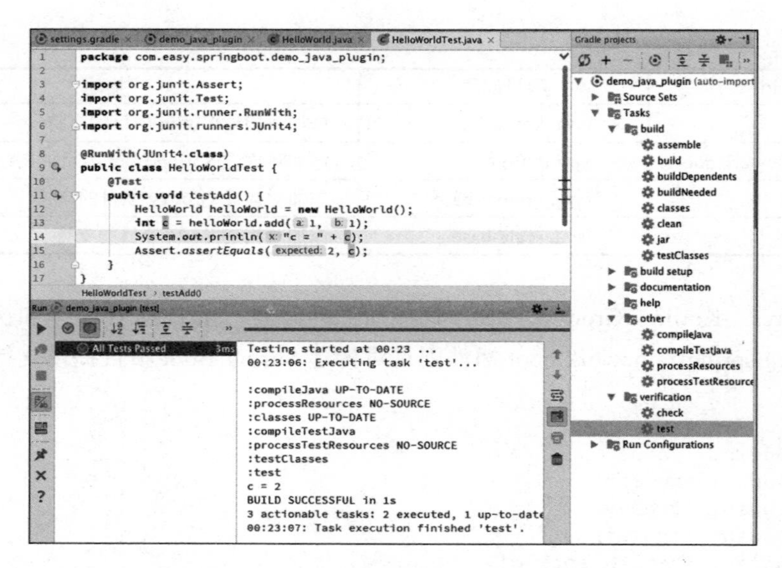

图 6-4 Gradle Java 插件 Tasks 列表

其中，HelloWorldTest 代码如下：

```
@RunWith(JUnit4.class)
public class HelloWorldTest {
    @Test
    public void testAdd() {
        HelloWorld helloWorld = new HelloWorld();
        int c = helloWorld.add(1, 1);
        System.out.println("c = " + c);
        Assert.assertEquals(2, c);
    }
}
```

执行 gradle test，可以看到控制台输出构建任务日志如下：

```
Testing started at 00:16 ...
00:16:12: Executing task 'test'...

:compileJava
:processResources NO-SOURCE
:classes
:compileTestJava
:processTestResources NO-SOURCE
:testClasses
:test
BUILD SUCCESSFUL in 4s
3 actionable tasks: 3 executed
00:16:16: Task execution finished 'test'.
```

执行 gradle build，控制台输出构建任务如下：

```
00:20:59: Executing task 'build'...

:compileJava UP-TO-DATE
:processResources NO-SOURCE
:classes UP-TO-DATE
:jar
:assemble
:compileTestJava UP-TO-DATE
:processTestResources NO-SOURCE
:testClasses UP-TO-DATE
:test UP-TO-DATE
:check UP-TO-DATE
:build

BUILD SUCCESSFUL in 0s
4 actionable tasks: 1 executed, 3 up-to-date
00:21:00: Task execution finished 'build'.
```

提示　更多关于 Java 插件的介绍，可参考：https://docs.gradle.org/4.4.1/userguide/java_plugin.html。

另外，我们在 build.gradle 中添加：

```
apply plugin: 'application'
```

即可使用 application 插件。

使用 application 插件可以配置 mainClassName：

```
mainClassName = 'com.easy.springboot.demo_java_plugin.HelloWorld'
```

完整的 build.gradle 配置文件内容如下：

```
group 'com.easy.springboot'
version '1.0-SNAPSHOT'

apply plugin: 'java'
apply plugin: 'application'

sourceCompatibility = 1.8

repositories {
    mavenCentral()
}

mainClassName = 'com.easy.springboot.demo_java_plugin.HelloWorld'

dependencies {
    testCompile group: 'junit', name: 'junit', version: '4.12'
}
```

其中，HelloWorld 代码如下：

```
package com.easy.springboot.demo_java_plugin;

public class HelloWorld {
    public static void main(String[] agrs) {
        System.out.println("Hello, World!");
    }

    public int add(int a, int b) {
        return a + b;
    }

}
```

这样，我们可以直接在命令行运行：

```
$ gradle run

> Task :run
Hello, World!


BUILD SUCCESSFUL in 1s
2 actionable tasks: 2 executed
```

本节示例工程代码地址为 https://github.com/KotlinSpringBoot/demo_java_plugin。
关于 application 插件的更多内容，可以参考：https://docs.gradle.org/4.4.1/userguide/application_plugin.html 。

6.4 项目实战

本节我们通过具体的项目实例介绍如何自己开发一个 Gradle 插件，该插件将会应用到我们后面的实际项目的开发中。使用的开发语言是 Kotlin、Groovy 和 Java。

项目说明：

- 插件名称：korGenerate。
- 功能概述：自动生成 entity、dao、service、controller 层的模板代码。
- 输入参数：korArgs、entityName 实体类名称。

6.4.1 创建项目

依次选择 File → New → Project，在 New Project 页面中，选择 Gradle，勾选 Java、Groovy、Kotlin（Java），点击 Next，如图 6-5 所示。

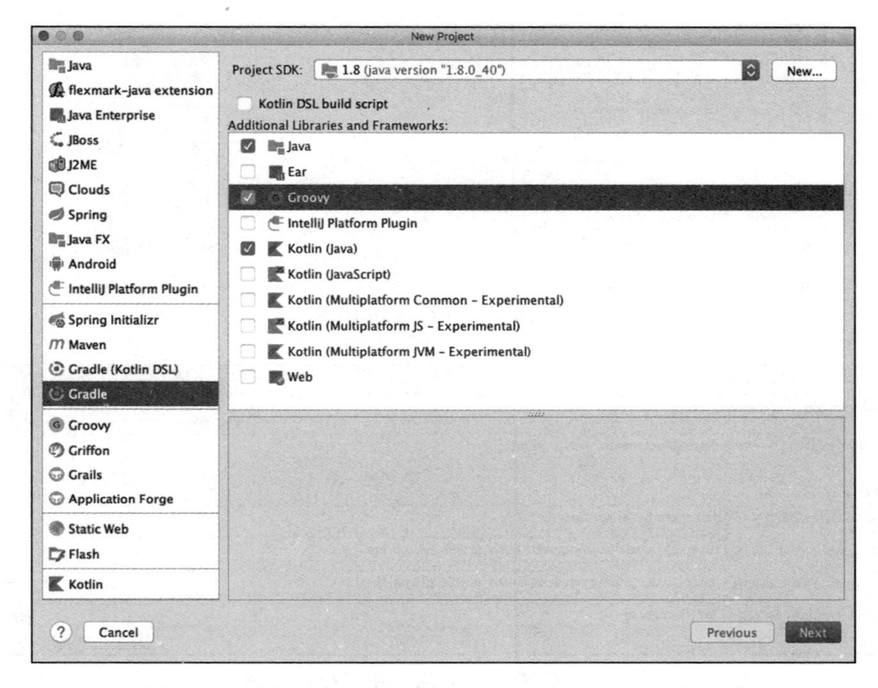

图 6-5　选择 Gradle，勾选 Java、Groovy、Kotlin（Java）

输入项目坐标信息，如图 6-6 所示。

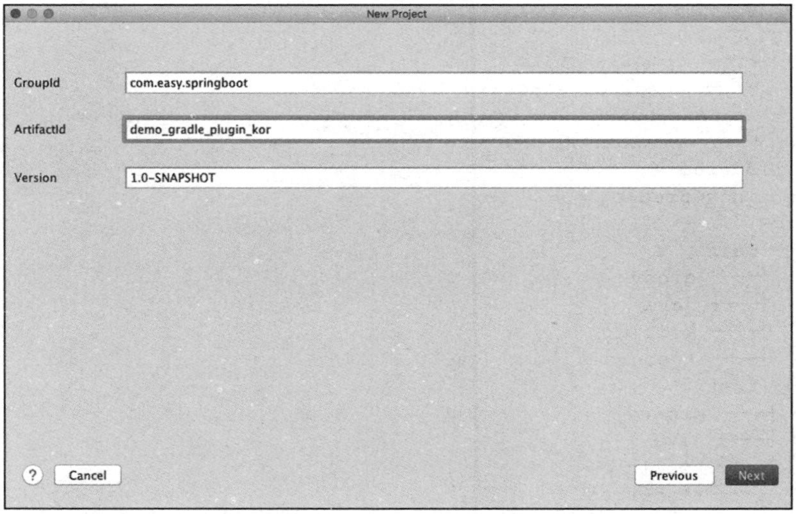

图 6-6　输入项目坐标信息

配置 Gradle 环境如图 6-7 所示。配置项目名称、项目存放目录，如图 6-8 所示。

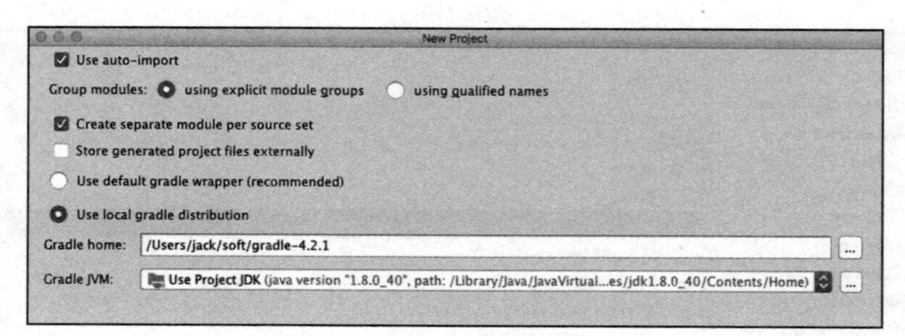

图 6-7 配置 Gradle 环境

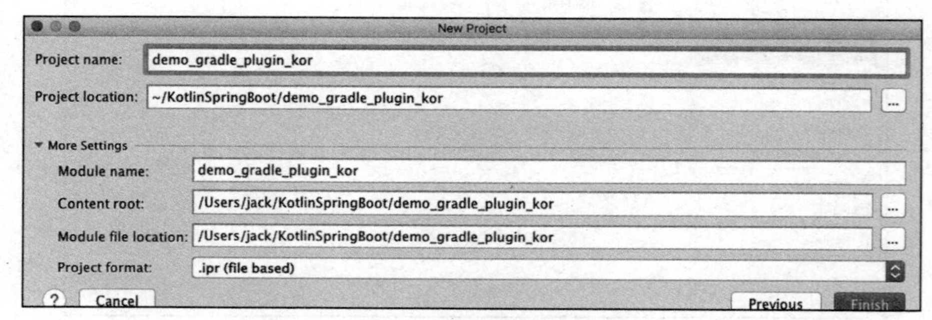

图 6-8 配置项目名称、项目存放目录

项目目录如下：

```
.
├── build
│   ├── kotlin
│   │   ├── compileKotlin
│   │   └── compileTestKotlin
│   └── kotlin-build
│       └── version.txt
├── build.gradle
├── settings.gradle
└── src
    ├── main
    │   ├── groovy
    │   ├── java
    │   ├── kotlin
    │   └── resources
    └── test
        ├── groovy
        ├── java
        ├── kotlin
        └── resources
```

其中 Gradle 配置文件 build.gradle 的内容如下：

```
buildscript {
    ext.kotlin_version = '1.2.10'
```

```
    repositories {
        mavenCentral()
    }
    dependencies {
        classpath "org.jetbrains.kotlin:kotlin-gradle-plugin:$kotlin_version"
    }
}

group 'com.easy.springboot'
version '1.0-SNAPSHOT'

apply plugin: 'groovy'
apply plugin: 'java'
apply plugin: 'kotlin'

sourceCompatibility = 1.8

repositories {
    mavenCentral()
}

dependencies {
    compile "org.jetbrains.kotlin:kotlin-stdlib-jdk8:$kotlin_version"
    compile 'org.codehaus.groovy:groovy-all:2.3.11'
    testCompile group: 'junit', name: 'junit', version: '4.12'
}

compileKotlin {
    kotlinOptions.jvmTarget = "1.8"
}
compileTestKotlin {
    kotlinOptions.jvmTarget = "1.8"
}
```

6.4.2　添加依赖

在 dependencies 中添加 gradleApi 依赖：

```
dependencies {
    compile "org.jetbrains.kotlin:kotlin-stdlib-jdk8:$kotlin_version"
    compile 'org.codehaus.groovy:groovy-all:2.3.11'
    testCompile group: 'junit', name: 'junit', version: '4.12'
    compile gradleApi()
}
```

其中，compile gradleApi() 是使用 Gradle 的 API 依赖。

6.4.3　配置上传本地 Maven 仓库

配置上传到 Maven 仓库，这里我们配置上传至本机的目录下：

```
apply plugin: 'maven'
```

```
uploadArchives {
    repositories {
        mavenDeployer {
            repository(url: uri('/Users/jack/.m2/repository'))
        }
    }
}
```

6.4.4 实现插件

我们使用 Groovy 来实现 Gradle 插件的执行逻辑。在 Groovy 源码目录下新建 package，如图 6-9 所示。

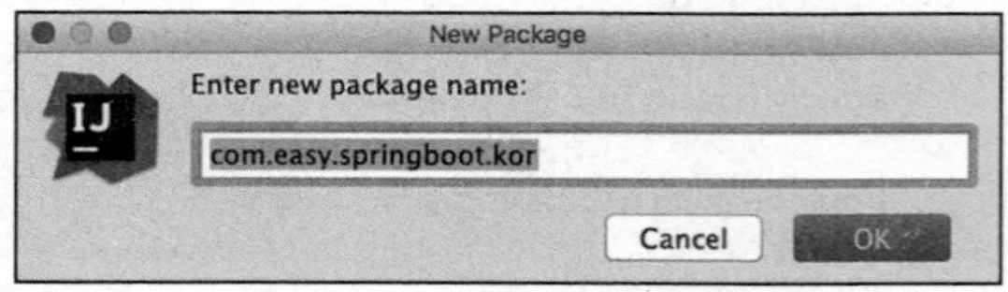

图 6-9 新建 package

在 package 下面新建 Groovy Class，如图 6-10 所示。

设置 Groovy 的类名称为 KorPlugin。类 KorPlugin 实现 org.gradle.api.Plugin 接口，IDEA 会自动提示实现 methods，如图 6-11 所示。

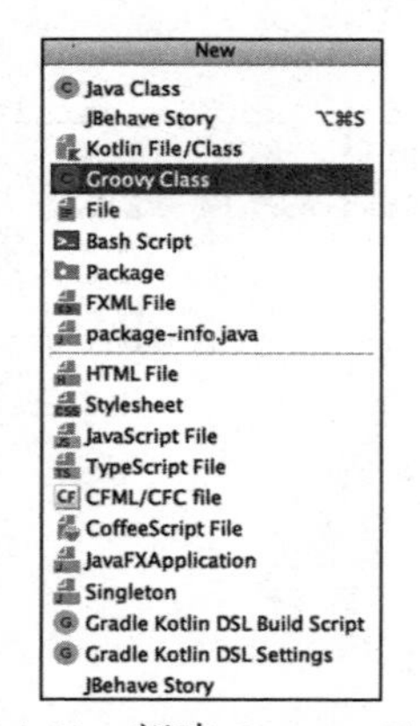

图 6-10 新建 Groovy Class

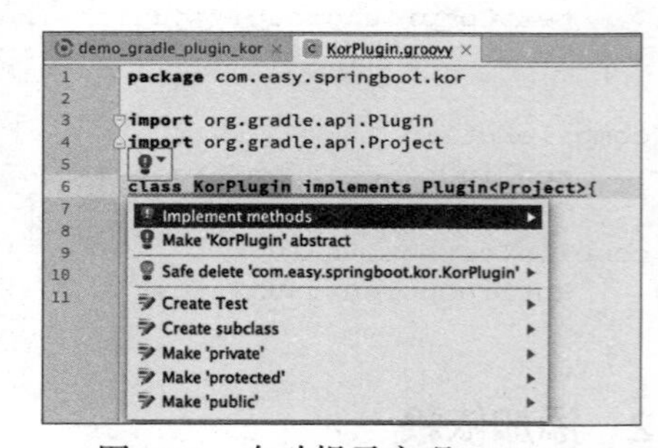

图 6-11 自动提示实现 methods

点击 Implement methods，选择要实现的方法，如图 6-12 所示。

点击 OK，自动生成的代码如下：

```
package com.easy.springboot.kor

import org.gradle.api.Plugin
import org.gradle.api.Project

class KorPlugin implements Plugin<Project>{

    @Override
    void apply(Project project) {

    }
}
```

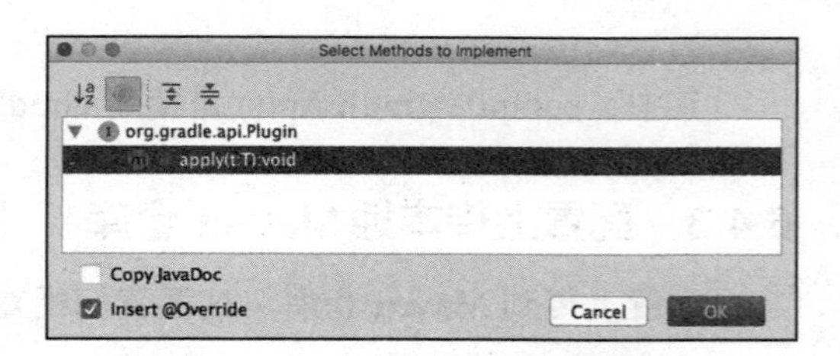

图 6-12 选择要实现的方法

KorPlugin.groovy 具体的实现代码是：

```
package com.easy.springboot.kor

import org.gradle.api.Plugin
import org.gradle.api.Project

class KorPlugin implements Plugin<Project> {

    @Override
    void apply(Project project) {
        String projectDir = project.projectDir.absolutePath

        project.extensions.create('korArgs', KorPluginExtension)

        project.task("korGenerate") << {
            println("Hello, Kor !")
            println("Group: $project.group")
            println("Name: $project.name")
            println("korArgs: $project.korArgs.entity")
            String packageName = "$project.group.$project.name"
            String entityName = project.korArgs.entity

            KorGenerateJava korGenerateJava = new KorGenerateJava()
            korGenerateJava.doGenerate(
                projectDir,
                packageName,
                entityName
            )
        }
    }
}

class KorPluginExtension {
    def entity = "Kor"
}
```

KorGenerateJava.java 实现代码如下：

```
package com.easy.springboot.kor;

public class KorGenerateJava {
    public void doGenerate(String projectDir, String packageName,String en-
        tityName){

        KorGenerateKotlin korGenerateKotlin = new KorGenerateKotlin();
        korGenerateKotlin.doGenerate(projectDir,packageName,entityName);

    }
}
```

我们使用 Java 代码作为连接 Kotlin 与 Groovy 代码的桥梁。

其中真正的业务逻辑实现代码在 KorGenerateKotlin.kt 中，完整的代码参考本节实例工程源码。

6.4.5 添加插件属性配置

在 src/main/resources/ 目录下新建 META-INF/gradle-plugins 目录，然后在该目录下添加插件属性配置文件，如图 6-13 所示。

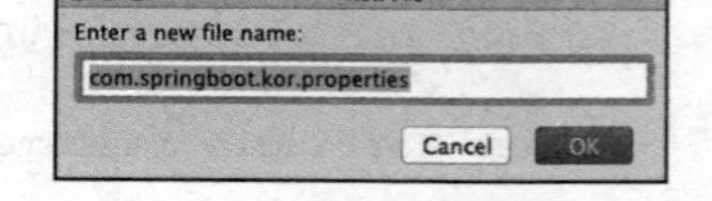

图 6-13 添加插件属性配置文件

其中，com.springboot.kor.properties 属性文件内容如下：

```
implementation-class=com.easy.springboot.kor.KorPlugin
```

完整的工程目录结构如下所示：

```
▼ src
  ▼ main
    ▼ groovy
      ▼ com
        ▼ easy
          ▼ springboot
            ▼ kor
              ▼ KorPlugin.groovy
                ▼ KorPlugin
                    apply(Project):void
                ▼ KorPluginExtension
                    entity:Object =
    ▼ java
      ▼ com
        ▼ easy
          ▼ springboot
            ▼ kor
              ▼ KorGenerateJava
                  doGenerate(String, String, String)
    ▼ kotlin
      ▼ com
        ▼ easy
          ▼ springboot
            ▼ kor
              ▶ KorGenerateKotlin
    ▼ resources
      ▼ META-INF
        ▼ gradle-plugins
            com.springboot.kor.properties
  ▶ test
  build.gradle
```

6.4.6 运行测试

直接运行编译打包上传本地 Maven 仓库 Executing task 'uploadArchives'：

```
gradle uploadArchives
```

会报错：

```
$ gradle --stacktrace compileGroovy
Caused by: groovy.lang.GroovyRuntimeException: Conflicting module ve-rsions.Module
    [groovy-all is loaded in version 2.4.12 and you are trying to load version 2.3.11
        ... 16 more


* Get more help at https://help.gradle.org
```

```
BUILD FAILED in 2s
3 actionable tasks: 1 executed, 2 up-to-date
```

通过报错日志内容我们可以看出，是因为 groovy-all 有多个版本冲突。我们发现在类路径下确实有 groovy-all-2.4.12 和 groovy-all-2.3.11，如下所示：

External Libraries
gradle-api-4.2.1.jar library root
gradle-installation-beacon-4.2.1.jar library root
groovy-all-2.4.12.jar library root
< 1.8 > /Library/Java/JavaVirtualMachines/jdk1.8.0_40/Conte
Gradle: junit:junit:4.12
Gradle: org.codehaus.groovy:groovy-all:2.3.11
Gradle: org.hamcrest:hamcrest-core:1.3
Gradle: org.jetbrains.kotlin:kotlin-stdlib:1.2.10
Gradle: org.jetbrains.kotlin:kotlin-stdlib-jdk7:1.2.10
Gradle: org.jetbrains.kotlin:kotlin-stdlib-jdk8:1.2.10
Gradle: org.jetbrains:annotations:13.0

原因分析：因为在 gradleApi() 中引入了 groovy-all 2.4.12，而我们又使用 IDEA 自动创建的 Groovy、Java、Kotlin 混合编程项目，添加了 groovy-all:2.3.11 依赖，导致冲突。我们把 groovy-all:2.3.11 依赖注释掉即可：

```
dependencies {
    compile "org.jetbrains.kotlin:kotlin-stdlib-jdk8:$kotlin_version"
    // compile 'org.codehaus.groovy:groovy-all:2.3.11'
    testCompile group: 'junit', name: 'junit', version: '4.12'
    compile gradleApi()
}
```

然后，重新执行 task 'uploadArchives'，如下所示：

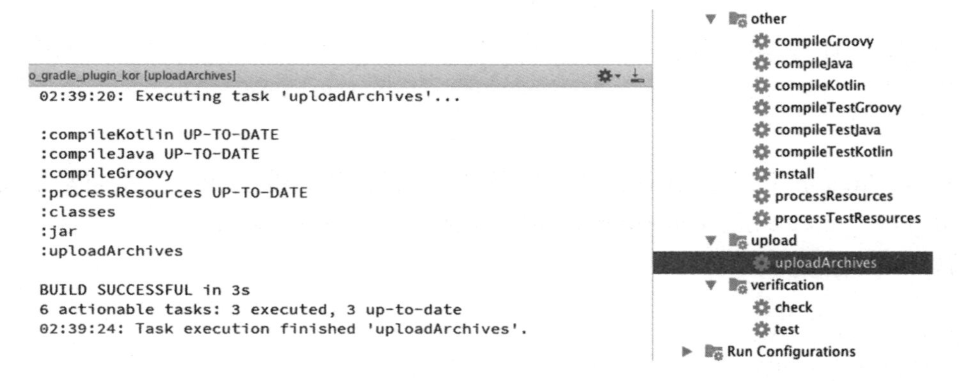

我们可以去本机的 Maven 仓库目录看到上传之后的插件 jar 包与 pom 信息：

```
~/.m2/repository/com/easy/springboot/demo_gradle_plugin_kor/1.0-SNAPSHOT
$ ls -al|awk '{print $9}'
...
maven-metadata-remote.xml
maven-metadata-remote.xml.sha1
maven-metadata.xml
```

```
maven-metadata.xml.md5
maven-metadata.xml.sha1
resolver-status.properties
```

插件项目实例工程源代码：https://github.com/KotlinSpringBoot/demo_gradle_plugin_kor

6.4.7 在项目中使用 kor 插件

在 Spring Boot 项目中使用 kor 插件的方法是直接添加 build 依赖和应用该插件，在 build.gradle 的配置文件内容如下：

```
buildscript {
    ext {
        kotlinVersion = '1.2.10'
        springBootVersion = '2.0.0.M7'
    }
    repositories {
        mavenLocal()              // 使用本地仓库
        …
    }
    dependencies {
        …
        classpath('com.easy.springboot:demo_gradle_plugin_kor:1.0-SNAPSHOT')
                                  // 添加插件依赖
    }
}

apply plugin: 'com.springboot.kor' //应用插件，名称对应到 com.springboot.kor.properties
                                   //属性文件名
…
// 自定义 kor 插件的输入参数
korArgs {
    entity = 'Kor'
}
```

其中，korArgs 是我们在 kor 插件开发代码中定义的配置参数。entity 对应自动生成实体类名、dao 层、service 层、controller 层的模板代码也会使用这个名称。

然后，命令行执行：

```
$ gradle korGenerate
我们可以看到控制台输出：
> Configure project :
The Task.leftShift(Closure) method has been deprecated and is scheduled to be removed
    in Gradle 5.0. Please use Task.doLast(Action) instead.
        at build_7dlaj3ua42ulmuink98ovyuau.run(/Users/jack/KotlinSpringBoot/demo_
            using_gradle_plugin/build.gradle:20)
            (Run with --stacktrace to get the full stack trace of this deprecation
                warning.)

> Task :korGenerate
```

```
Hello, Kor !
Group: com.easy.springboot
Name: demo_using_gradle_plugin
korArgs: Kor
Write Text:
 …
@Entity
class Kor {
    @Id
    @GeneratedValue(strategy = GenerationType.IDENTITY)
    var id: Long = -1
    var gmtCreate = Date()
    var gmtModify = Date()
    var isDeleted = 0
}
…
BUILD SUCCESSFUL in 1s
1 actionable task: 1 executed
```

然后，在我们的项目中自动生成了 entity/dao/service/controller 包，其中的自动生成的代码目录结构如下所示：

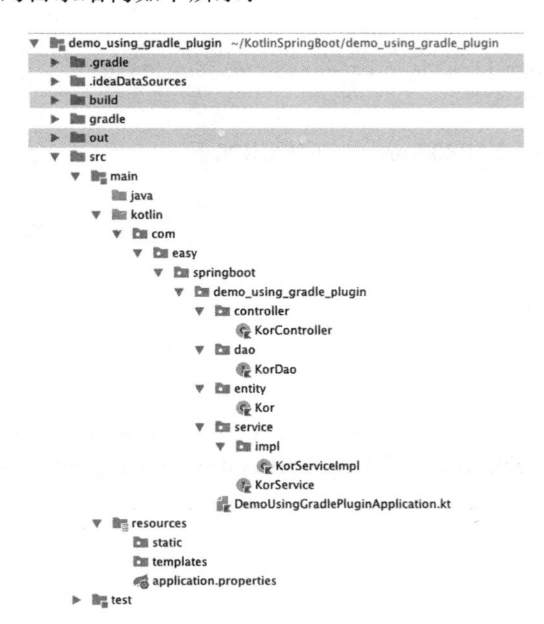

在 application.properties 中配置数据库连接：

```
spring.datasource.url=jdbc:mysql://localhost:3306/kor?useUnicode=true&character-
    Encoding=UTF8&useSSL=false
spring.datasource.username=root
spring.datasource.password=root
spring.datasource.driverClassName=com.mysql.jdbc.Driver
spring.jpa.database=MYSQL
spring.jpa.show-sql=true
```

```
# Hibernate ddl auto (create, create-drop, update)
spring.jpa.hibernate.ddl-auto=create-drop
spring.jpa.properties.hibernate.dialect=org.hibernate.dialect.MySQL5Dialect
```

在 Spring Boot 应用入口类 DemoUsingGradlePluginApplication.kt 中初始化测试数据，代码如下：

```
@SpringBootApplication
class DemoUsingGradlePluginApplication

fun main(args: Array<String>) {
    SpringApplicationBuilder().initializers(
            beans {
                bean {
                    ApplicationRunner {
                        /** 初始化测试数据 */
                        val korDao = ref<KorDao>()
                        val kor = Kor()
                        korDao.save(kor)
                    }
                }
            }
    ).sources(DemoUsingGradlePluginApplication::class.java).run(*args)
}
```

重新启动应用，浏览器输入 http://127.0.0.1:8080/kor/，响应输出如下：

```
[
  {
    "id": 1,
    "gmtCreate": "2018-01-17T14:33:43.000+0000",
    "gmtModify": "2018-01-17T14:33:43.000+0000"
  }
]
```

使用 kor 插件的示例工程源代码：https://github.com/KotlinSpringBoot/demo_using_gradle_plugin

6.5 本章小结

在软件开发的过程中，会有很多重复性的手工劳动，例如构建、打包、新建源码文件等操作。如果我们能够通过自己开发工具、插件的方式来实现这些操作的自动化，想必会大大提升工作效率。如 Ruby on Rails、Grails、React、Spring Roo 等框架都专门提供了快速开发项目的脚手架工具。

在本章中，我们就开发了一个 Gradle kor 插件，通过输入实体类名自动生成了通常所需要 entity、dao、service、controller 层的模板代码，大大节省了新建源码文件的手工劳动。这个思想值得借鉴。

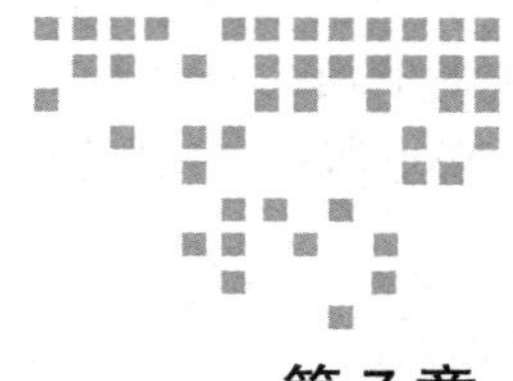

第 7 章 Chapter 7

使用 Spring MVC 开发 Web 应用

Spring 框架提供了两大核心特性：依赖注入（DI）和面向切面编程（AOP），以解耦业务逻辑层和其他各层的耦合，它将面向接口的编程思想贯穿整个应用开发过程。

本章我们先总体上介绍一下 Spring MVC，然后给出基于 Kotlin 编程语言并集成 Spring Boot 和 Spring MVC 进行服务器端 Web 应用程序的开发实战案例。

7.1 Spring MVC 简介

Spring MVC 是构建在 Servlet API 上的原生 Web 框架，并从一开始就包含在 Spring 框架中。正式名称为“ Spring Web MVC”这个名称来自其源模块 webmvc，我们现在通常将其称为“ Spring MVC”。与 Spring MVC 并行，Spring 5.0 引入了一个响应式编程框架，其名称为 Spring WebFlux。我们先介绍 Spring MVC。后面章节会介绍 Spring WebFlux。

7.1.1 Servlet 概述

Servlet 技术是 Java Web 应用的底层技术。1996 年，Sun 公司发布了 Servlet 和 JSP 规范，以替代 CGI（Common Gateway Interface，公共网关接口）技术。CGI 技术为每个请求创建进程，而创建进程会耗费大量资源。而在 Servlet 中，每个请求由一个轻量级的 Java 线程处理（而不是重量级的操作系统进程）。而 Servlet 程序运行在 JVM 进程中。Servlet 继承了 Java 的优秀特性（如垃圾回收、异常处理、Java 安全管理器等），Servlet 不太容易会出现内存管理问题和内存泄漏故障。这使得基于 Servlet 的 Web 应用程序更加安全、健壮。

Web 服务器使用轻量级线程调用 Servlet，因此，通过使用 Java 的多线程功能，Servlet

可以同时处理多个客户端请求。而 CGI 必须为每个客户端请求启动一个新的进程，轻量级线程与进程对资源的消耗完全不在一个量级。因此 Servlet 是真正高效且可伸缩的。对来自客户端的请求，Web 服务器使用轻量级线程调用 Servlet 的过程如图 7-1 所示。

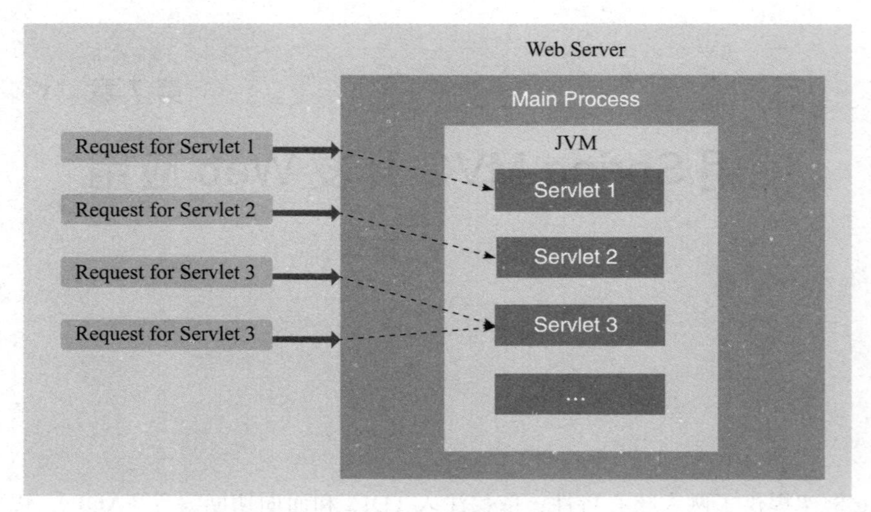

图 7-1 Web 服务器使用轻量级线程调用 Servlet 的过程

在通信量大的服务器上，Java Servlet 的优点在于执行速度更快于 CGI 程序。各个用户请求被激活成单个程序中的一个线程，而无需创建单独的进程，这意味着服务器端处理请求的系统开销将明显降低。

Servlet API 主要由两个 Java 包组成：javax.servlet 和 javax.servlet.http。直接从源代码的角度看，Servlet 就是一个实现了 javax.servlet.Servlet 接口的类。常用的实现了 Servlet 接口的类有：javax.servlet.GenericServlet 和 javax.servlet.http.HttpServlet 等。

HttpServlet 类是 GenericServlet 类的子类。HttpServlet 类为 Serlvet 接口提供了与 HTTP 协议相关的通用实现，在开发的 Java Web 应用中，自定义的 Servlet 类一般都扩展自 HttpServlet 类。

在开始解释 Servlet 如何工作之前，让我们先熟悉以下 3 个术语：

- **Web Server 服务器**：它可以处理客户端发送的 HTTP 请求，并使用 HTTP 响应对请求进行响应。
- **Web Application 应用程序**：就是我们通常所说的服务端的 Java Web 应用。它是 Servlet 的集合。
- **Web Container 容器**：也称为 Servlet 容器和 Servlet 引擎（例如 Tomcat、Jetty 等）。它是与 Servlet 交互的 Web 服务器的一部分。Web 容器是管理 Servlet 生命周期的 Web 服务器的主要组件。

Servlet 运行于支持 Java 的应用服务器中。从原理上讲，Servlet 可以响应任何类型的请求，但绝大多数情况下 Servlet 只用来扩展基于 HTTP 协议的 Web 服务器。Servlet 容器的请求执行过程如图 7-2 所示。

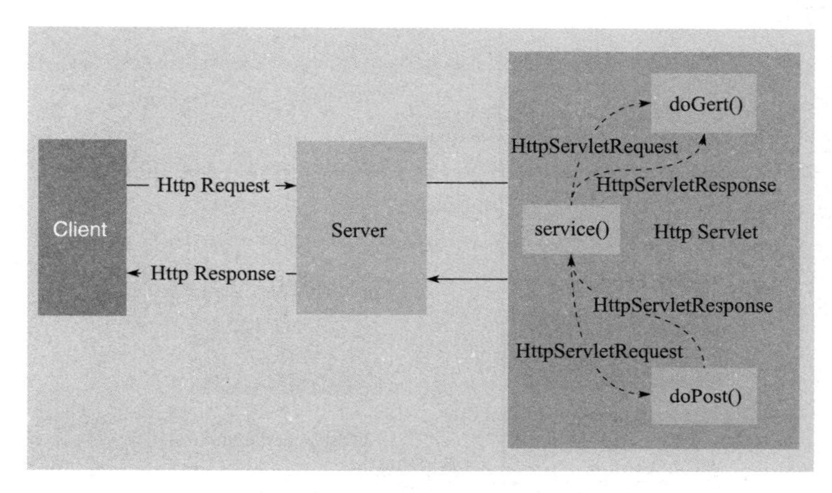

图 7-2 Servlet 的请求过程

在 Servlet 中，使用传统 CGI 程序很难完成的任务可以轻松地完成。例如，Servlet 能够直接和 Web 服务器交互，而普通的 CGI 程序不能。Servlet 还能够在各个程序之间共享数据，使得数据库连接池之类的功能很容易实现。这些使得 Servlet 成为 CGI 之后的 Web 服务端开发的主流标准。

7.1.2 MVC 简介

MVC 是模型（model）—视图（view）—控制器（controller）的缩写，一种解耦软件组件、提高代码复用性的框架模式。

MVC 是 XeroxPARC 在 20 世纪 80 年代为编程语言 Smalltalk — 80 发明的一种软件设计模式，已被广泛使用。后来成为 JavaEE 平台的基本架构模式。

MVC 分层有助于开发大型的、复杂的应用程序，可以进行分层开发和测试。例如，我们可以同时进行视图层、控制器逻辑和业务逻辑（模型层）的开发和测试。

MVC 的概念最早由 TrygveReenskaug（挪威奥斯陆大学教授）在 1978 年提出（参考论文：https://www.duo.uio.no/bitstream/handle/10852/9621/Reenskaug-MVC.pdf），并应用在 Smalltalk 系统中。MVC 三层架构如图 7-3 所示。

其中，MVC 各层分别介绍如下。

- Models（模型层），处理核心业务（数据）逻辑，模型对象负责在数据库中存取数据。这里的“数据”不仅限于数据本身，还包括处理数据的逻辑。
- Views（视图层），用于展示数据，通常视图依据模型数据创建。

- Controllers（控制器层），用于处理用户输入请求和响应输出：从视图读取数据，控制用户输入，并向模型发送数据。Controller 是在 Model 和 View 之间双向传递数据的中间协调者。

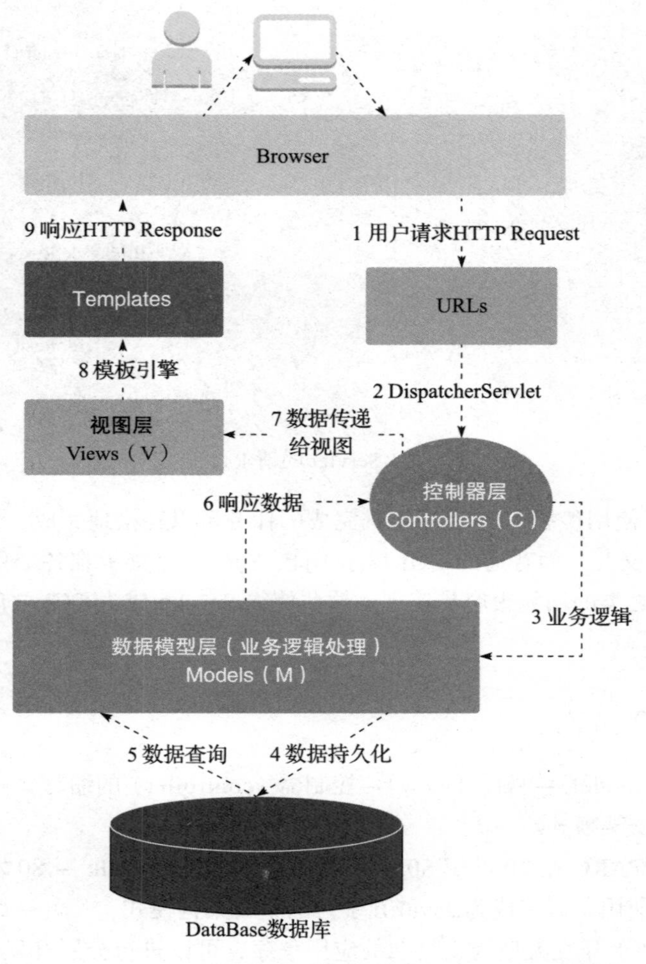

图 7-3 MVC 三层架构

7.1.3 Spring、Spring MVC 与 Spring Boot 2.0

引用一张来自 Spring 官网的一张极具代表性的 Spring 5.0 框架体系架构，如图 7-4 所示。

本章介绍的是右边的 Servlet Stack 部分。Servlet API 是同步阻塞 I/O 的架构模式。从这张图我们可以清楚地了解到 Spring Boot 2.0 与 Spring MVC 在系统整体架构中的位置。基于 Spring MVC 构建的系统架构大致如图 7-5 所示。

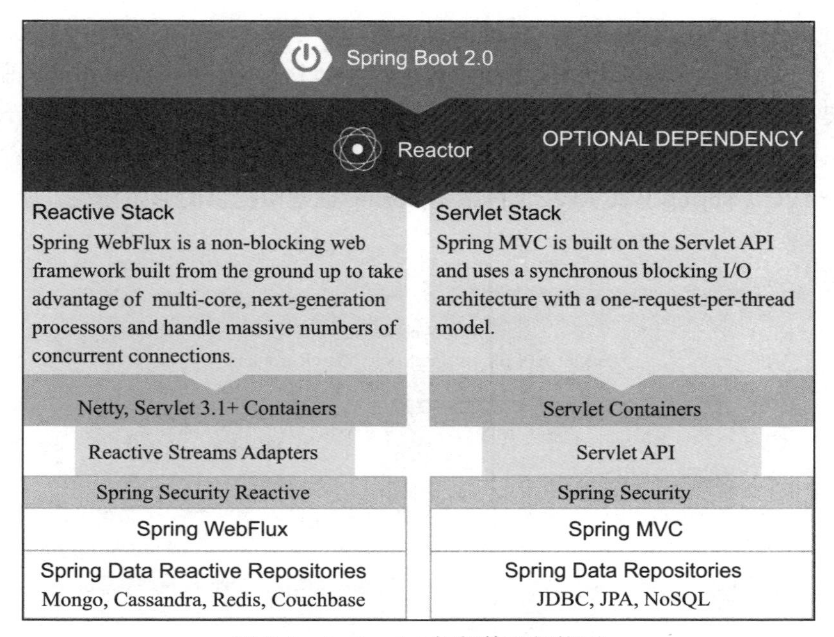

图 7-4　Spring 5.0 框架体系架构图

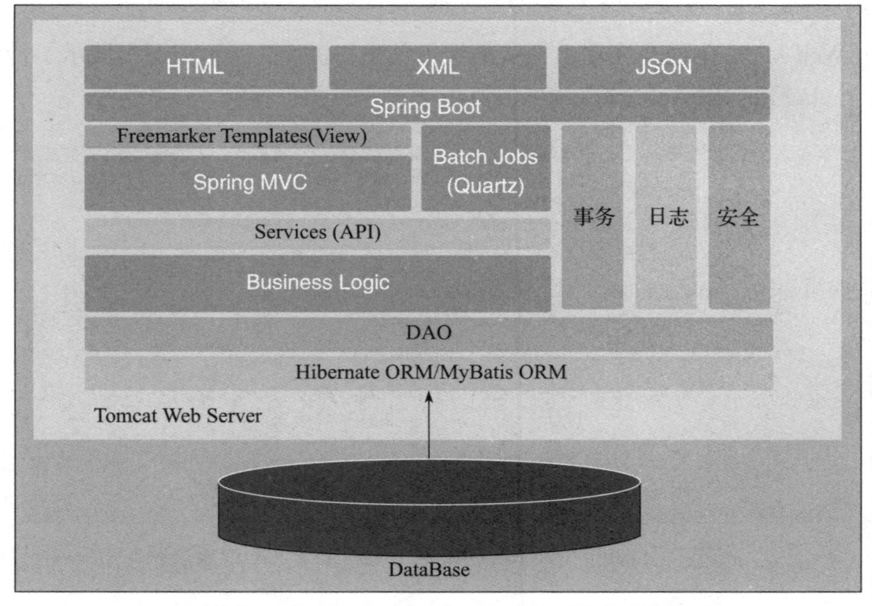

图 7-5　基于 Spring MVC 构建的系统架构图

7.1.4　Spring MVC 框架

Spring MVC 框架通过策略接口实现了高度可配置、可扩展。例如，Spring MVC 可以集成多种视图技术：JSP、Velocity、Freemarker 等。

Spring MVC 采用了约定优于配置的契约式编程风格。Spring MVC 是请求驱动的轻量级 Web 框架，使用了 MVC 架构模式的思想。基于请求驱动指的就是使用“请求 – 响应”模型。另外还有基于事件驱动的 Web 框架，如 Tapestry、JSF 等。

Spring MVC（Spring Web MVC）框架是 Spring 框架中提供的原生模块，所在的位置如图 7-6 所示。

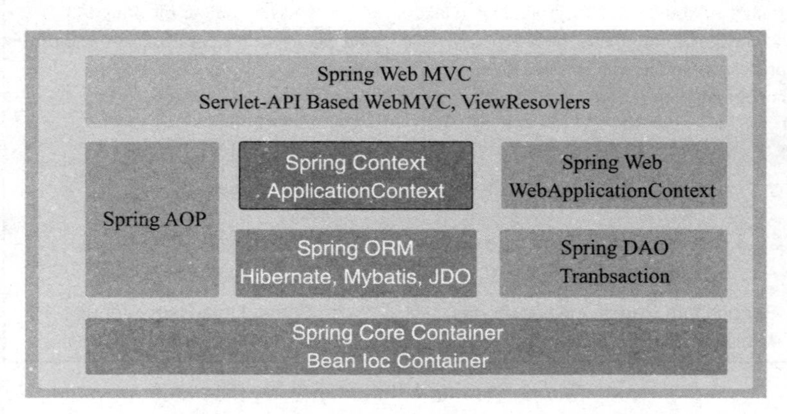

图 7-6 Spring Web MVC 框架是 Spring 框架中的原生模块

Spring Web MVC 中两个应用上下文的继承关系如图 7-7 所示。图中展示了传统 Spring MVC 中的 root/child WebApplicationContext 继承关系。

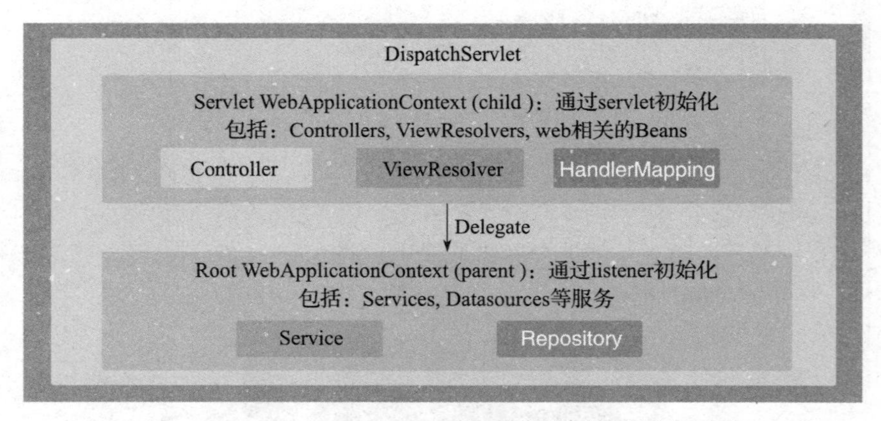

图 7-7 Spring Web MVC 中两个应用上下文的继承关系

在 Servlet WebApplicationContext（child）应用上下文中加载包含 Web 组件的 Bean，如：控制器、视图解析器和处理映射器的 Bean 等。

另外的 Root WebApplicationContext 是由 org.springframework.web.context.ContextLoader 创建的，其中包含应用的业务层、数据库层等 Bean。

root WebApplicationContext 里的 Bean 可以在不同的 child WebApplicationContext 里共享，而不同的 child WebApplicationContext 里的 Bean 区不干扰，这本来是个很好的设计。但是实际上有会不少的问题，例如：如果有两个 WebApplicationContext，常常会被误用导致重复构造 Bean、Bean 无法注入等问题。

对一些特殊问题场景下的 Bean，比如全局的 AOP 处理类，如果先在 root WebApplication-Context 里初始化了，那么 child WebApplicationContext 里的初始化的 Bean 就处理不到。如果在 child WebApplicationContext 里初始化，在 root WebApplicationContext 里的类就没有办法注入了。

区分哪些 Bean 放在 root/child 很麻烦，如果不小心容易搞错。

一劳永逸的解决办法：Bean 都由 Root WebApplicationContext 加载。在 Spring Boot 里默认情况下只有一个 Root WebApplicationContext，所以在 Spring Boot 里问题就简单了很多。

Spring MVC 处理一个 HTTP 请求的流程如图 7-8 所示。

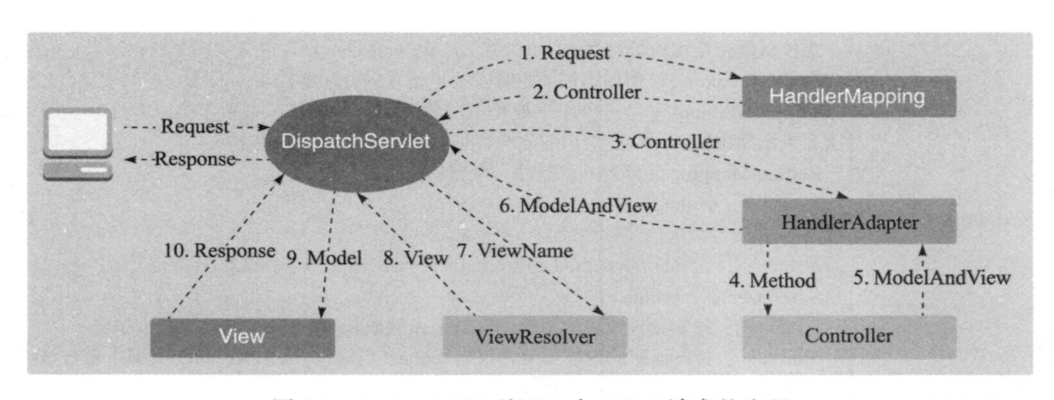

图 7-8　Spring MVC 处理一个 HTTP 请求的流程

整个过程详细介绍如下：

1）用户发送请求至前端控制器 DispatcherServlet。DispatcherServlet 收到请求调用 Handler-Mapping 处理器映射器。

2）处理器映射器根据请求 URL 找到具体的 Controller 处理器返回给 DispatcherServlet。

3）DispatcherServlet 通过 HandlerAdapter 处理器适配器调用 Controller 处理请求。

4）执行 Controller 处理器方法。

5）Controller 执行完成返回 ModelAndView。

6）HandlerAdapter 将 controller 执行结果 ModelAndView 返回给 DispatcherServlet。

7）DispatcherServlet 将 ModelAndView 的 ViewName 传给 ViewReslover 视图解析器。

8）ViewReslover 解析后返回具体 View。

9）DispatcherServlet 传递 Model 数据至 View，对 View 进行渲染视图（即将模型数据填充至视图中）。

10）DispatcherServlet 响应用户。

提示 更加详尽的原理可以参考 org.springframework.web.servlet. DispatcherServlet 类的源代码实现。更多关于 Spring Web MVC 的介绍参考官方文档：https://docs.spring.io/spring/docs/current/spring-framework-reference/web.html。

7.2 Spring MVC 常用注解

Spring MVC 中的常用注解如表 7-1 所示。

表 7-1 Spring MVC 中的常用注解

注解	功能说明
@Controller	负责注册一个控制器 Bean 到 Spring 上下文中。在 Spring MVC 中，控制器 Controller 负责处理由 DispatcherServlet 分发的请求，把用户请求的数据经过业务处理层处理之后封装成一个 Model，然后再把该 Model 返回给对应的 View 进行展示
@RequestMapping	为控制器指定可以处理哪些 URL 请求。@Controller 只是定义了一个控制器类，而使用 @RequestMapping 注解的方法才是真正处理请求的处理器。 @RequestMapping 是一个用来处理请求地址映射的注解，可用于类或方法上。用于类上，表示类中的所有响应请求的方法都是以该地址作为父路径。 RequestMapping 注解有 6 个属性，可分成三类进行说明。 1）value，method value：指定请求的实际地址。 method：指定请求的 method 类型，GET、POST、PUT、DELETE 等。 2）consumes，produces consumes：指定处理请求的提交内容类型，例如 application/json、text/html。 produces：指定返回的内容类型，仅当 request 请求头中的（Accept）类型中包含该指定类型才返回。
@RequestMapping	3）params，headers params：指定 request 中必须包含某些参数值是，才让该方法处理。 headers：指定 request 中必须包含某些指定的 header 值，才让该方法处理请求。
@RequestBody	该注解用于读取 Request 请求的 body 部分数据，使用 HandlerAdapter 配置的 HttpMessageConverter 进行解析，然后把 HttpMessageConverter 返回的对象数据绑定到 Controller 中方法的参数上
@ResponseBody	用于将 Controller 的方法返回的对象，通过 HttpMessageConverter 转换为指定格式，然后写入到 Response 对象的 body 数据区。当返回的数据不是 HTML 页面，而是其他某种格式的数据时（如 JSON、XML 等）使用
@ModelAttribute	该注解有两个用法，一个是作用于方法上，另一个是作用于参数上。 当作用于方法上时：通常用来在处理 @RequestMapping 之前为请求绑定需要从后台查询的 model； 当作用于参数上时：把相应名称的值绑定到注解的参数 Bean 上；要绑定的值通常来源于下面 3 种途径： • @SessionAttributes 标注的 attribute 对象上 • @ModelAttribute 作用于方法上时指定的方法返回 model 对象 • 上述两种情况都没有时，new 一个需要绑定的 Bean 对象，然后把 request 中按名称对应的方式把值绑定到 Bean 中

（续）

注解	功能说明
@SessionAttributes	将值放到 session 作用域中，作用在 class 上面
@RequestParam	使用 @RequestParam 标注处理方法入参，可以把请求参数传递给请求方法
@PathVariable	绑定 URL 路径占位符到入参
@ExceptionHandler	注解到方法上，出现异常时会执行该方法
@ControllerAdvice	使一个 Contoller 成为全局的异常处理类，类中用 @ExceptionHandler 方法注解的方法可以处理所有 Controller 发生的异常

@RequestHeader	把 Request 请求 header 部分的值绑定到方法的参数上。代码示例：
	`@RequestMapping("/showHeaderInfo")`
	`public void displayHeaderInfo(`
	`@RequestHeader("Accept-Encoding") String encoding,`
	`    @RequestHeader("Keep-Alive") long keepAlive)  {`
	`}`
	上面的代码，把 request header 部分的 Accept-Encoding 的值绑定到参数 encoding 上了，Keep-Alive header 的值绑定到参数 keepAlive 上

@CookieValue	把 Request header 中关于 cookie 的值绑定到方法的参数上。代码示例如下：
	`@RequestMapping("/displayHeaderInfo.do")`
	`public void displayHeaderInfo(@CookieValue("JSESSIONID") String`
	`    cookie) {`
	`}`
	即把 JSESSIONID 的值绑定到参数 cookie 上

关于这些注解的具体使用，我们将在后面的项目实例中讲到。

7.3　项目实战：使用 FreeMarker 模板引擎

本节介绍 Spring Boot 集成 FreeMarker 模板引擎进行视图层的开发。

7.3.1　FreeMarker 简介

FreeMarker 是一款模板引擎，模板编写语法为 FreeMarker Template Language（FTL）。FreeMarker 最初的设计是用于在 MVC 模式的 Web 开发框架中生成 HTML 页面，它没有被绑定到 Servlet 或 HTML 或任意 Web 相关的东西上，它也可以用于非 Web 应用环境中。来自官方文档（https://freemarker.apache.org/）的工作原理图如图 7-9 所示。

这种方式就是典型的 MVC（模型 – 视图 – 控制器）模式。更多关于 FreeMarker 的模板语言内容参考：https://sourceforge.net/projects/freemarker/files/chinese-manual/

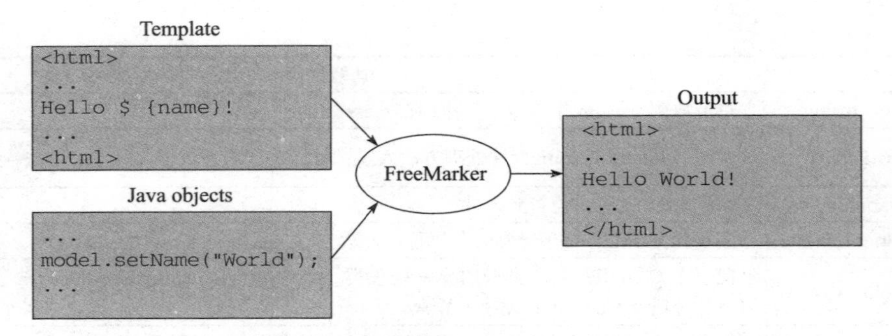

图 7-9 FreeMarker 工作原理图

7.3.2 实现一个分页查询页面

本小节我们使用 FreeMarker 模板引擎实现一个分页查询页面。

1. 工程搭建

首先，使用 Spring Initializr 创建项目，如图 7-10 所示。

Generate a Gradle Project with Kotlin and Spring Boot 2.0.0 M7

Project Metadata

Artifact coordinates

Group: com.easy.springboot

Artifact: demo_freemarker_template

Name: demo_freemarker_template

Description: Demo project for Spring Boot

Package Name: com.easy.springboot.demo_freemarker_template

Packaging: Jar

Java Version: 8

Too many options? Switch back to the simple version.

Dependencies

Add Spring Boot Starters and dependencies to your application

Search for dependencies: Web, Security, JPA, Actuator, Devtools...

Selected Dependencies: Web, Freemarker, MySQL, JPA

Generate Project

图 7-10 使用 Spring Initializr 创建项目

数据库层使用 JPA、MySQL。创建完的项目中 build.gradle 配置文件中的 compile('org.springframework.boot:spring-boot-starter-freemarker') 就是 Spring Boot 集成 FreeMarker 的起步依赖。

2. 后台分页接口

实现的后台分页接口是：http://127.0.0.1:8008/api/category/search?pageSize=10&pageNo=0&searchText= 数学 &type=1，响应的数据结构如下：

```
{
  - content: […
  +   - {
            id: 1,
            gmtCreate: "2017-12-29T14:24:58.000+0000",
            gmtModify: "2017-12-29T14:24:58.000+0000",
            name: "数学",
            detail: "",
            type: 1,
            code: "110"
        },
      +-{...},
      +-{...},
      +-{...},
      +-{...},
      +-{...},
      +-{...},
      +-{...},
      +-{...},
      +-{...}
    ],
  +-pageable: {...},
    totalPages: 80,
    totalElements: 800,
    last: false,
    size: 10,
    numberOfElements: 10,
    number: 0,
  +-sort: {...},
    first: true
}
```


其中，CategoryDao.kt 代码如下：

```
interface CategoryDao : JpaRepository<Category, Long> {
    @Query("select a from #{#entityName} a where type = :type and a.name like%:sea-
        rchText%")
    fun page(@Param("searchText") searchText: String,
             @Param("type") type: Int,
             pageable: Pageable): Page<Category>

}
```

搜索词是根据 Category 的 name 模糊匹配查询。

CategoryController.kt 关键代码如下：

```
@RestController
@RequestMapping("/api/category")
class CategoryController {

    @Autowired lateinit var CategoryDao: CategoryDao
    ...
    @GetMapping(value = ["/search"])
    fun page(
            @RequestParam(value = "pageNo", defaultValue = "0") pageNo: Int,
                 // Spring DataJPA 的分页默认第一页是: pageNo = 0
            @RequestParam(value = "pageSize", defaultValue = "10") pageSize:Int,
            @RequestParam(value = "searchText", defaultValue = "") searchText:
                String,
            @RequestParam(value = "type", defaultValue = "1") type: Int
    ): Page<Category> {
        return CategoryDao.page(
                searchText,
                type,
```

```
                PageRequest.of(pageNo, pageSize))
    }
}
```

路由 RouterController.kt 代码如下：

```
@Controller
class RouterController {

    @GetMapping(value = ["/category/list/{type}"])
    fun category(@PathVariable(value = "type") type: Int, model: Model): String {
        model["type"] = type
        return "category/list"
    }

    @GetMapping(value = ["/", ""])
    fun index(model: Model): String {
        model["type"] = 1
        return "category/list"
    }

}
```

在展示 list 页面的时候，我们使用了路径参数注解 @PathVariable 从路径中获取参数 type 的值。

3. 视图模板代码

对应的 FreeMarker 的视图模板代码目录结构如下所示。

```
▼ resources
  ▼ static
    ▼ app
        category_list.js
    ▶ bower_components
  ▼ templates
    ▼ category
        list.ftl
    ▼ layout
        foot.ftl
        head.ftl
        nav.ftl
  application.properties
  application-daily.properties
  application-dev.properties
  application-prod.properties
  logback-spring.xml
  全球行业分类.data
  图书分类.data
  学科分类.data
```

head.ftl、list.ftl 等前端视图代码参考工程源代码。

其中，head.ftl 中的 <#include 'nav.ftl'> 表示在 head.ftl 模板文件中插入 nav.ftl 模板文件的内容。这里的 include 指令的语法格式如下：

```
<#include path> 或
<#include path options>
```

path 参数表示的是要包含文件的路径，path 参数可以是如 "foo.ftl" 和 "../foo.ftl" 一样的相对路径，或者是如 "/foo.ftl" 这样的绝对路径。

path 是一个字符串表达式。path 不一定是一个固定的字符串，它也可以是像 profile.baseDir + "/menu.ftl" 这样的表达式。options 选项有如表 7-2 所示。

表 7-2　options 选项说明

options 的值	说明
parse	如果为 true，那么被包含的文件将会当作 FTL 来解析，否则整个文件将被视为简单文本（也就是说不会在其中查找 FreeMarker 的结构）。这个选项默认值为 true
encoding	一般情况下，被包含文件从包含它的文件继承编码方式（实际就是字符集），除非你用这个选项来指定编码方式。合法的名字有：ISO-8859-2、UTF-8、Shift_JIS、Big5、EUC-KR、GB2312。编码名称要和 java.io.InputStreamReader 中支持的一致
ignore_missing	当为 true 时，模板引用为空时会压制住错误，<#include...> 不会输出任何东西。当为 false 时，如果模板不存在，那么模板处理就会发生错误并停止。这个选项默认值为 false

代码示例如下：

```
<#include"/common/navbar.html" parse=false encoding="Shift_JIS" ignore_missing=true>
```

我们还可以用一个星号（*）来匹配路径，匹配 " 当前目录或其他任意其父目录 "。例如下面的代码指令：

```
<#include "*/footer.ftl">
```

那么 FreeMarker 引擎就会在下面的位置上按这个顺序查找模板：

```
/foo/bar/footer.ftl
/foo/footer.ftl
/footer.ftl
```

4. 前端代码

表格分页 JavaScript 的代码在 category_list.js 中。其中，使用了 jquery.js、bootstrap.js、bootstrap-table.js。分页的 JavaScript 代码如下：

```
var App = {
    initBootstrapTable: function () {
        var columns = [] // 表格的每一列
        columns.push(
            {
                title: "ID",
                field: "id",
                align: 'left',
                valign: 'middle',
                formatter: function (value, row, index) {
                    return value;
                }
            },
            {
                title: " 名称 ",
                field: "name",
```

```
                align: 'left',
                valign: 'middle',
                formatter: function (value, row, index) {
                    return value;
                }
            },
            {
                title: "简介",
                field: "detail",
                align: 'left',
                valign: 'middle',
                formatter: function (value, row, index) {
                    return value;
                }
            },
            {
                title: "编码",
                field: "code",
                align: 'left',
                valign: 'middle',
                formatter: function (value, row, index) {
                    return value;
                }
            }
        )

        $.extend($.fn.bootstrapTable.defaults, $.fn.bootstrapTable.locales['zhCN'])
                // 表格的文案使用中文
        var searchText = $('.search').find('input').val()
        $('#App').bootstrapTable({
            url: '/api/category/search',
            sidePagination: "server",
            queryParamsType: 'pageNo,pageSize,type',
            method: 'get',
            striped: true,
                        //是否显示行间隔色
            cache: false,
                        //是否使用缓存，默认为true，所以一般情况下需要设置一下这个属性（*）
            pagination: true,
                        //是否显示分页（*）
            paginationPreText: '上一页',
            paginationNextText: '下一页',
            search: true,
            searchText: searchText,
            searchAlign: 'right',
            searchOnEnterKey: false,
            trimOnSearch: true,
            pageNumber: 1,
                        //初始化加载第一页，默认第一页
            pageSize: 10,
                        //每页的记录行数（*）
            pageList: [10, 20, 50, 100, 200],
                        // 可选的每页数据
```

```
            totalField: 'totalElements',
                        // 所有记录 count
            dataField: 'content',
                        // 后端 json 对应的表格 List 数据的 key
            columns: columns,
            smartDisplay: true,
            queryParams: function (params) {
                // 处理查询参数
                return {
                    pageSize: params.pageSize,
                    pageNo: params.pageNumber - 1,
                    sortName: params.sortName,
                    sortOrder: params.sortOrder,
                    searchText: params.searchText,
                    type:$('#type').val()
                }
            },
            responseHandler: function (response) {
                console.log(response)
                // 类似 Filter，处理响应数据；记得要返回 response
                return response
            }
        })
    },

    init: function () {
        App.initBootstrapTable()
    }
}

$(App.init())
```

5. 运行测试

启动应用，浏览器输入 http://127.0.0.1:8008/category/list/1，输出页面如图 7-11 所示。

搜索

ID	名称	简介	编码
1	数学		110
2	数学史		11011
3	数理逻辑与数学基础	包括演绎逻辑学（亦称符号逻辑学）；证明论（亦称元数学）；递归论；模型论；公理集合论；数学基础；数理逻辑与数学基础其他学科	11014
4	数论	包括初等数论；解析数论；代数数论；超越数论；丢番图逼近；数的几何；概率数论；计算数论；数论其他学科	11017
5	代数学	线性代数；群论；域论；李群；李代数；Kac-Moody代数；环论（包括交换环与交换代数,结合环与结合代数,非结合环与非结合代数等）；模论；格论；泛代数理论；范畴论；同调代数；代数K理论；微分代数；代数编码理论；代数学其他学科	11021
6	代数几何学		11024
7	几何学	几何学基础；欧氏几何学；非欧几何学(包括黎曼几何学等)；球面几何学；向量和张量分析；仿射几何学；射影几何学；微分几何学；分数维几何；计算几何学；几何学其他学科	11027
8	拓扑学	点集拓扑学；代数拓扑学；同伦论；低维拓扑学；同调论；维数论；格上拓扑学；纤维丛论；几何拓扑学；奇点理论；微分拓扑学；拓扑学其他学科	11031
9	数学分析	微分学；积分学；级数论；数学分析其他学科	11034
10	非标准分析		11037

显示第 1 到第 10 条记录，总共 800 条记录 每页显示 10 条记录　　上一页 1 2 3 4 5 ... 80 下一页

图 7-11　表格后端分页

点击页码、上一页、下一页，可以看到表格的翻页效果。搜索框输入“计算机”，表格的查询结果分页如图 7-12 所示。

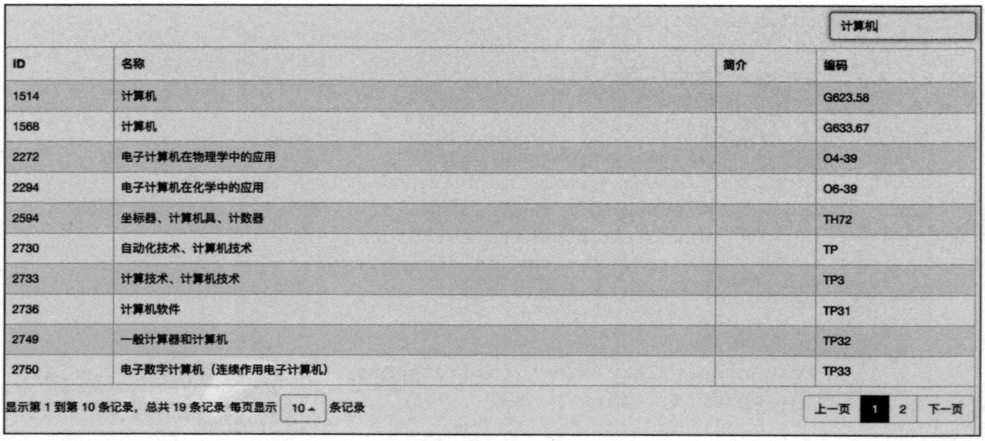

计算机

ID	名称	简介	编码
1514	计算机		G623.58
1568	计算机		G633.67
2272	电子计算机在物理学中的应用		O4-39
2294	电子计算机在化学中的应用		O6-39
2594	坐标器、计算机具、计数器		TH72
2730	自动化技术、计算机技术		TP
2733	计算技术、计算机技术		TP3
2736	计算机软件		TP31
2749	一般计算器和计算机		TP32
2750	电子数字计算机（连续作用电子计算机）		TP33

显示第 1 到第 10 条记录，总共 19 条记录 每页显示 10 条记录　　上一页 1 2 下一页

图 7-12　搜索结果的分页

7.4 实现文件下载

下面是一个简单的实现文件下载功能的代码：

```
package com.easy.springboot.demo_download_file

import org.springframework.core.io.FileSystemResource
import org.springframework.core.io.InputStreamResource
import org.springframework.http.HttpHeaders
import org.springframework.http.MediaType
import org.springframework.http.ResponseEntity
import org.springframework.stereotype.Controller
import org.springframework.web.bind.annotation.RequestMapping
import org.springframework.web.bind.annotation.RequestMethod

@Controller
class DownloadController {
    @RequestMapping(value = ["/download"], method = [RequestMethod.GET])
    fun downloadFile(log: String): ResponseEntity<InputStreamResource> {
        val filePath = "/Users/jack/logs/" + log
        val file = FileSystemResource(filePath)
        val headers = HttpHeaders()
        headers.add("Cache-Control", "no-cache, no-store, must-revalidate");
        headers.add("Content-Disposition", String.format("attachment; filename=\"%s\"",
            file.getFilename()))
        headers.add("Pragma", "no-cache")
        headers.add("Expires", "0")
```

```
        return ResponseEntity
                .ok()
                .headers(headers)
                .contentLength(file.contentLength())
                .contentType(MediaType.parseMediaType("application/octet-stream"))
                .body(InputStreamResource(file.getInputStream()))
    }

}
```

其中，contentType() 方法中设置文件类型为 application/octet-stream，从而以流的形式下载文件，这样可以实现任意格式的文件下载。

浏览器访问 http://127.0.0.1:8080/download?log=demo_logging，可以实现本机 /Users/jack/logs/demo_logging 文件的下载。

提示　本小节示例工程源代码 https://github.com/EasySpringBoot/demo_download_file

7.5　本章小结

Spring MVC 是 Spring 框架自己提供的 Web 框架，基于 Spring 框架、Servlet，采用了“模型 – 视图 – 控制器”（MVC）架构模式实现。Spring MVC 提供了丰富的注解和方便的功能特性，使得我们可以构建灵活和松耦合的服务端 Web 应用程序。

在 Spring 体系中，Spring 框架核心是“引擎”，而 Spring MVC 是基于 Spring 的 MVC 框架，Spring Boot 是一套快速开发框架。Kotlin 则是助力快速编码开发的强大语言，可以与 Spring 框架无缝集成使用。基于 Kotlin + Spring Boot 的技术栈无疑是 Java 企业级应用服务端开发的极佳选择。

提示　本章项目示例源代码地址为 https://github.com/KotlinSpringBoot/demo_freemarker_template。

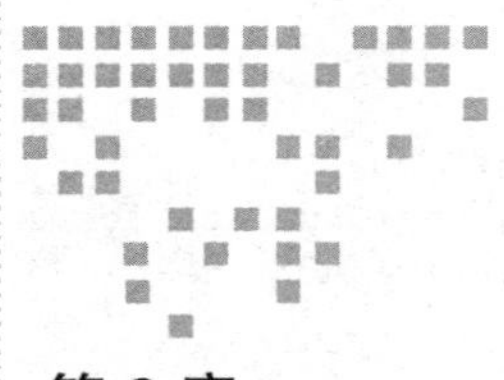

Chapter 8 第8章

Spring Boot自定义Web MVC配置

Spring Boot不仅提供相当简单实用的自动配置功能，而且开放了非常自由灵活的配置类。例如，在使用Spring MVC框架开发Java Web项目的时候，我们可以通过WebMvcConfigurationSupport等配置类来实现自由灵活的定制。本章就详细介绍在Spring Boot项目中怎样自定义Web MVC配置。

8.1 Web MVC配置简介

Spring MVC为我们提供了一个WebMvcConfigurationSupport类和一个注解@EnableWebMvc以帮助我们减少配置Bean的声明。WebMvcConfigurationSupport提供的常用配置方法如表8-1所示。

表8-1 WebMvcConfigurationSupport中常用的配置方法

配置方法	说明
addCorsMappings()	配置跨域路径映射
addFormatters()	配置格式化器
addInterceptors()	配置拦截器
addViewControllers()	配置视图控制器映射

另外，WebMvcConfigurerAdapter也可以配置MVC。WebMvcConfigurerAdapter是一个实现了WebMvcConfigurer接口的抽象类。对于这两种方式，Spring Boot都提供了默认的配置类，如果想具体了解配置可以去阅读一下源码。

下面我们简单说明一下如何自定义一个继承自WebMvcConfigurationSupport的配置类

WebMvcConfig。

首先，使用 @Configuration 将 WebMvcConfig 类标注为 Spring 配置类，代码如下：

```
@Configuration
public class WebMvcConfig: WebMvcConfigurationSupport() {
    //覆盖写配置方法
    …
}
```

然后，通过重写以下配置方法：addCorsMappings()、addFormatters()、addInterceptors()、addViewControllers() 实现自定义的配置。

当我们使用 Java Config 方式来启用 Spring MVC 配置，还需要启用 WebMvc 的配置，我们可以将 @EnableWebMvc 添加到 SpringBootApplication 入口类上，代码如下：

```
@SpringBootApplication
@ServletComponentScan(basePackages = ["com.easy.springboot.demo_spring_mvc.filter"])
@EnableAutoConfiguration(exclude = [ErrorMvcAutoConfiguration::class])
@EnableWebMvc    // 启用 Spring MVC 配置
open class DemoSpringMvcApplication

fun main(args: Array<String>) {
    SpringApplicationBuilder().initializers(
        beans {
            bean {
                ApplicationRunner {
                    initUser()
                    initCategory()
                }
            }
        }
    ).sources(DemoSpringMvcApplication::class.java).run(*args)
}
```

这样我们就可以实现自定义的 WebMvc 配置了。

下面我们来具体介绍一下 WebMvc 配置中的静态资源、拦截器、跨域、视图控制器、消息转换器、数据格式化器、视图解析器等配置方法。

8.1.1 静态资源配置

Spring Boot 中默认的静态资源配置，是把类路径下的 /static、/public、/resources 和 /META-INF/resources 文件夹的静态文件直接映射为 /**。我们可以通过覆盖写 addResourceHandlers 来定制静态资源路径映射，使用注册类 ResourceHandlerRegistry 添加相应的 ResourceHandler。示例代码如下：

```
override fun addResourceHandlers(registry: ResourceHandlerRegistry) {
    super.addResourceHandlers(registry)
    registry.addResourceHandler("/app/**")
            .addResourceLocations("classpath:/static/app/")
```

```
    registry.addResourceHandler("/bower_components/**")
            .addResourceLocations("classpath:/static/bower_components/")

}
```

上面的代码把来自浏览器的静态资源请求 /app/**、/bower_components/** 映射到 Web 工程中的类路径下面的 /static/app/、/static/bower_components/ 目录。

8.1.2 拦截器配置

通过覆盖写 addInterceptors() 方法，使用 InterceptorRegistry 注册器来添加拦截器 HandlerInterceptor。代码示例如下：

```
override fun addInterceptors(registry: InterceptorRegistry) {
    super.addInterceptors(registry)
    //注册自定义拦截器，添加拦截路径和排除拦截路径
    registry.addInterceptor(loginSessionHandlerInterceptor).
            addPathPatterns("/**").
            excludePathPatterns(
                    "/index",
                    "/login",
                    "/doLogin",
                    "/logout",
                    "/register",
                    "/doRegister",
                    "/**/*.js",
                    "/**/*.css",
                    "/**/*.css.map",
                    "/**/*.jpeg",
                    "/**/*.ico",
                    "/**/*.jpg",
                    "/**/*.png",
                    "/**/*.woff",
                    "/**/*.woff2"
            )
}
```

默认拦截所有请求 /**，白名单在 excludePathPatterns 中配置。其中，loginSessionHandlerInterceptor 是实现了 HandlerInterceptor 的拦截器。完整代码参考示例工程 LoginSessionHandlerInterceptor.kt。

8.1.3 跨域配置

通过重写 addCorsMappings 方法实现跨域配置的支持，使用 CorsRegistry 注册类添加路径映射，代码示例如下：

```
override fun addCorsMappings(registry: CorsRegistry) {
    super.addCorsMappings(registry)
```

```
    registry.addMapping("/**")
            .allowedOrigins("*")
            .allowedMethods("PUT,POST,GET,DELETE,OPTIONS")
            .allowedHeaders("*")
}
```

其中

- addMapping：配置允许跨域的路径。
- allowedMethods：配置允许访问该跨域资源服务器的请求方法，如：POST、GET、PUT、DELETE等。
- allowedOrigins：配置允许访问的跨域资源的请求域名。
- allowedHeaders：配置允许请求header的访问，如：X-TOKEN。

8.1.4 视图控制器配置

通过重写addViewControllers方法，使用ViewControllerRegistry实现视图控制器配置，代码示例如下：

```
override fun addViewControllers(registry: ViewControllerRegistry) {
    super.addViewControllers(registry)

    registry.addViewController("/").setViewName("/index")
    registry.addViewController("/index").setViewName("/index")
    registry.addViewController("/about").setViewName("/about")
    registry.addViewController("/error/403").setViewName("/error/403")
    registry.addViewController("/error/500").setViewName("/error/500")
}
```

上面的代码实现与下面的代码在逻辑上是一样的：

```
@GetMapping(value = ["/","/index"])
fun index(): String {
    return "index"
}

@GetMapping(value = ["/about"])
fun about(): String {
    return "about"
}

@GetMapping(value = ["/error/403"])
fun error_403(): String {
    return "error/403"
}

@GetMapping(value = ["/error/500"])
fun error_500(): String {
    return "error/500"
}
```

8.1.5 消息转换器配置

HttpMessageConverters（消息转换器）是在 HttpMessageConvertersAutoConfiguration 类中自动注册的。与 HttpMessageConverters（消息转换器）相关的类如下：

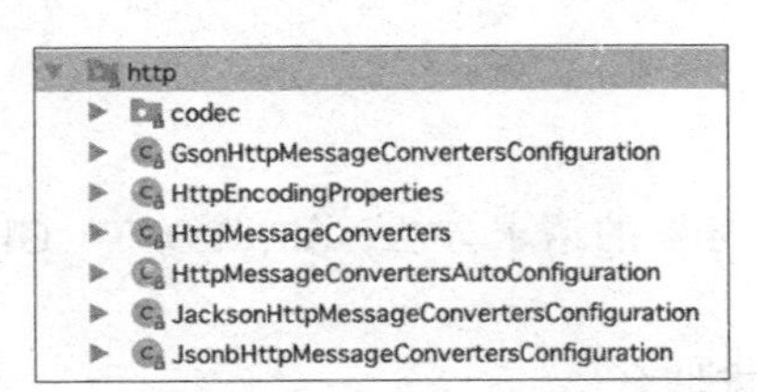

StringHttpMessageConverter 是 Spring Boot 默认自动配置的 HttpMessageConverter。除了默认的 StringHttpMessageConverter，在 HttpMessageConvertersAutoConfiguration 的自动配置类里还使用了 @Import 注解，引入了 JacksonHttpMessageConvertersConfiguration 和 GsonHttpMessageConverterConfiguration。自动配置的逻辑如下：

1）若 jackson 的 jar 包在类路径上，则 Spring Boot 通过 JacksonHttpMessageConverterConfiguration 增加 MappingJackson2HttpMessage Converter 和 MappingJackson2XmlHttpMessageConverter。

2）若 gson 的 jar 包在类路径上，则 Spring Boot 通过 GsonHttpMessageConverterConfiguration 增加 GsonHttpMessageConverter。

在 Spring Boot 中，如果要新增自定义的 HttpMessageConverter，则只需定义一个自己的 HttpMessageConverters 的 Bean，然后在此 Bean 中注册自定义 HttpMessageConverter 即可。通过覆盖重写 configureMessageConverters 方法来配置消息转换器。代码示例如下：

```
override fun configureMessageConverters(converters: MutableList<HttpMessage-
    Converter<*>>) {
    super.configureMessageConverters(converters)
    //创建 fastjson 消息转换器 : FastJsonHttpMessageConverter
    val fastConverter = FastJsonHttpMessageConverter()
    //创建 FastJsonConfig 配置类
    val fastJsonConfig = FastJsonConfig()
    //定制过滤 JSON 返回
    fastJsonConfig.setSerializerFeatures(
            SerializerFeature.WriteNullNumberAsZero,
            SerializerFeature.DisableCircularReferenceDetect,
            SerializerFeature.WriteMapNullValue,
            SerializerFeature.WriteNullStringAsEmpty
    )
    fastConverter.setFastJsonConfig(fastJsonConfig)
    //将 fastConverter 添加到视图消息转换器列表内
    converters.add(fastConverter)
}
```

8.1.6 数据格式化器配置

通过覆盖重写 addFormatters 方法来添加数据格式化器。Spring MVC 接收 HTTP 请求会把请求参数自动绑定映射到 Controller 请求参数上。Spring 中没有默认配置将字符串转换为日期类型。这个时候，可以通过添加一个 DateFormatter 类来实现自动转换。

例如，把来自 Controller 请求参数中 String 类型的日期数据格式化为 Date 日期类型，代码如下：

```
override fun addFormatters(registry: FormatterRegistry) {
    super.addFormatters(registry)
    registry.addFormatter(DateFormatter("yyyy-MM-dd"))
}
```

8.1.7 视图解析器配置

Spring MVC 提供了一个特殊的视图解析器协调类 ContentNegotiatingViewResolver，它通过代理给不同的 ViewResolver 来处理不同的视图。通过覆盖重写 configureViewResolvers() 方法来配置视图解析器。下面我们来配置一个 FreeMarker 的视图解析器，代码如下：

```
override fun configureViewResolvers(registry: ViewResolverRegistry) {
    super.configureViewResolvers(registry)
    registry.viewResolver(freeMarkerViewResolver())
}
```

其中，freeMarkerViewResolver() 是 FreeMarker 视图解析器配置，代码如下：

```
/**
 *
 * FreeMarker 视图解析器配置
 * 配置了 @Bean 注解，该注解会将方法返回值加入 Spring Ioc 容器。
 * @return
 */
@Bean
open fun freeMarkerViewResolver(): FreeMarkerViewResolver {
    val viewResolver = FreeMarkerViewResolver()
    // freemarker 本身配置了 templateLoaderPath 而在 viewResolver 中不需要配置 prefix，且路
      径前缀必须配置在 templateLoaderPath 中
    viewResolver.setPrefix("")
    viewResolver.setSuffix(".ftl")
    viewResolver.isCache = false
    viewResolver.setContentType("text/html;charset=UTF-8")
    viewResolver.setRequestContextAttribute("requestContext")
    // 为模板调用时，调用 request 对象的变量名
    viewResolver.order = 0
    viewResolver.setExposeRequestAttributes(true);
    viewResolver.setExposeSessionAttributes(true);
    return viewResolver
}
```

另外，我们还需要声明一个 FreeMarkerConfigurer Bean，这个类是 FreeMarker 配置类。我们还可以在这里添加自定义的内置变量，以便在前端 ftl 模板代码中使用。代码如下：

```
@Bean
open fun freemarkerConfig(): FreeMarkerConfigurer {
    val freemarkerConfig = FreeMarkerConfigurer()
    freemarkerConfig.setDefaultEncoding("UTF-8")
    freemarkerConfig.setTemplateLoaderPath("classpath:/templates/")
    var configuration: Configuration? = null
    try {
        configuration = freemarkerConfig.createConfiguration()
        configuration.defaultEncoding = "UTF-8"
    } catch (e: IOException) {
        log.error("freemarker 配置 bean, IO 异常 : {}", e)
    } catch (e: TemplateException) {
        log.error("freemarker 配置 bean, TemplateException 异常 : {}", e)
    }

    val freemarkerVars = mutableMapOf<String, Any>()
    freemarkerVars["rootContextPath"] = environment.getProperty("root.context.path")
    freemarkerConfig.setFreemarkerVariables(freemarkerVars)
    return freemarkerConfig
}
```

其中，freemarkerConfig.setTemplateLoaderPath("classpath:/templates/") 设置 ftl 模板文件的加载路径。在 FreeMarker 中添加内置变量 rootContextPath，代码是：

```
freemarkerVars["rootContextPath"] = environment.getProperty("root.context.path")
freemarkerConfig.setFreemarkerVariables(freemarkerVars)
```

在前端 ftl 模板代码中使用的例子是：

```
<script src="${rootContextPath}/app/login.js"></script>
<#include 'layout/foot.ftl'>
```

8.2 全局异常处理

要解决 controller 层的异常问题，当然不能在每个处理请求的方法中加上各种 try catch 的异常处理代码，那样会使代码相当烦琐。Spring MVC 中全局异常捕获处理的解决方案通常有两种方式：

1）使用 @ControllerAdvice + @ExceptionHandler 进行全局的 Controller 层异常处理。

2）实现 org.springframework.web.servlet.HandlerExceptionResolver 接口中的。resolveException() 方法。

下面我们分别通过实例来介绍。

8.2.1 使用 @ControllerAdvice 和 @ExceptionHandler 注解

在 Java Web 系统的开发中，不管是 dao 层、service 层还是 controller 层，都有可能抛出异常。在 Spring MVC 中，我们将所有类型的异常处理从各个单独的方法中解耦出来，进行异常信息的统一处理和维护。下面举例说明，如何使用 @ControllerAdvice 将 Controller 层的异常和数据校验的异常进行统一处理。

1. 定义统一异常处理类

首先，定义一个使用 @ControllerAdvice 标注的 WikiExceptionHandler 类，再声明一个 @ExceptionHandler 标注的方法 defaultErrorHandler，代码如下：

```
@ControllerAdvice
class WikiExceptionHandler {
    var log = LoggerFactory.getLogger(WikiExceptionHandler::class.java)

    @ExceptionHandler(value = ArithmeticException::class)
    fun defaultErrorHandler(req: HttpServletRequest, e: Exception): ModelAndView {
        log.error("WikiExceptionHandler ===> {}", e.message)
        e.printStackTrace()
        //这里可根据不同异常引起的类做不同处理方式
        val exceptionName = ClassUtils.getShortName(e.javaClass)
        log.info("WikiExceptionHandler ===> {}", exceptionName)
        val mav = ModelAndView()
        mav.addObject("stackTrace", e.stackTrace)
        mav.addObject("errorMessage", e.message)
        mav.addObject("url", req.requestURL)
        mav.viewName = "forward:/error/500"
        return mav
    }

}
```

其中，@ExceptionHandler(value = ArithmeticException::class) 标注 defaultErrorHandler() 方法捕获的异常是 ArithmeticException。当捕获到 ArithmeticException 异常时，会进入该方法 defaultErrorHandler() 中的逻辑：把异常信息放入 model，跳转 /error/500 请求 URL。

2. 异常信息展示

对应的请求视图映射在一个继承 WebMvcConfigurationSupport 类的 WebMvcConfig 类的 addViewControllers 方法中。代码如下：

```
override fun addViewControllers(registry: ViewControllerRegistry) {
    super.addViewControllers(registry)

    registry.addViewController("/").setViewName("/index")
    registry.addViewController("/index").setViewName("/index")
    registry.addViewController("/about").setViewName("/about")
    registry.addViewController("/error/403").setViewName("/error/403")
    registry.addViewController("/error/500").setViewName("/error/500")
}
```

500.ftl 视图模板代码如下：

```
<#include '../layout/head.ftl'>
```

```
<div class="container">
    <h1>Exception:</h1>
    <h3>${url!}</h3>
    <h3>${errorMessage!}</h3>
    <code>
        <#list stackTrace! as line>
            ${line}
        </#list>
    </code>
</div>
<#include '../layout/foot.ftl'>
```

3. 测试异常条件

在 RouterController 类中编写一个抛出 ArithmeticException 异常的测试逻辑，代码如下：

```
@Controller
class RouterController {

    /**
     * 测试全局异常处理:
     * @ControllerAdvice
     * class WikiExceptionHandler
     */
    @GetMapping(value = ["/test/ArithmeticException"])
    fun testArithmeticException(): String {
        val x = 1 / 0 // 抛出 ArithmeticException
        return "exception"
    }
    ...
}
```

4. 运行测试

启动应用，浏览器访问：http://127.0.0.1:9000/wiki/test/ArithmeticException，可以看到页面跳转错误信息展示页面，如图 8-1 所示。

@ControllerAdvice 也还可以结合 @InitBinder、@ModelAttribute 等注解一起使用。

8.2.2 实现 HandlerExceptionResolver 接口

Spring MVC 还提供了一个 HandlerExceptionResolver 接口，可用于统一异常处理。

1. HandlerExceptionResolver 接口

HandlerExceptionResolver 接口的定义如下：

```
public interface HandlerExceptionResolver {

    @Nullable
    ModelAndView resolveException(HttpServletRequest request, HttpServletResponse
        response, @Nullable Object handler, Exception ex);

}
```

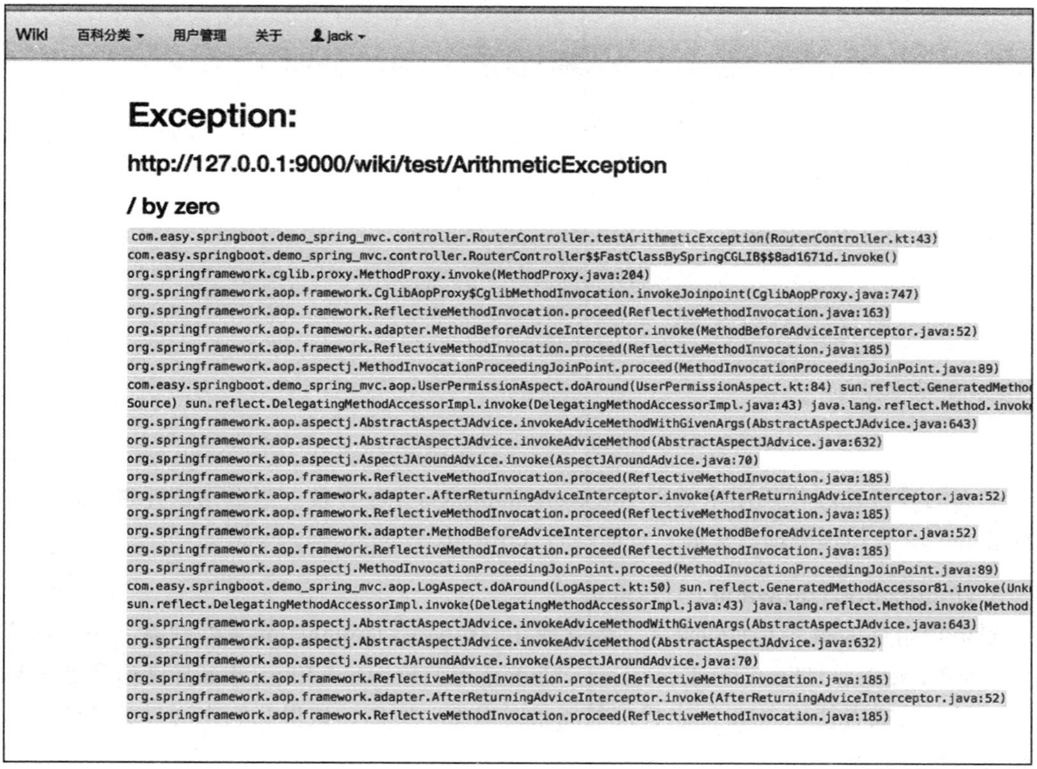

Wiki　百科分类 ▾　用户管理　关于　jack ▾

Exception:

http://127.0.0.1:9000/wiki/test/ArithmeticException

/ by zero

```
com.easy.springboot.demo_spring_mvc.controller.RouterController.testArithmeticException(RouterController.kt:43)
com.easy.springboot.demo_spring_mvc.controller.RouterController$$FastClassBySpringCGLIB$$8ad1671d.invoke()
org.springframework.cglib.proxy.MethodProxy.invoke(MethodProxy.java:204)
org.springframework.aop.framework.CglibAopProxy$CglibMethodInvocation.invokeJoinpoint(CglibAopProxy.java:747)
org.springframework.aop.framework.ReflectiveMethodInvocation.proceed(ReflectiveMethodInvocation.java:163)
org.springframework.aop.framework.adapter.MethodBeforeAdviceInterceptor.invoke(MethodBeforeAdviceInterceptor.java:52)
org.springframework.aop.framework.ReflectiveMethodInvocation.proceed(ReflectiveMethodInvocation.java:185)
org.springframework.aop.aspectj.MethodInvocationProceedingJoinPoint.proceed(MethodInvocationProceedingJoinPoint.java:89)
com.easy.springboot.demo_spring_mvc.aop.UserPermissionAspect.doAround(UserPermissionAspect.kt:84) sun.reflect.GeneratedMetho
Source) sun.reflect.DelegatingMethodAccessorImpl.invoke(DelegatingMethodAccessorImpl.java:43) java.lang.reflect.Method.invok
org.springframework.aop.aspectj.AbstractAspectJAdvice.invokeAdviceMethodWithGivenArgs(AbstractAspectJAdvice.java:643)
org.springframework.aop.aspectj.AbstractAspectJAdvice.invokeAdviceMethod(AbstractAspectJAdvice.java:632)
org.springframework.aop.aspectj.AspectJAroundAdvice.invoke(AspectJAroundAdvice.java:70)
org.springframework.aop.framework.ReflectiveMethodInvocation.proceed(ReflectiveMethodInvocation.java:185)
org.springframework.aop.framework.adapter.AfterReturningAdviceInterceptor.invoke(AfterReturningAdviceInterceptor.java:52)
org.springframework.aop.framework.ReflectiveMethodInvocation.proceed(ReflectiveMethodInvocation.java:185)
org.springframework.aop.framework.adapter.MethodBeforeAdviceInterceptor.invoke(MethodBeforeAdviceInterceptor.java:52)
org.springframework.aop.framework.ReflectiveMethodInvocation.proceed(ReflectiveMethodInvocation.java:185)
org.springframework.aop.aspectj.MethodInvocationProceedingJoinPoint.proceed(MethodInvocationProceedingJoinPoint.java:89)
com.easy.springboot.demo_spring_mvc.aop.LogAspect.doAround(LogAspect.kt:50) sun.reflect.GeneratedMethodAccessor81.invoke(Unk
sun.reflect.DelegatingMethodAccessorImpl.invoke(DelegatingMethodAccessorImpl.java:43) java.lang.reflect.Method.invoke(Method
org.springframework.aop.aspectj.AbstractAspectJAdvice.invokeAdviceMethodWithGivenArgs(AbstractAspectJAdvice.java:643)
org.springframework.aop.aspectj.AbstractAspectJAdvice.invokeAdviceMethod(AbstractAspectJAdvice.java:632)
org.springframework.aop.aspectj.AspectJAroundAdvice.invoke(AspectJAroundAdvice.java:70)
org.springframework.aop.framework.ReflectiveMethodInvocation.proceed(ReflectiveMethodInvocation.java:185)
org.springframework.aop.framework.adapter.AfterReturningAdviceInterceptor.invoke(AfterReturningAdviceInterceptor.java:52)
org.springframework.aop.framework.ReflectiveMethodInvocation.proceed(ReflectiveMethodInvocation.java:185)
```

图 8-1　/test/ArithmeticException 错误页面

HandlerExceptionResolver 接口中定义了一个 resolveException 方法，用于处理 Controller 中的异常。“Exception ex”参数即 Controller 抛出的异常。返回值类型是 ModelAndView，可以通过这个返回值来设置异常时显示的页面。

2. 异常处理实例

下面给出一个实现 HandlerExceptionResolver 接口的异常处理实例。

首先定义一个使用 @Component 标注的类 WikiHandlerExceptionResolver，该类实现 HandlerExceptionResolver 接口中的 resolveException() 方法，代码如下：

```
@Component
class WikiHandlerExceptionResolver : HandlerExceptionResolver {
    var log = LoggerFactory.getLogger(WikiHandlerExceptionResolver::class.java)
    override fun resolveException(httpServletRequest: HttpServletRequest, httpServlet-
        Response: HttpServletResponse, o: Any, e: Exception): ModelAndView?{
        var ex: Exception = Exception()
        // 先处理 UndeclaredThrowableException
        if (e is UndeclaredThrowableException) {
            ex = e.undeclaredThrowable as Exception
```

```
        }
        //这里可根据不同异常引起的类做不同处理方式
        val exceptionName = ClassUtils.getShortName(ex::class.java)
        if ("NoPermissionException" == exceptionName) {
            log.info("WikiHandlerExceptionResolver NoPermissionException ===> {}",
                exceptionName)
            e.printStackTrace()
            //向前台返回错误信息
            val model = HashMap<String, Any?>()
            model.put("stackTrace", e.stackTrace)
            model.put("errorMessage", e.message)
            model.put("url", httpServletRequest.requestURL)
            return ModelAndView("forward:/error/403", model)
        }
        return null
    }
}
```

上面的异常处理代码用来处理 NoPermissionException 异常。当遇到该异常，把异常信息放进 model 中传递到前台页面，页面跳转 /error/403。

3. UndeclaredThrowableException

如果我们直接测试上面的异常，会发现在 resolveException() 异常处理器方法中捕获到的异常是 UndeclaredThrowableException。UndeclaredThrowableException 异常通常是在 RPC 接口调用场景或者使用 JDK 动态代理的场景时发生。所以，这里我们先通过类型转换来获取真实的异常对象：

```
if (e is UndeclaredThrowableException) {
    ex = e.undeclaredThrowable as Exception
}
```

至此，WikiHandlerExceptionResolver 就可以处理 Controller 抛出的 NoPermissionException 异常了。

4. 异常逻辑测试

我们在 RouterController 中显示地抛出一个 NoPermissionException 异常：

```
@Controller
class RouterController {
    @GetMapping(value = ["/test/NoPermissionException"])
    fun testNoPermissionException(): String {
        throw NoPermissionException("没有权限")
    }
}
```

5. 运行测试

启动应用，在浏览器中输入 http://127.0.0.1:9000/wiki/test/NoPermissionException，可以看到系统跳转到无权限页面，如图 8-2 所示。

Wiki　百科分类 ▾　用户管理　关于　jack ▾

No Permission

Exception:

http://127.0.0.1:9000/wiki/test/NoPermissionException

```
com.easy.springboot.demo_spring_mvc.controller.RouterController$$EnhancerBySpringCGLIB$$1e34f1df.testNoPermissionException()
sun.reflect.NativeMethodAccessorImpl.invoke0(Native Method) sun.reflect.NativeMethodAccessorImpl.invoke(NativeMethodAccessorImpl.java:62)
sun.reflect.DelegatingMethodAccessorImpl.invoke(DelegatingMethodAccessorImpl.java:43) java.lang.reflect.Method.invoke(Method.java:497)
org.springframework.web.method.support.InvocableHandlerMethod.doInvoke(InvocableHandlerMethod.java:209)
org.springframework.web.method.support.InvocableHandlerMethod.invokeForRequest(InvocableHandlerMethod.java:136)
org.springframework.web.servlet.mvc.method.annotation.ServletInvocableHandlerMethod.invokeAndHandle(ServletInvocableHandlerMethod.java:102)
org.springframework.web.servlet.mvc.method.annotation.RequestMappingHandlerAdapter.invokeHandlerMethod(RequestMappingHandlerAdapter.java:871)
org.springframework.web.servlet.mvc.method.annotation.RequestMappingHandlerAdapter.handleInternal(RequestMappingHandlerAdapter.java:777)
org.springframework.web.servlet.mvc.method.AbstractHandlerMethodAdapter.handle(AbstractHandlerMethodAdapter.java:87)
org.springframework.web.servlet.DispatcherServlet.doDispatch(DispatcherServlet.java:991)
org.springframework.web.servlet.DispatcherServlet.doService(DispatcherServlet.java:925)
org.springframework.web.servlet.FrameworkServlet.processRequest(FrameworkServlet.java:978)
org.springframework.web.servlet.FrameworkServlet.doGet(FrameworkServlet.java:870) javax.servlet.http.HttpServlet.service(HttpServlet.java:635)
org.springframework.web.servlet.FrameworkServlet.service(FrameworkServlet.java:855) javax.servlet.http.HttpServlet.service(HttpServlet.java:742)
org.apache.catalina.core.ApplicationFilterChain.internalDoFilter(ApplicationFilterChain.java:231)
org.apache.catalina.core.ApplicationFilterChain.doFilter(ApplicationFilterChain.java:166)
org.apache.tomcat.websocket.server.WsFilter.doFilter(WsFilter.java:52)
org.apache.catalina.core.ApplicationFilterChain.internalDoFilter(ApplicationFilterChain.java:193)
org.apache.catalina.core.ApplicationFilterChain.doFilter(ApplicationFilterChain.java:166)
com.easy.springboot.demo_spring_mvc.filter.AuthenticationOncePerRequestFilter.doFilterInternal(AuthenticationOncePerRequestFilter.kt:26)
org.springframework.web.filter.OncePerRequestFilter.doFilter(OncePerRequestFilter.java:107)
org.apache.catalina.core.ApplicationFilterChain.internalDoFilter(ApplicationFilterChain.java:193)
org.apache.catalina.core.ApplicationFilterChain.doFilter(ApplicationFilterChain.java:166)
org.springframework.web.filter.CharacterEncodingFilter.doFilterInternal(CharacterEncodingFilter.java:200)
org.springframework.web.filter.OncePerRequestFilter.doFilter(OncePerRequestFilter.java:107)
org.apache.catalina.core.ApplicationFilterChain.internalDoFilter(ApplicationFilterChain.java:193)
```

图 8-2　无权限页面

8.3　定制 Web 容器

Spring Boot 支持嵌入式容器（例如：Apache Tomcat、Eclipse Jetty、RedHat Undertow），默认启动嵌入式 Tomcat 容器。

关于 Spring Boot 中 Web 服务器的配置在 application.properties 中的属性以 server.* 开头。要全部精通这些属性的配置，可谓是一项“大工程”。这个列表里的属性 Spring Boot 都提供了默认配置。不过，如果我们需要自定义一些常用的属性，可以直接在 application.properties 配置文件中设置。另外，还可以通过自定义 WebServerFactoryCustomizer 接口来实现代码层面上的定制。

在 Spring Boot 1.x 中，我们通过 EmbeddedServletContainerCustomizer 接口调优 Tomcat 自定义配置。而在 Spring Boot 2.0 中，通过 WebServerFactoryCustomizer 接口定制。Spring Boot 内置的 ServletWebServerFactory 配置类是：

```
org.springframework.boot.autoconfigure.web.servlet.DefaultServletWebServerFactory-
    Customizer
```

该类基于 Servlet 容器 Web 服务器的配置。另外，DefaultReactiveWebServerCustomizer 基于 Reactive 响应式 Web 服务器默认配置。Spring Boot 内置的 Web 容器有 Undertow、Jetty、Tomcat、Netty 等。Spring Boot 对这三个容器分别进行了实现。这些实现类如表 8-2 所示。

表 8-2 各种 WebServer 实现类

实现类	说明
TomcatWebServer	使用 TomcatServletWebServerFactory、TomcatReactiveWebServerFactory 分别创建一个基于 Servlet 和 Reactive 栈的 Tomcat Web Server
JettyWebServer	使用 JettyServletWebServerFactory、JettyReactiveWebServerFactory 分别创建一个基于 Servlet 和 Reactive 栈的 Jetty Web Server
NettyWebServer	使用 NettyReactiveWebServerFactory 创建一个基于 Reactive 栈的 Netty Web Server
UndertowWebServer	使用 UndertowServletWebServerFactory、UndertowReactiveWebServerFactory 创建一个基于 Servlet 和 Reactive 栈的 Undertow Web Server

例如，我们通过 WebServerFactoryCustomizer 接口来定制一个 TomcatServletWebServer，端口号 9000，uri 编码 UTF-8，Session 过期时间为 10 分钟。代码如下：

```
@Component、
class WikiWebServerFactoryCustomizer : WebServerFactoryCustomizer<TomcatServle-
    tWebServerFactory> {
    @Autowired lateinit var environment: Environment

    override fun customize(server: TomcatServletWebServerFactory) {
        server.port = 9000      //端口号
        server.contextPath = environment.getProperty("root.context.path")
        server.uriEncoding = Charset.forName("UTF-8")
        server.sessionTimeout = Duration.ofMinutes(10L)
                                //session 过期时间
    }
}
```

其中，environment.getProperty("root.context.path") 获取 Web 应用的上下文根路径。其值是在 application.properties 配置文件中自定义属性 root.context.path 的值：

```
root.context.path=/wiki
```

这里自定义的 WikiWebServerFactoryCustomizer Bean 由 WebServerFactoryCustomizerBeanPostProcessor 类调用。代码片段如下：

```
private Collection<WebServerFactoryCustomizer<?>> getWebServerFactoryCusto-
    mizerBeans() {
    return (Collection) this.beanFactory
            .getBeansOfType(WebServerFactoryCustomizer.class, false, false).values();
}
```

BeanFactoryPostProcessor 是在 Spring 容器加载了 Bean 的定义文件之后，在 Bean 实例化之前执行。

8.4 定制 Spring Boot 应用程序启动 Banner

在启用 Spring Boot 应用程序的时候，我们通常都会看到下面的启动 Banner：

```
  .   ____          _            __ _ _
 /\\ / ___'_ __ _ _(_)_ __  __ _ \ \ \ \
( ( )\___ | '_ | '_| | '_ \/ _` | \ \ \ \
 \\/  ___)| |_)| | | | | || (_| |  ) ) ) )
  '  |____| .__|_| |_|_| |_\__, | / / / /
 =========|_|==============|___/=/_/_/_/
 :: Spring Boot ::        (v2.0.0.M7)

2018-01-28 14:27:16.345  INFO 6670 --- [           main] c.k.k.KsbWithSecurityApplicationKt
2018-01-28 14:27:16.347  INFO 6670 --- [           main] c.k.k.KsbWithSecurityApplicationKt
2018-01-28 14:27:16.705  INFO 6670 --- [           main] ConfigServletWebServerApplicationContext
2018-01-28 14:27:24.913  INFO 6670 --- [           main] trationDelegate$BeanPostProcessorChecker
2018-01-28 14:27:25.170  INFO 6670 --- [           main] trationDelegate$BeanPostProcessorChecker
2018-01-28 14:27:25.199  INFO 6670 --- [           main] trationDelegate$BeanPostProcessorChecker
2018-01-28 14:27:25.225  INFO 6670 --- [           main] trationDelegate$BeanPostProcessorChecker
```

其实，这段 Banner 是在 Spring Boot 中默认配置的，配置类是 SpringBootBanner。如果想通过 SpringApplication 打印自己的 Banner，Spring Boot 中使用 SpringApplicationBannerPrinter 这个类来实现支持 TextBanner（banner.txt）、ImageBanner（banner.gif/ banner.jpg/ banner.png）两种 Banner 展示。

下面我们在项目的 src/main/resources 目录下添加 banner 文件：

其中，banner.png 图片如图 8-3 所示。

图 8-3 banner.png

banner.txt 的内容如下：

```
${AnsiColor.GREEN}
```

```
  _____            _               ____              _     ___   ___
 / ____|          (_)             |  _ \            | |   |__ \ / _ \
| (___  _ __  _ __ _ _ __   __ _  | |_) | ___   ___ | |_     ) || | | |
 \___ \| '_ \| '__| | '_ \ / _` | |  _ < / _ \ / _ \| __|   / / | | | |
 ____) | |_) | |  | | | | | (_| | | |_) | (_) | (_) | |_   / /_ | |_| |
|_____/| .__/|_|  |_|_| |_|\__, | |____/ \___/ \___/ \__| |____(_)___/
       | |                  __/ |
       |_|                 |___/

${AnsiColor.BRIGHT_RED}
Application : ${spring.application.name}
Web Context Path : ${root.context.path}
Spring Boot Version: ${spring-boot.version}${spring-boot.formatted-version}
```

我们可以在 banner.txt 中直接使用 application.properties 配置文件中的属性。

启动程序，将看到如下的启动界面：

```
&&&&&&&&&&&&&&&&&&&&&&&&&oooooooooooooooooooooooo
&&&&&&&&&&&&&&&&&&&&&&&oooooooooooooooooooooooo
&&&&&&&&&&&&&&&&&&&&&oooooooooooooooooooooooo
&&&&&&&&&&&&&&&&&&&oooooooooooooooooooooooo
&&&&&&&&&&&&&&&&&oooooooooooooooooooooooo
&&&&&&&&&&&&&&&oooooooooooooooooooooooo oooooooooooooooooooo
&&&&&&&&&&&&&oooooooooooooooooooooooo  oooooooooooooooooooooo
&&&&&&&&&&&oooooooooooooooooooooooo   ooooooooooo**ooooooooooo*
&&&&&&&&o&oooooooooooooooooooooo   *ooooooo  ooo  *oo  oooooooo
&&&&&&&&&&&&oooooooooooooooooo    ooooooo    ooo  *oo   ooooooo
&&&&&&&&&&&&&&&oooooooooooooo    ooooooo   ooooo  *oooo   ooooooo
&&&&&&&&&&&&&&&&&&ooooooooo    ooooooo   oooooo  *ooooo   ooooooo
&o&&&&&&&&&&&&&&&&&&ooooo     oooooooo   oooooo  *ooooo   oooooooo
&&&&&&&&&&&&&&&&&&&&&&o&&&&     :ooooooo   oooooooooooooo   oooooooo
&&&&&&&&&&&&&&&&&&&&&o&&&&&&&&    .ooooooo   oooooooooooo   oooooooo
&&&&&&&&&&&&&&&&&&&o&&&&&&&&&&&&    ooooooo    oooooooo    ooooooo
&&&&&&&&&&&&&&&&&&&&&&&&&&&&&&&&&&   oooooooo:            .oooooooo
&&&&&&&&&&&&&&&&&&&&&&&&&&&&&&&&&&&&  oooooooooooooooooooooooooo
&&&&&&&&&&&&&&&&&&&&&&&&&&&&&&&&&&&&&& oooooooooooooooooooooooo
&&&&&&&&&&&&&&&&&&&&&&&&&&&&&&&&&&&&&&&&oooooooooooooooooooooo
&&&&&&&&&&&&&&&&&&&&&&&&&&&&&&&&&&&&&&&&&&ooooooooooooooooooo
&&&&&&&&&&&&&&&&&&&&&&&&&&&&&&&&&&&&&&&&&&&&
&&&&o&&&&&&&&&&&&&&&&&&&&&&&&&&&&&&&&&&&&&&&&&
&&&&&&&&&&&&&&&&&&&&&&&&&&&&&&&&&&&&&&&&&&&&&&&&
&&&&&&&&&&&&&&&&&&&&&&&&&&&&&&&&&&&&&&&&&&&&&&&&&&
```

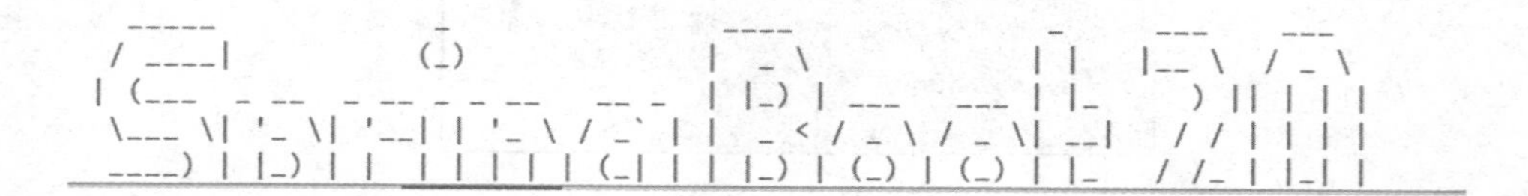

提示 本节示例工程源代码位于 https://github.com/KotlinSpringBoot/demo_spring_mvc/tree/WebMvcConfigurer。

8.5 自定义注册 Servlet、Filter 和 Listener

Spring Boot 中已经移除了 web.xml 文件，如果需要注册添加 Servlet、Filter、Listener 为 Spring Bean，在 Spring Boot 中有两种方式：

- Servlet 3.0 API 中的注解 @WebServlet、@WebListener、@WebFilter 用来配置。
- Spring Boot JavaConfig 注解配置 Bean 的方式进行配置。

本节介绍在 Spring Boot 应用中注册 Servlet、Filter 和 Listener 的方法。

首先使用 http://start.spring.io/ 创建项目、导入 IDEA 中，等待构建完毕，得到一个标准的 Gradle Spring Boot 项目。下面我们就来分别介绍如何注册 Servlet、Filter 和 Listener。

8.5.1 注册 Servlet

在包 com.easy.springboot.demo1_add_servlet 下面添加 HelloServlet.kt 类，如下代码所示：

```
@WebServlet(urlPatterns = ["/HelloServlet"])
class HelloServlet : HttpServlet() {

    override fun doGet(req: HttpServletRequest, resp: HttpServletResponse) {
        doPost(req, resp)
    }

    override fun doPost(req: HttpServletRequest, resp: HttpServletResponse) {
        println("HelloServlet doGet: req.requestURI = ${req.requestURI}")

        var cookieStr = ""
        req.cookies.forEach {
            cookieStr += "${it.name}=${it.value};"
        }

        var headers = hashMapOf<String, String>()
        println("Request Headers:")
        req.headerNames.iterator().forEach {
            println("${it} : ${req.getHeader(it)}")
            headers[it] = req.getHeader(it)
        }

        val om = ObjectMapper()
        val resultJson = om.writeValueAsString(
                User(
                        username = "你好，World",
                        id = "10000000001",
                        cookie = cookieStr,
                        headers = headers
                )
        )
        resp.contentType = "application/json;charset=UTF-8"
        val out = resp.writer
        println("HelloServlet doPost ===> ${resultJson}")
```

```
        out.println(resultJson)
    }

    data class User(
            var username: String,
            var id: String,
            var cookie: String,
            var headers: Map<String, String>
    )

}
```

我们声明了一个继承自 HttpServlet 的 HelloServlet 类，并覆写了 doGet 和 doPost 方法。在 doPost 方法中，我们打印出了请求的 Headers 和 Cookies。

其中，@WebServlet 注解标注在继承了 HttpServlet 类的 HelloServlet 之上：

```
@WebServlet(urlPatterns = ["/HelloServlet"])
```

该注解的作用等价于在 web.xml 中配置的该 servlet 的 <servlet-mapping> 元素中 <url-pattern> 的配置：

```
<servlet>
<!-- 类名 -->
<servlet-name> HelloServlet </servlet-name>
<!-- 所在的包 -->
<servlet-class> com.easy.springboot.demo_servlet_filter_listener. HelloServlet</
    servlet-class>
</servlet>
<servlet-mapping>
<servlet-name> HelloServlet </servlet-name>
<!-- 访问的 URL 路径地址 -->
<url-pattern> /HelloServlet </url-pattern>
</servlet-mapping>
```

写完 Servlet 之后，要在 Spring Boot 入口类添加 @ServletComponentScan 注解，告诉 Spring Boot 去扫描使用下面注解注册的 Servlet、Filter、Listener：

- @WebFilter 注册配置 Filter 类。
- @WebServlet 注册配置 Servlet 类。
- @WebListener 注册配置 Listener 类。

代码如下：

```
package com.easy.springboot.demo_servlet_filter_listener

import org.springframework.boot.autoconfigure.EnableAutoConfiguration
import org.springframework.boot.autoconfigure.SpringBootApplication
import org.springframework.boot.autoconfigure.web.servlet.error.ErrorMvcAuto-
    Configuration
import org.springframework.boot.runApplication
import org.springframework.boot.web.servlet.ServletComponentScan
```

```
@SpringBootApplication
@ServletComponentScan
class DemoServletFilterListenerApplication

fun main(args: Array<String>) {
    runApplication<DemoServletFilterListenerApplication>(*args)
}
```

启用应用，在浏览器中输入 http://127.0.0.1:8080/HelloServlet，响应输出如下所示：


```
{
   username: "你好，World",
   id: "10000000001",
   cookie:
   "cna=dHZeEm9PhWkCAbR/ddciOmfw;value_USER_COOKIE=22EF2F7E92890682EE9E89
   CN;emplId=110138;swork-info_USER_COOKIE=35D28621689B126096ADC1D7FA54AC
   info_SSO_TOKEN=E96439AE1182B8C2C1545434B5802E8D512E154E0F3810695208F19
 - headers: {
       accept-language: "zh-CN,zh;q=0.9,en;q=0.8",
       cookie: "cna=dHZeEm9PhWkCAbR/ddciOmfw;
       value_USER_COOKIE=22EF2F7E92890682EE9E89AFB115047DE9EC4BB31EB14D64
       SSO_LANG=zh-CN; emplId=110138; swork-
       info_USER_COOKIE=35D28621689B126096ADC1D7FA54ACD661D5FB9F36D0C8AB1
       swork-info_SSO_TOKEN=E96439AE1182B8C2C1545434B5802E8D512E154E0F381
       host: "127.0.0.1:8080",
       upgrade-insecure-requests: "1",
       connection: "keep-alive",
       cache-control: "max-age=0",
       accept-encoding: "gzip, deflate, br",
       user-agent: "Mozilla/5.0 (Macintosh; Intel Mac OS X 10_12_1) Apple
       accept: "text/html,application/xhtml+xml,application/xml;q=0.9,ima
   }
}
```


后台日志输出如下所示：

```
2018-01-20 22:57:45.847  INFO 32118 --- [ost-startStop-1] o.s.b.w.servlet.
    ServletRegistrationBean  : Mapping servlet: 'dispatcherServlet' to [/]
2018-01-20 22:57:45.849  INFO 32118 --- [ost-startStop-1] o.s.b.w.servlet.
    ServletRegistrationBean  : Mapping servlet: 'com.easy.springboot.demo_
    servlet_filter_listener.HelloServlet' to [/HelloServlet]
2018-01-20 22:57:45.858  INFO 32118 --- [ost-startStop-1] o.s.b.w.servlet.
    FilterRegistrationBean  : Mapping filter: 'characterEncodingFilter' to: [/*]
2018-01-20 22:57:45.859  INFO 32118 --- [ost-startStop-1] o.s.b.w.servlet.
    FilterRegistrationBean  : Mapping filter: 'hiddenHttpMethodFilter' to: [/*]
2018-01-20 22:57:45.860  INFO 32118 --- [ost-startStop-1] o.s.b.w.servlet.
    FilterRegistrationBean  : Mapping filter: 'httpPutFormContentFilter' to: [/*]
2018-01-20 22:57:45.860  INFO 32118 --- [ost-startStop-1] o.s.b.w.servlet.
    FilterRegistrationBean  : Mapping filter: 'requestContextFilter' to: [/*]
```

可以看出，我们的 HelloServlet(映射请求 URL 路径到 /HelloServlet)Bean 由 ServletRegistrationBean 完成注册初始化。而 Spring MVC 默认的指派 Servlet Bean dispatcherServlet（URL 请求映射到根路径 /）也是由 ServletRegistrationBean 完成注册初始化。

8.5.2 注册 Filter

可将 Filter 认为是 Servlet 的一种“变种”，它主要用于对用户请求进行预处理，也可以对 HttpServletResponse 进行后处理，是典型的职责链模式。

Filter 与 Servlet 的区别在于，它不能直接向用户生成响应。在 HttpServletRequest 到达 Servlet 之前，Filter 拦截客户的 HttpServletRequest。根据需要检查 HttpServletRequest，也可以修改 HttpServletRequest 请求头和数据。

在 HttpServletResponse 到达客户端之前，Filter 拦截 HttpServletResponse。根据需要检查 HttpServletResponse，也可以修改 HttpServletResponse 响应头和数据。

创建一个 Kotlin 类，如图 8-4 所示。新建 HelloFilter.kt 类，如图 8-5 所示。

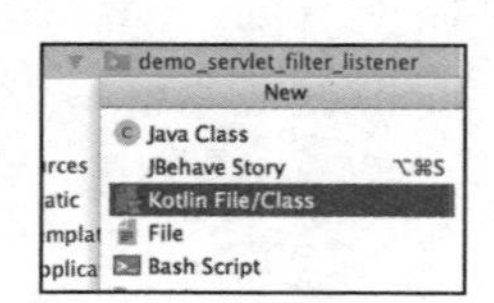

图 8-4 右击新建 Kotlin 类

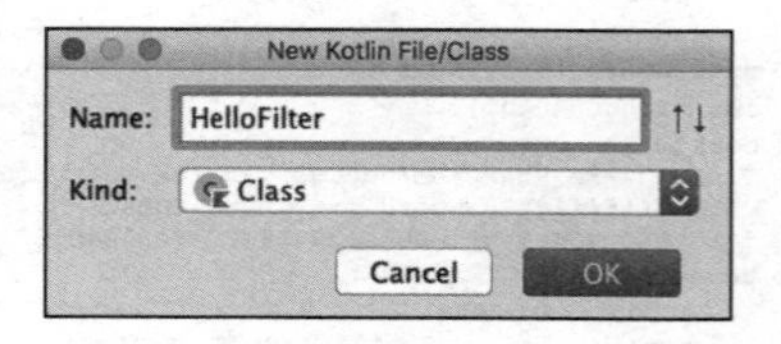

图 8-5 新建 Kotlin 类的界面

实现 javax.servlet.Filter 接口，如图 8-6 所示。

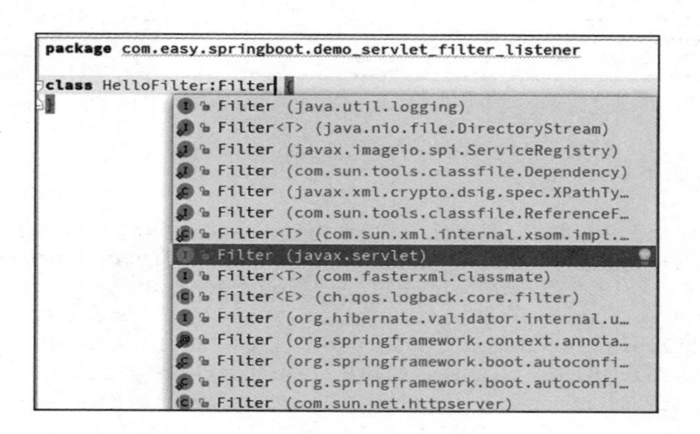

图 8-6 实现 javax. servlet. Filter 接口

使用 IDEA 的自动提示，生成代码功能，点击 Implement members，如图 8-7 所示。选择要实现的方法，如图 8-8 所示。

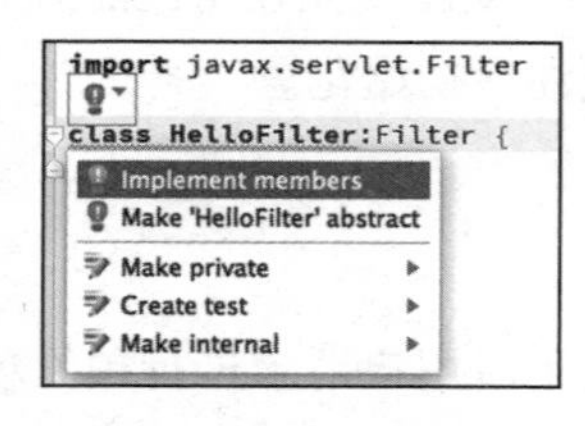

图 8-7 生成代码功能

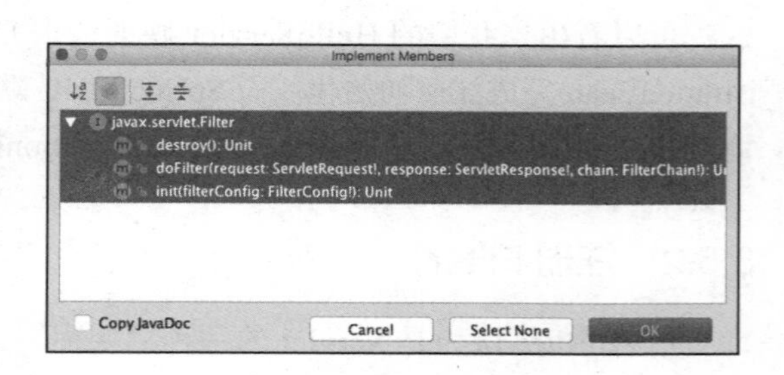

图 8-8 选择需要实现的方法

点击 OK，自动生成的代码如下所示：

```
package com.easy.springboot.demo_servlet_filter_listener

import javax.servlet.*

class HelloFilter:Filter {
    override fun destroy() {
    }

    override fun doFilter(request: ServletRequest?, response: ServletResponse?,chain:
        FilterChain?) {
    }

    override fun init(filterConfig: FilterConfig?) {
    }
}
```

我们实现 doFilter() 方法，拦截 /HelloServlet，在响应头中添加 Token 这个 Key：

```
res.setHeader("Token", token)
```

具体的实现代码如下：

```
@WebFilter(urlPatterns = ["/HelloServlet"])
class HelloFilter : Filter {
    override fun destroy() {
        println("===> 进入 HelloFilter 类 destroy 方法 ")
    }

    override fun doFilter(request: ServletRequest, response: ServletResponse, chain:
        FilterChain) {
        println("===> 进入 HelloFilter doFilter")

        chain.doFilter(request, response) // 处理请求跟响应的分界线

        println("===> chain.doFilter 后执行 setToken(response) 方法 ")
        setToken(response)
    }

    private fun setToken(response: ServletResponse) {
        val res = response as HttpServletResponse
        val token = MD5Util.md5("salt${System.currentTimeMillis()}")
        res.setHeader("Token", token)
    }

    override fun init(filterConfig: FilterConfig) {
        println("===> HelloFilter 类: ${filterConfig.filterName}")
        println("===> 进入 HelloFilter 类 init 方法 ")
    }
}
```

HelloFilter 实现了 doFilter() 方法完成对用户请求进行预处理（也可实现对服务器响应

进行后处理）。处理请求与响应的分界线是 chain.doFilter()，执行该方法之前，即对用户请求进行预处理；执行该方法之后，即对服务器响应进行后处理。在 HelloServlet 执行完逻辑，输出响应之后，又进入 chain.doFilter 之后的代码，执行 setToken（response）方法，对响应结果进行处理。

启动应用，在浏览器中输入：http://127.0.0.1:8080/HelloServlet，打开浏览器的 Console 后台，可以看到响应头中的 Token 信息：

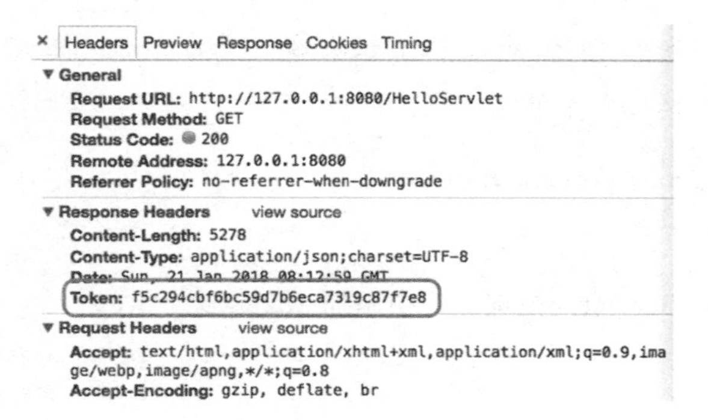

为了验证拦截的效果，我们写一个 /index 请求测试一下响应头中没有 Token 这个 Key 的场景。IndexController.kt 代码如下：

```
@RestController
class IndexController {
    @GetMapping("/index")
    fun index(): String {
        return "INDEX"
    }
}
```

打开 http://127.0.0.1:8080/index，响应头中则没有 Token 信息：

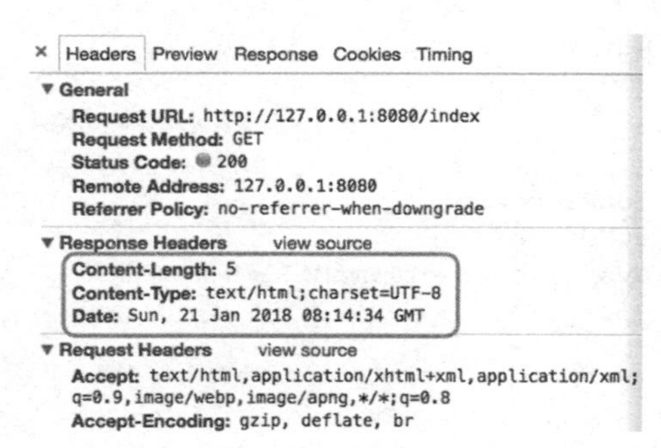

应用控制台日志信息如下：

```
INFO 34487 --- [ost-startStop-1] o.s.b.w.servlet.FilterRegistrationBean    :Mapping
    filter: 'com.easy.springboot.demo_servlet_filter_listener.HelloFilter' to urls:
    [/HelloServlet]
```

```
===> HelloFilter 类: com.easy.springboot.demo_servlet_filter_listener.HelloFilter
===> 进入 HelloFilter 类 init 方法
……
===> 进入 HelloFilter doFilter
===> 进入 HelloServlet doGet: req.requestURI = /HelloServlet
===> 进入 HelloServlet doPost: req.requestURI = /HelloServlet
HelloServlet 响应输出 ===> {"username":"你好, World","id":"10000000001","cookie":"cna=
    dHZeEm9PhWkCAbR/ddciOmfw;…
===> chain.doFilter 后执行 setToken(response) 方法
```

从日志可以看出，/HelloServlet 请求先进入 HelloFilter 过滤器，然后进入 HelloServlet 类中的 doGet、doPost 方法。HelloFilter 过滤器的 init 方法在 FilterRegistrationBean 注册完 Bean 之后就执行，而且仅执行一次。

关闭应用，我们可以看到控制台日志输出：

```
2018-01-21 16:28:40.528  INFO 34487 --- [           main] d.DemoServletFilterLis-
    tenerApplicationKt : Started DemoServletFilterListenerApplicationKt in4.443 seconds
    (JVM running for 5.359)
===> 进入 HelloFilter doFilter
===> 进入 HelloServlet doGet: req.requestURI = /HelloServlet
===> 进入 HelloServlet doPost: req.requestURI = /HelloServlet
HelloServlet resultJson ===> {"username":"你好,World","id":"10000000001","cookie":"cna=
    dHZeEm9PhWkCAbR/ddciOmfw;value_US…
…
===> 进入 HelloFilter 类 destroy 方法
```

当 AnnotationConfigServletWebServerApplicationContext 容器销毁的时候，进入 HelloFilter 类 destroy 方法。

Filter 接口中的方法总结如下：

- init()：类似于 Servlet 生命周期中的 init() 方法，用于初始化一些关于 Filter 接口的参数；只在 Servlet 启动时调用一次。
- doFilter()：类似于 Servlete 生命周期中的 service() 方法，该方法用于存放过滤器的业务逻辑实现代码。
- destory()：类似于 Servlet 生命周期中的 destroy() 方法，当 Servlet 容器销毁前调用该方法。

在开发中能用到 Filter 的场景主要有如下几种：

- 禁用浏览器的缓存（缓存处理）。
- 解决中文乱码的问题。
- 检查用户是否登录来管理用户权限。
- 用户授权的 Filter：Filter 负责检查用户请求，根据请求过滤用户非法请求。
- 日志 Filter：详细记录某些特殊的用户请求。
- 负责解码的 Filter：包括对非标准编码的请求解码。
- 其他特殊场景等。

8.5.3 注册 Listener

Servlet 的监听器 Listener 是实现了 javax.servlet.ServletContextListener 接口的服务器端程序，随 Web 应用的启动而启动，只初始化一次，随 Web 应用的停止而销毁。其主要作用是：做一些初始化的内容添加工作，设置一些基本的内容，比如一些参数或者一些固定的对象等。

添加 HelloListener.kt 类，代码如下：

```
@WebListener
class HelloListener: ServletContextListener {
    override fun contextInitialized(sce: ServletContextEvent) {
        println("===> HelloListener contextInitialized")
    }

    override fun contextDestroyed(sce: ServletContextEvent) {
        println("===> HelloListener contextDestroyed")
    }
}
```

重启应用程序，可以看到控制台日志输出：

```
...
Mapping filter: 'com.easy.springboot.demo_servlet_filter_listener.HelloFilter'to
   urls: [/HelloServlet]
===> HelloListener contextInitialized
===> HelloFilter 类: com.easy.springboot.demo_servlet_filter_listener.HelloFilter
===> 进入 HelloFilter 类 init 方法
```

HelloListener contextInitialized 在 Filter init 执行之前初始化。使用注解 @WebListener 标注一个 Listener 接口的类，如下所示：

```
javax.servlet.http.HttpSessionAttributeListener
javax.servlet.http.HttpSessionListener
javax.servlet.ServletContextAttributeListener
javax.servlet.ServletContextListener
javax.servlet.ServletRequestAttributeListener
javax.servlet.ServletRequestListener
javax.servlet.http.HttpSessionIdListener
```

可以将其分为三类，具体如下：

- 与 ServletContext 有关的 Listener 接口。包括：ServletContextListener、ServletContextAttributeListener。
- 与 HttpSession 有关的 Listener 接口。包括：HttpSessionListener、HttpSessionIdListener、HttpSessionAttributeListener。
- 与 ServletRequest 有关的 Listener 接口，包括：ServletRequestListener、ServletRequestAttributeListener。

关闭应用程序，控制台输出日志如下所示：

```
===> chain.doFilter 后执行 setToken(response) 方法
...
===> 进入 HelloFilter 类 destroy 方法
===> HelloListener contextDestroyed
```

HelloListener contextDestroyed 方法在 HelloFilter 类 destroy 方法执行之后执行。

在 Servlet 3.0 之前我们都是使用 web.xml 进行配置，需要增加 Servlet、Filter 或者 Listener 都是在 web.xml 增加相应的配置，使用起来比较烦琐。在 Spring Boot 中我们可以使用几个简单的注解即可完成注册 Servlet、Filter、Listener。

本节示例工程源代码位于 https://github.com/KotlinSpringBoot/demo_servlet_filter_listener。

8.6　本章小结

通常情况下，我们并不需要重新定义 Spring Boot 中的默认配置。但是，Spring Boot 在提供了一套默认的配置方案值之外，仍然完美支持灵活定制配置我们的应用。

在 Servlet 3.0 之前我们都是使用 web.xml 进行配置中，需要增加 Servlet、Filter 或者 Listener 时都是在 web.xml 增加相应的配置，使用起来比较烦琐。在 Spring Boot 中只需使用几个简单的注解即可完成注册 Servlet、Filter、Listener，非常极简。

Chapter 9 第9章

Spring Boot 中的 AOP 编程

本章中，我们首先介绍一下 AOP 编程的内容，然后基于 Spring Boot + Spring MVC，使用 AOP + Filter 来实现一个简单的用户登录鉴权与权限控制系统，同时，完成一个用户注册的功能。通过本章的学习，你将会深入理解 AOP 与 Filter 在 Spring Boot 实际项目中的应用，为后面深入理解 Spring Security 框架打下基础。

9.1 Spring Boot 与 AOP

本节介绍 Spring Boot 中的 AOP 编程，并给出实现一个日志切面的项目实战案例。

9.1.1 AOP 简介

我们知道面向对象编程（OOP）的特点是继承、多态和封装，而封装就要求将功能分散到不同的对象中去，这在软件设计中往往称为职责分配。而面向切面编程（Aspect Oriented Programming，AOP）则是通过预编译方式和运行期动态代理实现核心业务逻辑之外的横切行为的统一维护的一种技术，例如日志记录、跟踪、优化和监控、事务、持久化、认证、权限管理、异常处理以及针对具体行业应用场景的横切行为等。AOP 是 OOP 的补充和扩展，也是 Spring 框架中的一个重要内容，是函数式编程的一种衍生范型。利用 AOP 可以对业务逻辑的各个部分进行隔离，从而使得业务逻辑各部分之间的耦合度降低，提高程序的复用性，同时提高了开发的效率——这也正是极简编程理念的体现。

面向切面编程（AOP）技术解剖开封装的对象内部，并将那些影响了多个类的公共行为封装到一个可重用模块，并将其命名为切面（Aspect）。抽象出“切面”主要解决的问题是：

减少系统的重复代码，降低模块之间的耦合度，提高系统的可操作性和可维护性。

AOP 的切面示意图如图 9-1 所示。

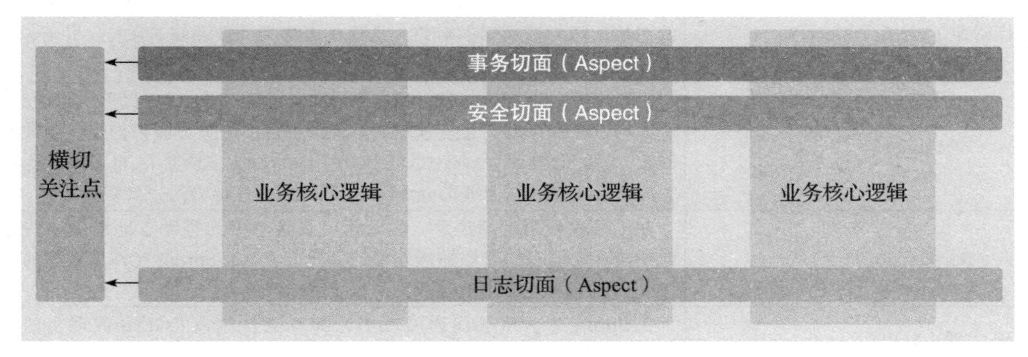

图 9-1 AOP 的切面示意图

使用“横切”技术，AOP 把软件系统分为两个部分：业务核心逻辑（纵向）关注点和横切关注点。

AOP 的作用在于分离系统中的各种关注点，将核心关注点和横切关注点分离开来。AOP 的核心概念如表 9-1 所示。

表 9-1 AOP 核心概念

名词	说明
横切关注点（Cross-cutting Concerns）	表示对哪些方法进行拦截，拦截后怎么处理
切面（Aspect）	@Aspect 将一个 java 类定义为切面类。类（Class）是对物体特征的抽象，切面（Aspect）是对横切关注点的抽象。在 Aspect 中会包含着一些 Pointcut 以及相应的 Advice
连接点（Joint Point）	被拦截到的点，例如被拦截的方法，对类成员的访问以及异常处理程序块的执行等等，它自身还可以嵌套其他 Joint Point
切入点（Pointcut）	使用 @Pointcut 定义一个切入点（规则表达式）。对连接点进行拦截定义。Pointcut 表示一组 joint point，这些 joint point 或是通过逻辑关系组合起来，或是通过通配、正则表达式等方式集中起来，它定义了相应的 Advice 将要发生的地方
通知（Advice）	所谓通知指的就是指拦截到连接点之后要执行的代码，通知分为前置、后置、异常、最终返回、环绕通知等 5 种： Advice 定义了在 Pointcut 里面定义的程序点具体要做的操作。通知类型分为： • 前置 @Before：在切入点开始处切入内容 • 后置 @After：在切入点结尾处切入内容 • 最终返回 @AfterReturning：在切入点 return 之后切入内容（返回值回调，可以用来对处理返回值做一些加工处理） • 环绕 @Around：在切入点前后切入内容，并自己控制何时执行切入点自身的内容 • 异常 @AfterThrowing：用来处理当切入内容部分抛出异常之后的处理逻辑

（续）

名词	说明
目标对象（Target Object）	目标对象表示需要被织入横切关注点的对象，即该对象是切入点选择的对象，需要被通知的对象，从而也可称为“被通知对象”；由于 Spring AOP 通过代理模式实现，从而这个对象永远是被代理对象
织入（Weaving）	织入是一个是将切面应用到目标对象从而创建出 AOP 代理对象的过程。织入可以在编译期、类装载期、运行期进行。组装方面来创建一个被通知对象。这可以在编译时完成（例如使用 AspectJ 编译器），也可以在运行时完成。Spring 和其他纯 Java AOP 框架一样，在运行时完成织入
引入（Introduction）	在不修改代码的前提下，引入可以在运行期为类动态地添加一些方法或字段，为已有的类添加额外新的字段或方法，Spring 允许引入新的接口（对应一个实现）到所有被代理对象（目标对象）
AOP 代理（AOP Proxy）	在 Spring 中，AOP 代理可以用 JDK 动态代理或 CGLIB 代理实现，而通过拦截器模型应用切面

9.1.2 Spring AOP 介绍

如果说依赖注入（DI，Dependency Injection）有助于应用对象之间的解耦，那么面向切面编程（AOP，Aspect Oriented Programing）则有助于横切关注点与它们所影响的对象之间的解耦。AOP 框架使用代理模式创建的对象，从而实现在连接点处插入通知（即应用切面），就是通过代理来对目标对象应用切面。

AOP 是 Spring 框架中的一个重要内容，在 Spring 中，AOP 代理可以用 JDK 动态代理或 CGLIB 代理 CglibAopProxy 实现。Spring 中 AOP 代理由 Spring 的 IOC 容器负责生成、管理，其依赖关系也由 IOC 容器负责管理。因此，AOP 代理可以直接使用容器中的 Bean 实例作为目标，这种关系可由 IOC 容器的依赖注入提供。

Spring 中创建代理的规则为：

- 默认使用 JDK 动态代理来创建 AOP 代理，这样就可以为任何接口实例创建代理。
- 当需要代理的类不是接口的时候，Spring 会切换为使用 CGLIB 代理，也可强制使用 CGLIB 代理。

9.1.3 实现一个简单的日志切面

进行 AOP 编程的关键就是定义切入点和定义增强处理，一旦定义了合适的切入点和增强处理，AOP 框架将自动生成 AOP 代理。日志代码往往横向地散布在所有对象层次中，而与它对应的对象的核心功能毫无关系对于其他类型的代码，如安全性、异常处理和透明的持续性也都是如此，这种散布在各处的无关的代码被称为横切（cross cutting），在 OOP 设计中，它导致了大量代码的重复，而不利于各个模块的重用。

本节通过使用 AOP 编程技术来实现一个简单的日志切面。

首先使用 IDEA 中集成的 Spring Initializr 创建 Spring Boot 项目，然后添加 AspectJ 依

赖。要进行AOP编程，还需要在build.gradle中添加AOP的起步依赖spring-boot-starter-aop。完整的配置参考示例工程源代码。

等待Gradle Sync完毕，可以看到AOP starter中引入了aspectjweaver的依赖。现在我们就可以通过@Aspect等注解方式进行AOP编程了。

1. 切面逻辑实现

写一个需要被拦截做日志记录的测试Http接口HelloAopController.kt，代码如下：

```
@RestController
class HelloAopController {

    @GetMapping("hello")
    fun hello(): World {
        return World(name = "AOP", id = "1002")
    }

    data class World(var name: String, var id: String)

}
```

浏览器输入：http://127.0.0.1:8080/hello，响应输出：

```
{
  "name": "AOP",
  "id": "1002"
}
```

下面我们实现针对/hello请求的日志记录。切面类的代码如下：

```
@Component
@Aspect
class LogAspect {
    private val LOG = LoggerFactory.getLogger(LogAspect::class.java)

    // 横切点
    @Pointcut("execution(public * com.easy.springboot.demo2_aop_logging.controller.
        *.*(..))")
    fun logPointCut() {
    }

    @Before("logPointCut()")
    @Throws(Throwable::class)
    fun doBefore(joinPoint: JoinPoint) {
        // 接收到请求，记录请求内容
        val attributes = RequestContextHolder.getRequestAttributes() as Servlet
            RequestAttributes
        val request = attributes.request

        // 记录下请求内容
        LOG.info("请求地址 : " + request.requestURL.toString())
        LOG.info("HTTP METHOD : " + request.method)
```

```
        LOG.info("IP : " + request.remoteAddr)
        LOG.info("CLASS_METHOD : " + joinPoint.getSignature().getDeclaringType-
            Name() + "."
                + joinPoint.getSignature().getName())
        LOG.info(" 参数 : " + Arrays.toString(joinPoint.getArgs()))

    }

    @AfterReturning(returning = "ret", pointcut = "logPointCut()")
                                // returning 的值和 doAfterReturning 的参数名一致
    @Throws(Throwable::class)
    fun doAfterReturning(ret: Any) {
        //处理完请求，返回内容
        LOG.info(" 返回值 : " + ret)
    }

    @Around("logPointCut()")
    @Throws(Throwable::class)
    fun doAround(pjp: ProceedingJoinPoint): Any {
        val startTime = System.currentTimeMillis()
        val ob = pjp.proceed()  //ob 为方法的返回值
        LOG.info(" 耗时 : " + (System.currentTimeMillis() - startTime))
        return ob
    }

}
```

2. Pointcut 表达式

在上面的切面类 LogAspect 的切入点 fun logPointCut() 函数上面标注的注解：

```
@Pointcut("execution(public * com.easy.springboot.demo2_aop_logging.controller.*.*(..))")
```

表示拦截这个 com.easy.springboot.demo2_aop_logging.controller 包下面的所有方法的执行。其中 execution 表达式表示方法执行的时候匹配。常用的 pointcuts 匹配类型如表 9-2 所示。

表 9-2　连接点（pointcuts）匹配类型

连接点（pointcuts）匹配类型	说明
execution(MethodSignature)	方法执行
call(MethodSignature)	方法调用
execution(ConstructorSignature)	构造器执行
call(ConstructorSignature)	构造器调用
staticinitialization(TypeSignature)	类初始化
get(FieldSignature)	属性读操作
set(FieldSignature)	属性写操作
handler(TypeSignature)	处理执行
initialization(ConstructorSignature)	对象初始化
preinitialization(ConstructorSignature)	对象预先初始化

而其中的 controller.*.*（..）代表所有子目录下的所有方法，最后括号里（..）的两个点代表所有参数。AspectJ 类型匹配的通配符语法如下：

- * 表示匹配任何数量字符。
- .. 表示匹配任何数量字符的重复，如在类型模式中匹配任何数量子包，而在方法参数模式中匹配任何数量参数。
- + 表示匹配指定类型的子类型，仅能作为后缀放在类型模式后边。

3. 运行测试

浏览器再次请求：http://127.0.0.1:8080/hello，我们可以看到后台的日志输出：

```
c.e.s.demo2_aop_logging.aop.LogAspect    : 请求地址 : http://127.0.0.1:8080/hello
c.e.s.demo2_aop_logging.aop.LogAspect     : HTTP METHOD : GET
c.e.s.demo2_aop_logging.aop.LogAspect     : IP : 127.0.0.1
c.e.s.demo2_aop_logging.aop.LogAspect     : CLASS_METHOD : com.easy.springboot.
    demo2_aop_logging.controller.HelloAopController.hello
c.e.s.demo2_aop_logging.aop.LogAspect     : 参数 : []
c.e.s.demo2_aop_logging.aop.LogAspect     : 耗时 : 10
c.e.s.demo2_aop_logging.aop.LogAspect    : 返回值 : World(name=AOP, id=1002)
```

AOP 是一种编程范式，是一种程序设计思想，与具体的计算机编程语言无关，所以不止是 Java，像 .Net 等其他编程语言也有 AOP 的实现方式。AOP 的思想理念就是将通用逻辑从业务逻辑中分离出来。

本节示例工程源代码：https://github.com/KotlinSpringBoot/demo2_aop_logging

9.2　项目实战：使用 AOP + Filter 实现登录鉴权与权限控制

本节通过使用 AOP + Filter 来实现系统中的用户登录鉴权与权限控制。

9.2.1　系统整体架构

本节项目所采用的系统技术栈如下：

- 前端：JS/CSS/HTML、jQuery、Bootstrap
- 视图引擎：FreeMarker
- Java Web 层框架：基于 Servlet 技术栈的 Spring MVC + Spring + Spring Boot
- AOP 框架：AspectJ
- 编程语言：Java、Kotlin
- ORM 层：Spring Data JPA + Hibernate
- 数据库：MySQL

系统技术架构图如图 9-2 所示。

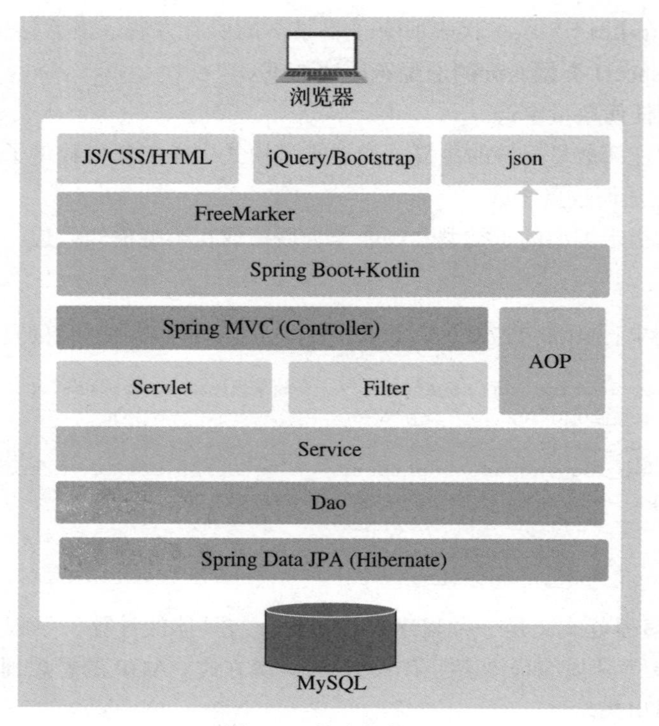

图 9-2 技术架构图

业务功能模块及其权限如表 9-3 所示。

表 9-3 业务功能模块及其权限

功能模块	请求路径	权限控制
首页	/index	任何用户
分类列表	/category/list/{type} /api/category/search	登录用户
用户管理	/user/list /api/user	管理员角色登录用户 ROLE_ADMIN
用户注册	/register /doRegister	任何用户
用户登录	/login /doLogin	任何用户
退出	/logout	任何用户
关于	/about	登录用户

9.2.2 创建工程

打开 IDEA，新建项目，使用 Spring Initialzr 来创建项目，选择起步依赖：Web、FreeMarker、JPA、MySQL、Aspects，点击 Next，如图 9-3 所示。

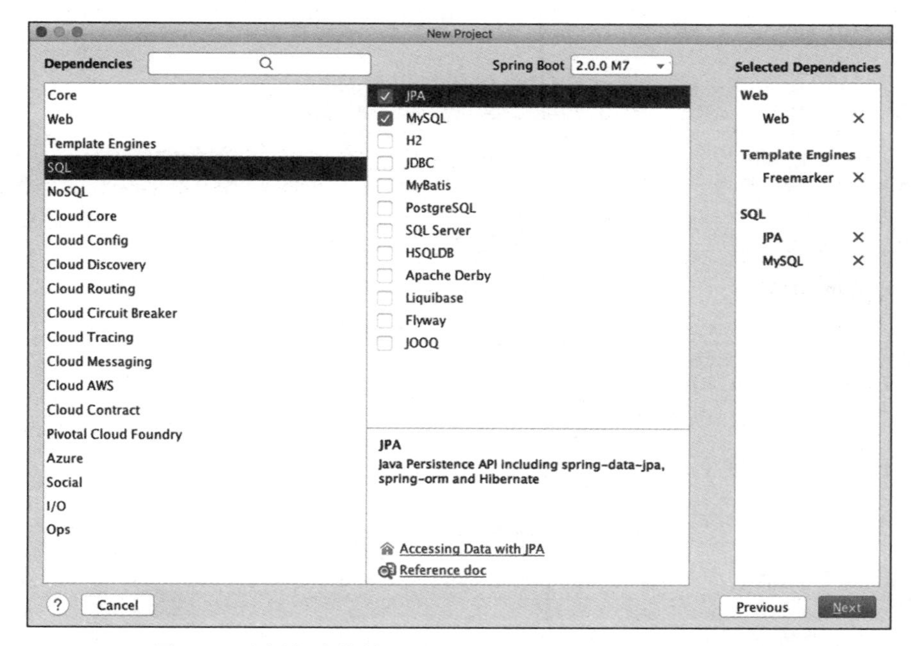

图 9-3　选择起步依赖：Web、FreeMarker、JPA、MySQL

设置项目名称，工程文件存放目录，点击 Finish，然后导入 Gradle 项目配置，点击 OK。

等待工程初始化完毕，我们将得到一个 Spring Boot 工程。该项目中我们使用在前面章节中开发的 Kor 插件，具体的配置参见在示例工程源代码中的 build.gradle 配置文件。

9.2.3　数据库表结构设计

用户数据存表为 user，角色存表为 role，用户 – 角色关联关系存入 user_roles 表，百科分类数据存表为 category。表结构分别如下。

user 表结构如下：

	Field	Type	Null	Key	Default	Extra
1	id	bigint(20)	NO	PRI	<null>	auto_increment
2	gmt_create	datetime	YES		<null>	
3	gmt_modify	datetime	YES		<null>	
4	is_deleted	int(11)	NO		<null>	
5	password	varchar(100)	YES		<null>	
6	username	varchar(100)	YES	UNI	<null>	

user 表数据如下：

	id	gmt_create	gmt_modify	is_deleted	password	username
1	1	2018-01-24 0…	2018-01-24 0…	0	ee11cbb19052e40b07a…	user
2	2	2018-01-24 0…	2018-01-24 0…	0	e10adc3949ba59abbe5…	jack
3	3	2018-01-24 0…	2018-01-24 0…	0	21232f297a57a5a7438…	admin

role 表结构如下：

	Field	Type	Null	Key	Default	Extra
1	id	bigint(20)	NO	PRI	<null>	auto_increment
2	gmt_create	datetime	YES		<null>	
3	gmt_modify	datetime	YES		<null>	
4	is_deleted	int(11)	NO		<null>	
5	role	varchar(255)	YES		<null>	

role 表数据如下：

	id	gmt_create	gmt_modify	is_deleted	role
1	1	2018-01-24 02:53:48	2018-01-24 02:53:48	0	ROLE_USER
2	2	2018-01-24 02:53:48	2018-01-24 02:53:48	0	ROLE_ADMIN

user_roles 表结构如下：

	Field	Type	Null	Key	Default	Extra
1	user_id	bigint(20)	NO	PRI	<null>	
2	roles_id	bigint(20)	NO	PRI	<null>	

user_roles 表数据如下：

	user_id	roles_id
1	1	1
2	2	1
3	2	2
4	3	1
5	3	2

category 表结构如下：

	Field	Type	Null	Key	Default	Extra
1	id	bigint(20)	NO	PRI	<null>	auto_increment
2	code	varchar(200)	YES	UNI	<null>	
3	detail	varchar(1000)	YES		<null>	
4	gmt_create	datetime	YES		<null>	
5	gmt_modify	datetime	YES		<null>	
6	is_deleted	int(11)	NO		<null>	
7	name	varchar(200)	YES	MUL	<null>	
8	type	int(11)	NO	MUL	<null>	

category 表数据如下：

	id	code	detail			is_deleted	name	type
1	1	110		-	-	0	数学	1
2	2	11011		-	-	0	数学史	1
3	3	11014	包括演绎逻辑学（亦称符号逻辑学）；证明论（亦称	-	-	0	数理逻辑与数学基础	1
4	4	11017	包括初等数论；解析数论；代数数论；超越数论；丢	-	-	0	数论	1
5	5	11021	线性代数；群论；域论；李群；李代数；Kac-Mood	-	-	0	代数学	1
6	6	11024		-	-	0	代数几何学	1
7	7	11027	几何学基础；欧氏几何学；非欧几何学(包括黎曼几何	-	-	0	几何学	1
8	8	11031	点集拓扑学；代数拓扑学；同伦论；低维拓扑学；同	-	-	0	拓扑学	1
9	9	11034	微分学；积分学；级数论；数学分析其他学科	-	-	0	数学分析	1
10	10	11037		-	-	0	非标准分析	1
11	11	11041	实变函数论；单复变函数论；多复变函数论；函数逼	-	-	0	函数论	1

9.2.4　用户登录逻辑

用户登录逻辑如图 9-4 所示。

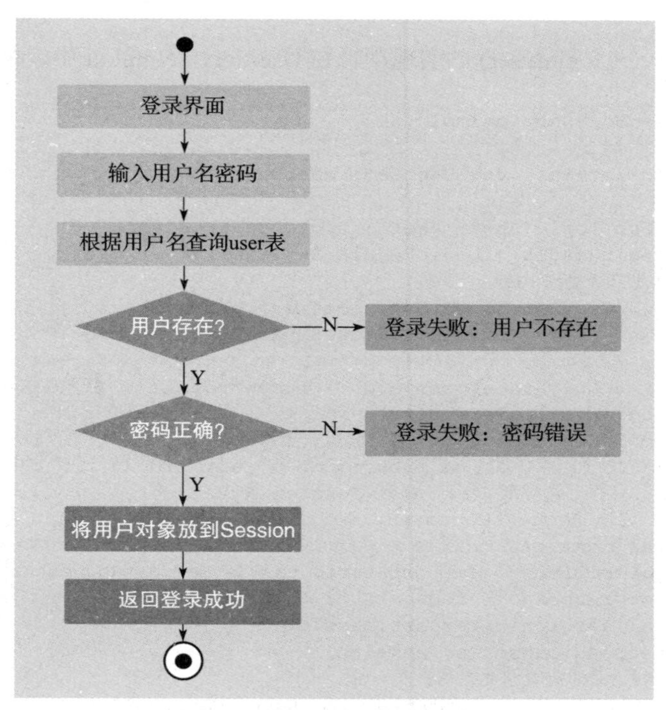

图 9-4　用户登录逻辑流程

关于用户登录的入口代码在 LoginController.kt 中。一个 /doLogin 的 POST 接口来处理登录请求，代码如下：

```
@Controller
class LoginController {
    @Autowired lateinit var UserService: UserService

    @PostMapping(value = ["/doLogin"])
    @ResponseBody
    fun doLogin(user: User): LoginResult<String> {
        return UserService.login(user)
    }

    @GetMapping(value = ["/login"])
    fun login(model: Model): String {
        return "login"
    }

    @GetMapping(value = ["/logout"])
    fun logout(session: HttpSession): String {
        session.removeAttribute(CommonContext.CURRENT_USER_CONTEXT)
```

```
        return "forward:/"
    }

}
```

其中，UserService.login(user) 的实现逻辑在 UserServiceImpl.kt 中，关键代码如下：

```
@Service class UserServiceImpl :
UserService {
    @Autowired lateinit var UserDao: UserDao

    override fun login(user: User): LoginResult<String> {
        val loginResult = LoginResult<String>()
        // 根据用户名查询 User 对象
        val u = UserDao.findByUsername(user.username)
        val password = MD5Util.md5(user.password)

        if (u != null && u.password != password) {// 1.用户存在 && 密码错误
            loginResult.isSuccess = false
            loginResult.msg = "登录失败：密码错误"
        } else if (u != null && u.password == password) {// 2.用户存在 && 密码正确
            // 3.将用户对象 User 放到 Session 中
            // 通过 RequestContextHolder 获取当前请求属性
            val requestAttributes = RequestContextHolder.currentRequestAttri-butes()
            val request = (requestAttributes as ServletRequestAttributes).request
            val response = requestAttributes.response
            val session = request.session
            setSessionUser(u, session)
            // 4.是否有登录之后的跳转 URL
            var toURL = session.getAttribute(CommonContext.LOGIN_REDIRECT_URL)as? String
            if (toURL == null) {
                toURL = "/"
            }
            // 5.返回登录成功
            loginResult.isSuccess = true
            loginResult.msg = "登录成功"
            loginResult.result = null
            loginResult.redirectUrl = toURL
        } else {
            // 5.直接返回用户登录失败
            loginResult.isSuccess = false
            loginResult.msg = "登录失败：用户不存在"
            loginResult.result = null
        }
        return loginResult
    }

    private fun setSessionUser(u: User, session: HttpSession) {
        session.setAttribute(CommonContext.CURRENT_USER_CONTEXT, u)
          //把当前登录用户放到 session 中
    }

}
```

其中，session.setAttribute（CommonContext.CURRENT_USER_CONTEXT, u）代码中的属性 CommonContext.CURRENT_USER_CONTEXT 的值是 "currentUser"。在前端的 ftl 模板文件中，展示当前登录用户的写法是：

```
${(Session["currentUser"].username)!}
```

前端模板代码的实例在 nav.ftl 中，完整的代码参考本章节的示例工程源代码。

9.2.5　登录态鉴权过滤器

下面来设计一个系统全局过滤器 AuthenticationFilter。用一个数组来存储过滤器白名单：包含这些名单中的 url 不需要过滤，直接通过。

```
val FILTER_PASS_URLS = arrayOf(
    "/index", "/login", "/doLogin", "/logout",
    "/register", "/doRegister",
    ".js", ".css", ".jpeg", ".ico", ".jpg", ".png", ".woff")
```

1. 用户鉴权逻辑

用户鉴权逻辑流程如图 9-5 所示。

AuthenticationFilter 完整的实现代码如下：

```
@WebFilter(urlPatterns = ["/*"])
class AuthenticationFilter : Filter {

    override fun destroy() {
        println("===> AuthenticationFilter destroy")
    }

     override fun doFilter(request: ServletRequest, response: ServletRespo-
nse, chain: FilterChain) {
        println("===> AuthenticationFilter doFilter")
        val requestURL = (request as HttpServletRequest).requestURL.toString()
        // 1. 判断是否需要鉴权
        if (isNeedAuth(requestURL, request)) {
            System.err.println("Auth RequestURL: ${requestURL}")
            // 2. 执行用户登录状态鉴权
            doAuthenticationFilter(request, response, chain)
            chain.doFilter(request, response)
        } else {
            println("Pass RequestURL: ${requestURL}")
            chain.doFilter(request, response)
        }
    }

    /**
     * 该请求 URL : 不在资源白名单 && 没有被过滤过
     */
    private fun isNeedAuth(requestURL: String, request: ServletRequest) =
        !isEscapeUrls(requestURL) && request.getAttribute(CommonContext.FILTERED
```

```
        _REQUEST) == null

    private fun doAuthenticationFilter(request: ServletRequest, response:
        ServletResponse, chain: FilterChain) {
        // 3.设置过滤标识，防止一次请求多次过滤
        request.setAttribute(CommonContext.FILTERED_REQUEST, true)
        val httpServletRequest = request as HttpServletRequest

        val sessionUser = getSessionUser(httpServletRequest)

        // 如果当前 session 中不存在该用户（ 用户未登录 ）
        if (sessionUser == null) {
            // 4.跳转登录页面
            redirectLogin(request, response)
        }
    }

    private fun redirectLogin(request: ServletRequest, response: Servlet-
        Response) {
        val httpServletRequest = request as HttpServletRequest
        var toURL = httpServletRequest.requestURL.toString()
        // 查询参数处理
        val queryString = httpServletRequest.queryString
        if (queryString != "") {
            toURL += "?$queryString"
        }
        // 将用户请求的 URL 存入 Session 中，用于登录成功之后跳转
        httpServletRequest.session.setAttribute(CommonContext.LOGIN_REDIRECT_URL,
            toURL)
        httpServletRequest.getRequestDispatcher("/login")
            .forward(request, response)
    }

    private fun isEscapeUrls(requestURI: String): Boolean {
        CommonContext.FILTER_PASS_URLS.iterator().forEach {
            if (requestURI.indexOf(it) >= 0) {
                return true
            }
        }
        return false
    }

    /**
     * 获取当前 Session 中是否有该用户
     */
    private fun getSessionUser(httpServletRequest: HttpServletRequest): User? {
        return httpServletRequest.session.getAttribute(CommonContext.CURRENT_USER_
            CONTEXT) as? User
    }

    override fun init(filterConfig: FilterConfig) {
        println("===> AuthenticationFilter init")
    }
}
```

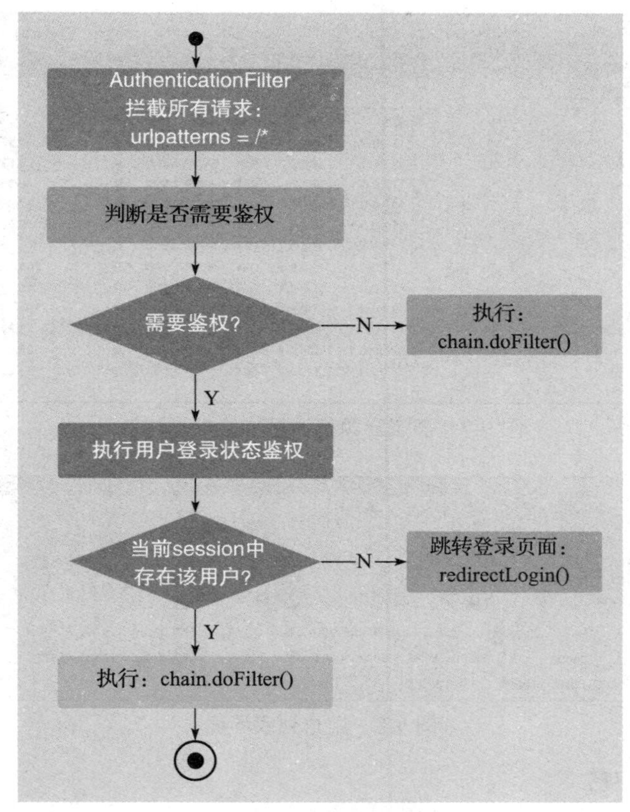

图 9-5　用户鉴权逻辑流程

2. 运行测试

浏览器输入：http://127.0.0.1:8008/user/list，系统跳转登录界面如图 9-6 所示。

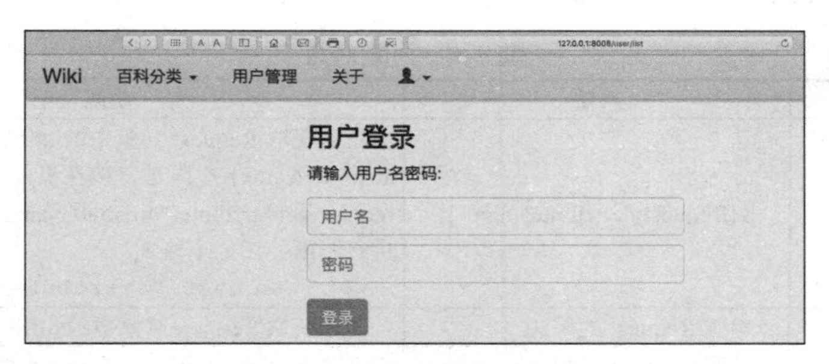

图 9-6　系统跳转登录界面

打开浏览器控制台，可以发现静态资源都通过了鉴权过滤器，如图 9-7 所示。

输入用户名、密码，我们可以看到登录成功，页面自动跳转回 http://127.0.0.1:8008/user/list，显示用户列表界面如图 9-8 所示。

名稱	網域	類型	方式	架構	狀態	已加入快取	大小
list	127.0.0.1	文件	GET	HTTP	200	否	3.92 KB
bootstrap-theme.css	127.0.0.1	樣式表	GET	HTTP	200	是	25.52 KB
bootstrap-table.css	127.0.0.1	樣式表	GET	HTTP	200	是	6.98 KB
bootstrap.css	127.0.0.1	樣式表	GET	HTTP	200	是	142.59 KB
pnotify.css	127.0.0.1	樣式表	GET	HTTP	200	是	2.60 KB
app.css	127.0.0.1	樣式表	GET	HTTP	200	是	133 B
jquery.js	127.0.0.1	工序指令	GET	HTTP	200	是	261.76 KB
bootstrap.js	127.0.0.1	工序指令	GET	HTTP	200	是	68.07 KB
bootstrap-table.js	127.0.0.1	工序指令	GET	HTTP	200	是	107.82 KB
bootstrap-table-zh-CN.js	127.0.0.1	工序指令	GET	HTTP	200	是	1.37 KB
pnotify.js	127.0.0.1	工序指令	GET	HTTP	200	是	28.98 KB
login.js	127.0.0.1	工序指令	GET	HTTP	200	是	862 B

图 9-7　静态资源都通过了鉴权过滤器

图 9-8　用户列表界面

3. 页面展示用户名

其中，用户名 jack 的显示是通过 FreeMarker 内置的 Session 对象去获取当前会话中的用户，属性 Key 是 "currentUser"：

```
${(Session["currentUser"].username)!}
```

FreeMarker 模板代码中可以直接使用的内置对象如表 9-4 所示。

表 9-4　内置对象

内置对象	用法	功能说明
Request	${Request["atrributeName"]}	用于获取 Request 对象中的 attribute 对象。${Request.atrributeName} 这样是直接在页面输出属性值，相当于 request.getAtrribute("atrributeName")；如果要对这个值进行判断使用如下格式：`<#if Request["atrributeName"]="xxx">`
Session	参照 Request 的用法	用于获取 Session 对象中的 attribute 对象
Application	参照 Request 的用法	获取 Application(ServletContext) 对象中的 attribute 对象
RequestParameters	${RequestParameters["atrributeName"]}	等同于 request.getParameter("atrributeName") 用于获取 Request 对象的 parameter 参数
Parameters	${Parameters["method"]}	属性获取，依次从 RequestParameters、Request、Session、Application 对象中获取对应属性参数，一旦获取，则不再向下查找

9.2.6　AOP 实现用户权限管理

本节使用 AOP 来实现页面权限的控制。

1. 功能简介

我们在系统中设计：用户管理页面只有管理员角色（ROLE_ADMIN）权限的用户才能访问。如果无权限，就跳转 No Permission 页面。用户 user 的角色中没有管理员角色，那么登录系统后，访问用户管理页面的效果如图 9-9 所示。

图 9-9　No Permission 页面

2. 实现权限控制

下面我们匹配请求中包含 "/user" 的请求，例如：/api/user，/user/list 等。这些请求只有管理员权限才能访问。管理员的角色是 USER_ADMIN。通过 AOP 切面中拦截方法的执行，在 CommonContext 中定义常量

```
val USER_ADMIN_PERM = "/user"
```

首先，我们从 Session 中获取当前用户对象，然后通过用户名查询该用户拥有的角色集合。如果该用户拥有 USER_ADMIN，就返回；否则进入无权限处理流程。代码如下：

```
if (requestURL.indexOf(CommonContext.USER_ADMIN_PERM) > 0) {
    val session = request.session
    val currentUser = session.getAttribute(CommonContext.CURRENT_USER_CONTEXT) as?
        User
    if (currentUser != null) {
        val roles = currentUser.roles
        roles.forEach {
            if (it.role == "ROLE_ADMIN") {
                return
            }
        }
    }
    processNoPermissionResponse(attributes)
}
```

其中，processNoPermissionResponse 主要用来处理没有权限的时候的交互。当请求是 Ajax 请求，返回 JSON 无权限信息；否则返回无权限页面：/error/403。

具体实现代码是：

```
private fun processNoPermissionResponse(attributes: ServletRequestAttributes) {
    if (isAjax(attributes)) {
        // 输出 JSON
        writeResponseJsonNoPermission(attributes)
    } else {
        // 跳转没有权限页面
        forwardNoPermissionResponse(attributes)
    }
```

```
}
private fun writeResponseJsonNoPermission(attributes: ServletRequestAttributes) {
    val response = attributes.response
    response.characterEncoding = "UTF-8"
    response.contentType = "application/json; charset=utf-8"
    val writer = response.writer
    val result = HashMap<String, Any>()
    result.put("code", "403")
    result.put("message", " 无权限 ")
    writer.write(ObjectMapper().writeValueAsString(result))
}

private fun forwardNoPermissionResponse(attributes: ServletRequestAttributes) {
    val request = attributes.request
    val response = attributes.response
    request.getRequestDispatcher("/error/403")
            .forward(request, response)
}

private fun isAjax(attributes: ServletRequestAttributes): Boolean {
    val request = attributes.request
    return "XMLHttpRequest".equals(request.getHeader("X-Requested-With"), ignore-
        Case = true)
}
```

切面逻辑的完整实现代码在实例工程的 UserPermissionAspect.kt 类中。

3. 无权限页面

我们在下面的这段代码中设置无权限跳转请求路径 /error/403：

```
request.getRequestDispatcher("/error/403")
        .forward(request, response)
```

其中，使用 RequestDispatcher 的 forward() 方法实现请求转发。

无权限路径请求接收路由：

```
@GetMapping(value = ["/error/403"])
fun error_403(): String {
    return "error/403"
}
```

无权限页面模板代码：

```
<#include '../layout/head.ftl'>
<div class="container">
    <h1>No Permission</h1>
</div>
<#include '../layout/foot.ftl'>
```

4. Pointcut 切入点配置

上面的切面代码中，切入点的配置是：

```
@Pointcut("execution(* com.easy.springboot.demo_spring_mvc.controller.*.*
    (..))")
fun userPermissionPointCut() {
}
```

其中，Pointcut 切入点 execution 配置表达式中

@Pointcut("execution(* com.easy.springboot.demo_spring_mvc.controller.*.*(..))")

关键的语法说明如表 9-5 所示。

表 9-5　Pointcut 切入点 execution 配置表达式语法

符号	含义
execution()	表达式的主体
com.easy 前的“”星号	匹配任意返回值的类型
com.easy.springboot.demo_spring_mvc.controller	AOP 拦截的包名
.*.*(..) 中的第 1 个“.*”	包下面的所有类
.*.*(..) 中的第 2 个“.*”	类中的所有方法
.*(..) 后面的 2 个点	前面的“.*”表示任何方法名，括号中的表示参数，两个点表示任何参数类型

5. 运行测试

使用 user 用户名登录，该用户没有管理员权限，访问 http://127.0.0.1:8008/user/list，跳转无权限页面，如图 9-10 所示。

使用 admin 用户登录，该用户拥有 ROLE_ADMIN 角色（管理员权限），可以正确访问到用户管理的列表页面。如图 9-11 所示。

图 9-10　跳转无权限页面

图 9-11　admin 用户可以正确访问到用户管理的列表页面

9.2.7　用户注册

本小节实现用户注册的功能。同时，还会讲到后端数据校验的相关内容。

用户注册界面如图 9-12 所示。

对应的 FreeMarker 模板代码是 register.ftl，前端的 JavaScript 代码是 register.js。处理 HTTP 接口 /doRegister POST 请求的后端 Controller 代码在 RegisterController.kt 中，代码如下：

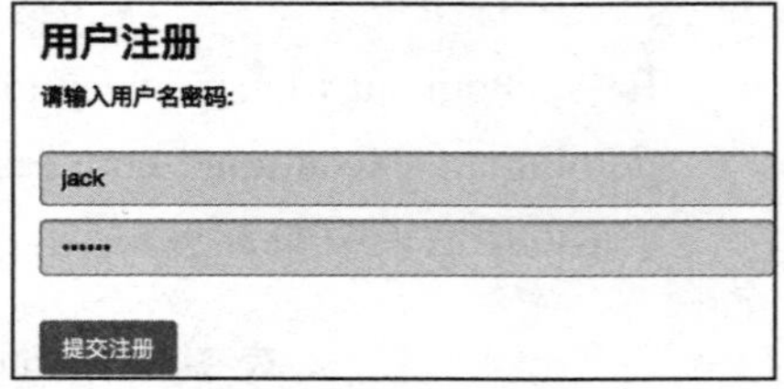

图 9-12 用户注册界面

```
@Controller
class RegisterController {
    @Autowired lateinit var UserService: UserService

    @PostMapping(value = ["/doRegister"])
    @ResponseBody
    fun doRegister(@Valid user: User,
        bindingResult: BindingResult):
        Register Result<String> {
         return UserService.register(user, bindingResult)
    }

    @GetMapping(value = ["/register"])
    fun register(): String {
        return "register"
    }
}
```

其中，处理用户注册的核心逻辑是 UserService.register() 方法，该方法中有数据校验的功能，下节介绍后端校验的实现。

9.2.8 数据后端校验

用户注册提交的表单数据通常都需要进行数据校验。数据校验是用户交互中必不可少的过程。一些比较好的用户交互可以先通过简单的前端 JavaScript 预校验，在真正提交表单的时候，在后台逻辑层还需要进行后端校验（因为前端 JavaScript 的校验，是可以绕过的）。

用户在前端页面上填写表单时，前端 JavaScript 程序会校验参数的合法性，当数据到了后端，为了防止恶意操作，保持程序的健壮性，后端同样需要对数据进行校验。最简单的做法是直接在业务方法里面进行判断，当判断成功之后再继续往下执行。但这样会使代码耦合、冗余。当多个地方需要校验时，就需要在每一个地方调用校验程序，导致代码很冗余，且不美观。

那么如何优雅地对参数进行校验呢？ JSR303 就是为了解决这个问题出现的。JSR303 是一套 Java Bean 参数校验的标准，它定义了很多常用的校验注解，我们可以直接将这些注解加在 Java Bean 的属性上面，就可以在需要校验的时候进行校验了。

在上面的 doRegister(@Valid user: User，bindingResult: BindingResult) 代码中，@Valid 注解用于校验 User 对象，所属包为 javax.validation.Valid。在 controller 层的方法要校验的参数添加 @Valid 注解，需要传入 BindingResult 对象，用于获取校验失败情况下的反馈信息。

使用 @Valid 注解校验对象，首先需要在实体类的相应字段上添加校验条件的注解，User 类字段的校验注解如下所示：

```
@Entity
class User {
    …
    @Column(unique = true)
    @NotEmpty(message = "用户名不能为空")
    @Size(max = 100, min = 1, message = "用户名长度在 1-20 之间")
    var username = ""
    @NotEmpty(message = "密码不能为空")
    @Size(max = 100, min = 10, message = "密码长度在 10-100 之间")
    var password = ""
    @ManyToMany(targetEntity = Role::class, fetch = FetchType.EAGER)
    lateinit var roles: Set<Role>
}
```

其中，@NotEmpty 用来进行非空校验，@Size 是字符串长度校验。

Bean Validation API 内置的验证约束注解在 jar 包 javax.validation:validation-api:2.0.0.Final 中，这些注解如下所示：

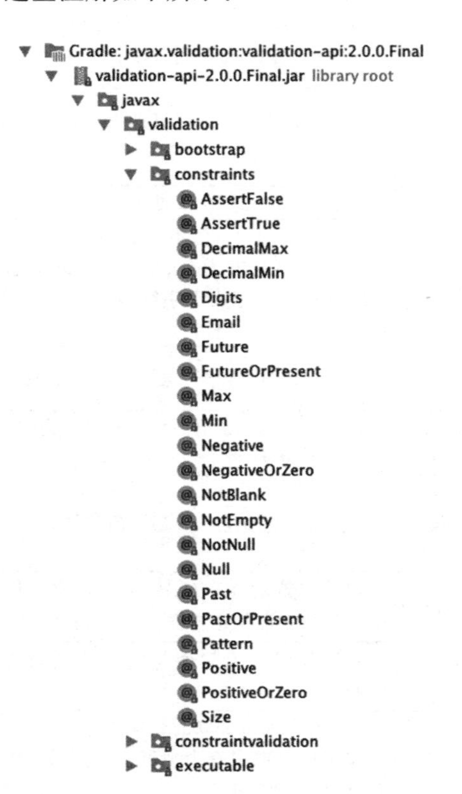

上面的这些注解简单说明如表 9-6 所示。

表 9-6 数据校验注解说明

验证注解	验证的数据类型	说明
@AssertFalse	Boolean, boolean	验证注解的元素值是 false
@AssertTrue	Boolean, boolean	验证注解的元素值是 true
@NotNull	任意类型	验证注解的元素值不是 null
@Null	任意类型	验证注解的元素值是 null
@Min(value= 值)	BigDecimal，BigInteger, byte, short, int, long，等任何 Number 或 CharSequence（存储的是数字）子类型	验证注解的元素值大于等于 @Min 指定的 value 值
@Max（value= 值）	和 @Min 要求一样	验证注解的元素值小于等于 @Max 指定的 value 值
@DecimalMin(value= 值)	和 @Min 要求一样	验证注解的元素值大于等于 @DecimalMin 指定的 value 值
@DecimalMax(value= 值)	和 @Min 要求一样	验证注解的元素值小于等于 @DecimalMax 指定的 value 值
@Digits(integer= 整数位数，fraction= 小数位数)	和 @Min 要求一样	验证注解的元素值的整数位数和小数位数上限
@Size（min= 下限，max= 上限）	字符串、Collection、Map、数组等	验证注解的元素值的在 min 和 max（包含）指定区间之内，如字符长度、集合大小
@Past	java.util.Date，java.util.Calendar；Joda Time 类库的日期类型	验证注解的元素值（日期类型）在当前时间之前
@Future	与 @Past 要求一样	验证注解的元素值（日期类型）在当前时间之后
@NotBlank	CharSequence 子类型	验证注解的元素值不为空（不为 null、去除首位空格后长度为 0），不同于 @NotEmpty，@NotBlank 只应用于字符串且在比较时会去除字符串的首位空格
@Length（min= 下限，max= 上限）	CharSequence 子类型	验证注解的元素值长度在 min 和 max 区间内
@NotEmpty	CharSequence 子类型、Collection、Map、数组	验证注解的元素值不为 null 且不为空（字符串长度不为 0、集合大小不为 0）
@Range（min= 最小值，max= 最大值）	BigDecimal, BigInteger, CharSequence, byte, short, int, long 等原子类型和包装类型	验证注解的元素值在最小值和最大值之间
@Email（regexp= 正则表达式，flag= 标志的模式）	CharSequence 子类型（如 String）	验证注解的元素值是 Email，也可以通过 regexp 和 flag 指定自定义的 email 格式
@Pattern（regexp= 正则表达式，flag= 标志的模式）	String，任何 CharSequence 的子类型	验证注解的元素值与指定的正则表达式匹配
@Valid	任何非原子类型	指定递归验证关联的对象；如用户对象中有个地址对象属性，如果想在验证用户对象时一起验证地址对象的话，在地址对象上加 @Valid 注解即可级联验证

虽然 JSR303 和 Hibernate Validtor 已经提供了很多校验注解，但是当面对复杂参数校验时，还是不能满足我们的要求，这时候也可以自定义校验注解。

在 BindingResult 对象中存储了数据校验的结果。BindingResult 类图结构如图 9-13 所示。

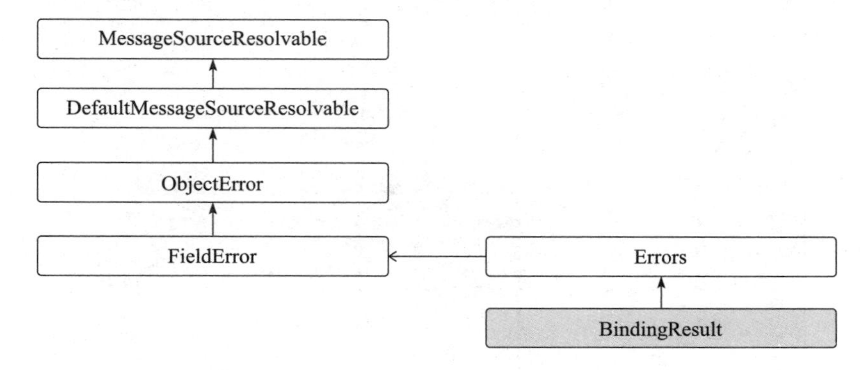

图 9-13 BindingResult 类图继承层次

获取 User 对象校验的信息如下：

```
override fun register(user: User, bindingResult: BindingResult):RegisterResult<
    String> {
    val RegisterResult = RegisterResult<String>()
    val sb = StringBuffer()
    if (bindingResult.hasErrors()) {
        bindingResult.fieldErrors.forEach {
            sb.append(it.defaultMessage).append(",")
        }
        RegisterResult.isSuccess = false
        RegisterResult.msg = sb.toString()
        return RegisterResult
    }
    ...
}
```

再加上用户名不能重复的校验：

```
val username = user.username
val u = UserDao.findByUsername(username)
if (u != null) {
    RegisterResult.isSuccess = false
    RegisterResult.msg = "用户名已存在"
    return RegisterResult
}
```

启用应用，进入用户注册界面，什么都不输入，直接点击“提交注册”，提示：“密码长度在 10 ～ 100 之间，用户名不能为空，用户名长度在 1 ～ 20 之间，密码不能为空”。

输入用户名 universsky、密码 123456，提示“密码长度在 10 ～ 100 之间”。

输入满足条件的密码注册，注册成功；使用注册的用户 universsky 登录系统，登录成功。页面成功跳转访问页面，如图 9-14 所示。

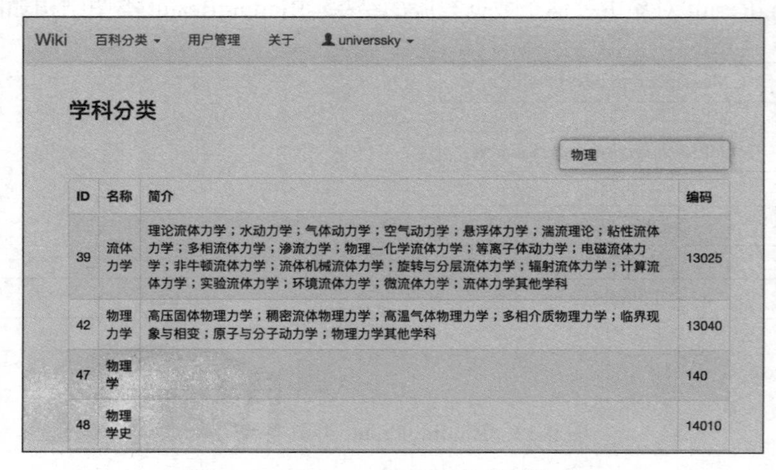

图 9-14　用户 universsky 登录系统页面

9.3 本章小结

本章使用 Spring Boot + Spring MVC 基于 Filter 过滤器和 AOP 切面实现了一个简单的用户登录注册、用户鉴权和权限控制的应用实例。通过本章的学习，我们可以深入了解 Filter 过滤器和 AOP 切面编程在编程实践中的应用。同时，还用实例介绍了 Spring MVC 中如何实现后端数据的校验功能。

下一章我们介绍 Spring Boot 集成 Security 实现系统的权限管理与安全。

提示　本章工程实例源代码地址为 https://github.com/KotlinSpringBoot/demo_spring_mvc

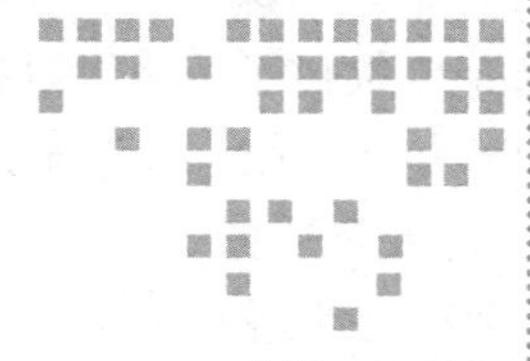

第 10 章 Chapter 10

Spring Boot 集成 Spring Security 安全开发

在企业级 Web 应用系统开发中，对系统的安全和权限控制通常是必需的。比如：没有访问权限的用户，不能访问系统页面。要实现访问控制的方法多种多样，可以通过过滤器、AOP、拦截器等实现，也可以直接使用框架实现，例如：Apache Shiro、Spring Security。很多成熟的大公司都会有一套完整的 SSO（单点登录）、ACL（Access Control List，权限访问控制列表）、UC(用户中心)系统，但是在开发中小型系统的时候，往往还是优先选择轻量级通用的框架解决方案。

Spring Security 就是一个 Spring 生态中关于安全方面的框架，它能够为基于 Spring 的企业应用系统提供声明式的安全访问控制解决方案。

本章使用 Spring Boot 集成 Spring Security 开发一个简单的自动化测试平台，其中，包括用户的登录鉴权、权限控制和一个简单的 HTTP 接口测试页面。本章通过该项目由浅入深地讲解 Spring Boot 集成 Spring Security 开发的相关知识。

10.1 Spring Security 简介

本节首先对 Spring Security 框架进行简要介绍。

Spring Security 的前身是 Acegi Security。Spring Security 是一种基于 Spring AOP 和 Servlet 过滤器 Filter 的安全框架，它提供全面的安全性解决方案，提供在 Web 请求和方法调用级别的用户鉴权和权限控制。图 10-1 展示了 Filter 在一个

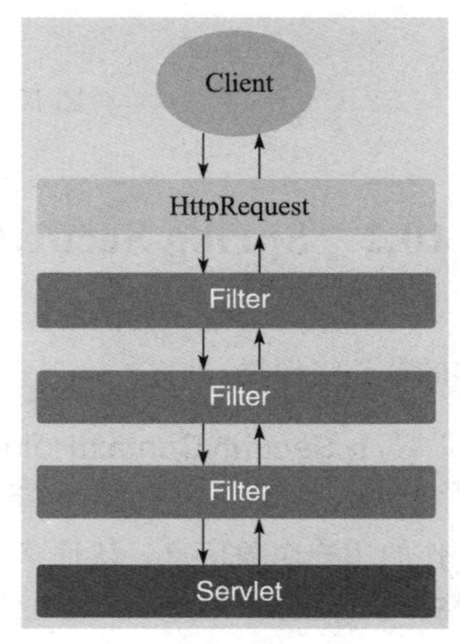

图 10-1 Filter 与安全框架

HTTP 请求过程中的执行流程。

Spring Security 提供了一组可以在 Spring 应用上下文中配置的 Bean，充分利用了 Spring IoC 容器和 AOP 功能，为应用系统提供声明式的安全访问控制功能，为基于 J2EE 企业应用软件提供了全面的安全服务。

Web 应用的安全性通常包括：用户认证（Authentication）和用户授权（Authorization）两个部分。分别阐述如下：

- **用户认证**：指的是验证某个用户是否为系统合法用户，也就是说用户能否访问该系统。用户认证一般要求用户提供用户名和密码。系统通过校验用户名和密码来完成认证过程。
- **用户授权**：指的是验证某个用户是否有权限执行某个操作。

Spring Security 提供了多种登录认证策略。例如典型的基于表单登录认证的流程如图 10-2 所示。

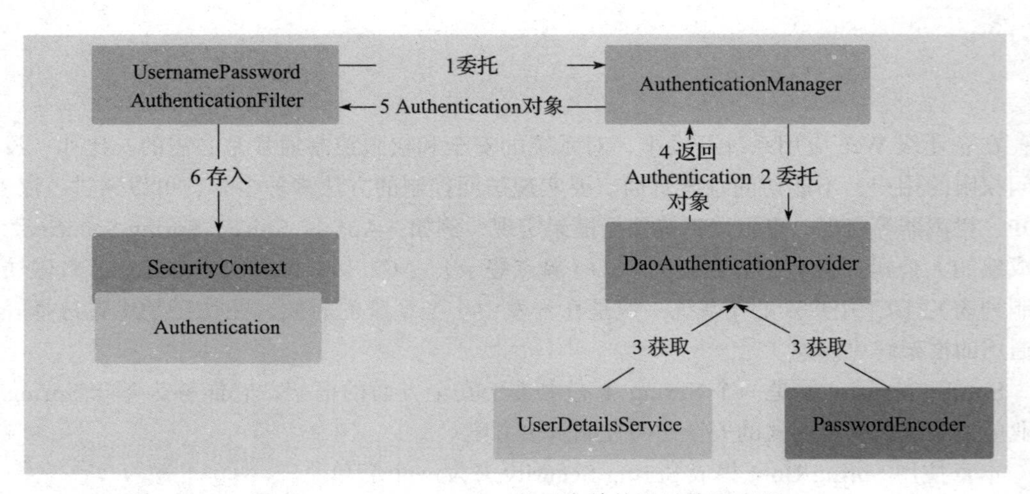

图 10-2 Spring Security 基于表单的登录认证流程

10.2 Spring Security 核心组件

在上一节中，我们已经了解到 Spring Security 基于表单认证的流程，本节介绍其中涉及的核心组件类。

1. SecurityContextHolder

SecurityContextHolder 用于存储安全上下文（Security Context）的信息。例如：当前操作的用户对象信息、认证状态、角色权限信息等，这些都保存在 SecurityContextHolder 中。SecurityContextHolder 默认使用 ThreadLocalSecurityContextHolderStrategy 类来存储认证信息。这个类定义如下：

```
final class ThreadLocalSecurityContextHolderStrategy implements
    SecurityContextHolderStrategy {
    private static final ThreadLocal<SecurityContext> contextHolder = new Thread-
        Local<SecurityContext>();
  ...
}
```

其中我们看到 static final ThreadLocal<SecurityContext> contextHolder 这个变量，使用 ThreadLocal 表示这是一种与线程绑定的策略。Spring Security 在用户登录时自动绑定认证信息到当前线程，在用户退出时清除当前线程的认证信息。

2. 获取当前用户的信息

因为身份信息是与线程绑定的，所以可以在程序的任何地方使用静态方法获取用户信息。典型的获取当前登录用户姓名的例子，使用 Java 代码如下所示：

```
Object principal = SecurityContextHolder.getContext().getAuthentication(). get-
    Principal();
if (principal instanceof UserDetails) {
    String username = ((UserDetails)principal).getUsername();
    } else {
    String username = principal.toString();
}
```

其中，getAuthentication() 返回认证信息，getPrincipal() 返回身份信息，UserDetails 是 Spring Security 对用户信息封装类。下面会详细介绍。

3. Authentication

Authentication（认证信息接口）是 org.springframework.security.core 包中的接口，继承自 Principal 类。Authentication 接口协议如下：

```
public interface Authentication extends Principal, Serializable {
    Collection<? extends GrantedAuthority> getAuthorities();
    Object getCredentials();
    Object getDetails();
    Object getPrincipal();
    boolean isAuthenticated();
     void setAuthenticated(boolean isAuthenticated) throws IllegalArgumentEx-
        ception;
}
```

接口中的方法参见表 10-1。

表 10-1　Authentication 接口中的方法说明

接口方法	功能说明
getAuthorities()	权限信息列表，默认是 GrantedAuthority 接口的一些实现类，通常是代表权限信息的一系列字符串
getCredentials()	密码信息，用户输入的密码字符串，在认证过后通常会被移除，用于保障安全

（续）

接口方法	功能说明
getDetails()	用户细节信息，Web 应用中的实现接口通常为 WebAuthenticationDetails，它记录了访问者的 ip 地址和 sessionId 的值
getPrincipal()	最重要的身份信息，大部分情况下返回的是 UserDetails 接口的实现类，也是框架中的常用接口之一。UserDetails 接口将在下面介绍

其中，Principal 是位于 java.security 包中的接口。通过 Authentication 接口，我们可以得到用户拥有的权限信息列表、密码、用户细节信息、用户身份信息、认证信息等。一个典型的 Authentication 数据示例如下：

```
{
  "SPRING_SECURITY_CONTEXT": {
    "authentication": {
      "authorities": [
        {
          "authority": "ROLE_ADMIN"
        },
        {
          "authority": "ROLE_USER"
        }
      ],
      "details": {
        "remoteAddress": "127.0.0.1",
        "sessionId": "F673D514413390BADE93ED21FF16A4A2"
      },
      "authenticated": true,
      "principal": {
        "password": null,
        "username": "admin",
        "authorities": [
          {
            "authority": "ROLE_ADMIN"
          },
          {
            "authority": "ROLE_USER"
          }
        ],
        "accountNonExpired": true,
        "accountNonLocked": true,
        "credentialsNonExpired": true,
        "enabled": true,
        "name": "admin"
      },
      "credentials": null,
      "name": "admin"
    }
  }
}
```

4. AuthenticationManager

AuthenticationManager（认证管理器）负责验证。认证成功后，AuthenticationManager 返回一个填充了用户认证信息（包括上面提到的权限信息、身份信息、细节信息等，但密码通常会被移除）的 Authentication 实例。SecurityContextHolder 安全上下文容器将填充了信息的 Authentication，通过如下方法

```
SecurityContextHolder.getContext().setAuthentication()
```

设置到 SecurityContextHolder 容器中。

AuthenticationManager 接口是认证相关的核心接口，也是发起认证的入口。AuthenticationManager 一般不直接认证，AuthenticationManager 接口的常用实现类 ProviderManager 内部会维护一个 List<AuthenticationProvider> 列表，存放多种认证方式，实际上这是委托（Delegate）模式的应用。

熟悉 Shiro 的朋友可以把 AuthenticationProvider 理解成 Realm。在默认策略下，只需要通过一个 AuthenticationProvider 的认证，即可被认为是登录成功。

下面是 ProviderManager 认证部分的关键源码：

```
public class ProviderManager implements AuthenticationManager, MessageSource-
    Aware, InitializingBean {
    // 维护一个 AuthenticationProvider 列表
    private List<AuthenticationProvider> providers = Collections.emptyList();
    // 认证逻辑
    public Authentication authenticate(Authentication authentication)
            throws AuthenticationException {
        Class<? extends Authentication> toTest = authentication.getClass();
        AuthenticationException lastException = null;
        Authentication result = null;
        // 遍历 Providers 列表依次认证
        for (AuthenticationProvider provider : getProviders()) {
            if (!provider.supports(toTest)) {
                continue;
            }
            try {
                result = provider.authenticate(authentication);
                if (result != null) {
                    copyDetails(authentication, result);
                    break;
                }
            }
            ...
            catch (AuthenticationException e) {
                lastException = e;
            }
        }
        // 如果有 Authentication 信息，则直接返回
        if (result != null) {
            if (eraseCredentialsAfterAuthentication
```

```
                    && (result instanceof CredentialsContainer)) {
                //移除密码
                ((CredentialsContainer) result).eraseCredentials();
            }
            //发布登录成功事件
            eventPublisher.publishAuthenticationSuccess(result);
            return result;
        }
        ...
        // 如果没有认证成功，包装异常信息
        if (lastException == null) {
            lastException = new ProviderNotFoundException(messages.getMessage(
                    "ProviderManager.providerNotFound",
                    new Object[] { toTest.getName() },
                    "No AuthenticationProvider found for {0}"));
        }
        prepareException(lastException, authentication);
        // 抛出异常
        throw lastException;
    }
}
```

ProviderManager 中的 List，会依照次序去认证，认证成功则立即返回，若认证失败则返回 null，下一个 AuthenticationProvider 会继续尝试认证，如果所有认证器都无法认证成功，则 ProviderManager 会抛出一个 ProviderNotFoundException 异常。

到这里，我们可以简单小结下：身份信息 Authentication 存放在容器 SecurityContext-Holder 中，身份认证管理器 AuthenticationManager 负责管理认证流程。真正进行认证的逻辑由 AuthenticationProvider 接口的具体实现提供。下面介绍最为常用的 DaoAuthenticationProvider。

5. DaoAuthenticationProvider

AuthenticationProvider（基于数据库的认证器）最常用的一个实现是 DaoAuthentication-Provider。顾名思义，Dao 正是数据访问层的缩写。用户前台提交了用户名和密码，而数据库中保存了用户名和密码，认证便是负责比对同一个用户名所提交的密码和数据库中保存的密码是否相同。DaoAuthenticationProvider 类的核心代码如下：

```
public class DaoAuthenticationProvider extends AbstractUserDetailsAuthentication-
    Provider {
    ...
    protected final UserDetails retrieveUser(String username,
            UsernamePasswordAuthenticationToken authentication)
            throws AuthenticationException {
        UserDetails loadedUser;

        try {
            loadedUser =
            this.getUserDetailsService()
                    .loadUserByUsername(username);
        }
```

```
        ...
        return loadedUser;
    }
    ...
}
```

retrieveUser() 方法返回一个 UserDetails 对象。在 Spring Security 中提交的用户名和密码，被封装成了 UsernamePasswordAuthenticationToken 对象，而根据用户名加载用户的任务则是交给了 UserDetailsService 去做。我们只需要实现一个 UserDetailsService 的接口实现即可完成基于数据库的用户名密码登录认证。具体项目实例我们将在下面的小节中详细介绍。

UsernamePasswordAuthenticationToken 和 UserDetails 密码的比对，由 additionalAuthenticationChecks 方法完成：

```
protected void additionalAuthenticationChecks(UserDetails userDetails,
        UsernamePasswordAuthenticationToken authentication)
        throws AuthenticationException {
    ...
    String presentedPassword = authentication.getCredentials().toString();
    if (!passwordEncoder.matches(presentedPassword, userDetails.getPassword())) {
        logger.debug("Authentication failed: password does not match stored
            value");
        throw new BadCredentialsException(messages.getMessage(
            "AbstractUserDetailsAuthenticationProvider.badCredentials",
            "Bad credentials"));
    }
}
```

如果这个 void additionalAuthenticationChecks() 方法没有抛异常，则认为比对成功。比对密码的过程，用到了 PasswordEncoder。

6. UserDetailsService

用户相关的信息是通过如下接口加载：

```
org.springframework.security.core.userdetails.UserDetailsService
```

该接口的唯一方法是：

```
loadUserByUsername(String username)
```

用来根据用户名加载相关的信息。这个方法的返回值是如下：

```
org.springframework.security.core.userdetails.UserDetails
```

其中包含了用户的信息，包括用户名、密码、权限、是否启用、是否被锁定、是否过期等。UserDetails 这个接口代表了最详细的用户信息，这个接口涵盖了一些必要的用户信息字段。

UserDetails 接口协议如下：

```
public interface UserDetails extends Serializable {
    Collection<? extends GrantedAuthority> getAuthorities();
    String getPassword();
    String getUsername();
    boolean isAccountNonExpired();
    boolean isAccountNonLocked();
    boolean isCredentialsNonExpired();
    boolean isEnabled();
}
```

其中最重要的是用户权限 getAuthorities()，由如下接口表示：

```
org.springframework.security.core.GrantedAuthority
```

UserDetails 接口和 Authentication 接口很类似，比如它们都拥有 username、authorities，两者对比参见表 10-2。

表 10-2 UserDetails 接口和 Authentication 接口对比

UserDetails 接口	Authentication 接口
Collection<? extends GrantedAuthority> getAuthorities()	Collection<? extends GrantedAuthority> getAuthorities()
String getUsername()	Object getDetails()
String getPassword()	Object getPrincipal()
isAccountNonExpired()	Object getCredentials()
isAccountNonLocked()	isAuthenticated()
isCredentialsNonExpired()	void setAuthenticated(boolean isAuthenticated)
isEnabled()	

不过，Authentication 的 getCredentials() 与 UserDetails 中的 getPassword() 是不同的。

getCredentials() 是用户提交的密码凭证，getPassword() 是用户正确的密码。认证器其实就是对这两者的比对。

其中，UserDetailsService 接口定义如下：

```
public interface UserDetailsService {
    UserDetails loadUserByUsername(String username) throws
            UsernameNotFoundException;
}
```

UserDetailsService 只负责从特定的地方（通常是数据库）加载用户信息。UserDetailsService 常见的实现类有从数据库加载用户信息的 JdbcDaoImpl 和从内存中加载用户信息的 InMemoryUserDetailsManager 等。

当然，我们也可以自己实现 UserDetailsService，在后面的项目实战的例子中，我们将会采用自己实现 UserDetailsService 接口的方式。

提示 更多关于 Spring Security 的内容可参考官方文档：https://docs.spring.io/spring-security/site/docs/5.0.0.RELEASE/reference/htmlsingle/

10.3　项目实战

本节由浅入深，逐步实现一个 Security 项目，包括默认认证用户名密码、内存用户名密码认证”，然后实现基于数据库的用户和角色权限控制，每个部分都给出了详尽的实例说明。

10.3.1　初阶 Security：默认认证用户名密码

首先创建集成 Security 的 Spring Boot 项目。选择 Spring Boot Starter ：Security、Web，如图 10-3 所示。

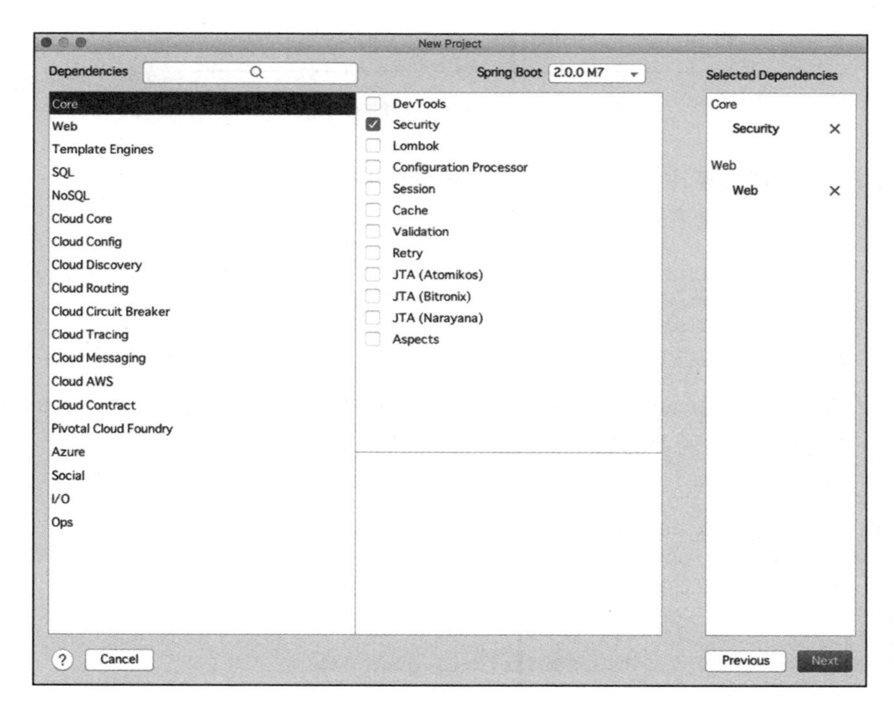

图 10-3　选择 Spring Boot Starter：Security、Web

导入 Gradle 项目配置。完成构建之后，我们可以看到项目配置文件 build.gradle 中的 Security 的相关依赖：

```
dependencies {
    compile('org.springframework.boot:spring-boot-starter-security')
    compile('org.springframework.boot:spring-boot-starter-web')
    …
    testCompile('org.springframework.boot:spring-boot-starter-test')
    testCompile('org.springframework.security:spring-security-test')
}
```

启动应用。浏览器打开页面：http://127.0.0.1:8080/，将看到一个登录表单页面，如图 10-4 所示。

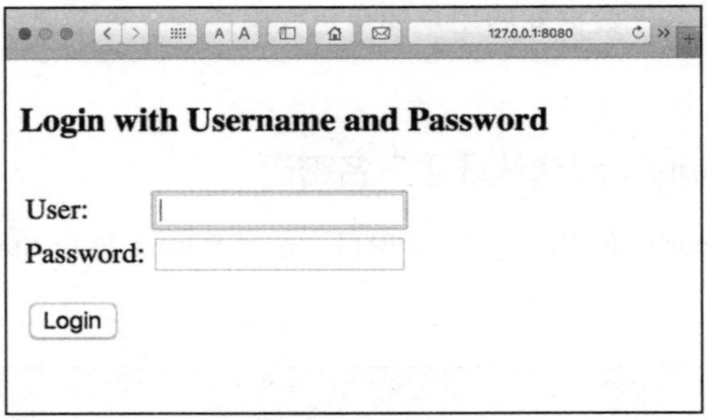

图 10-4 Security 内置的登录表单

其中用户名、密码是什么呢？让我们去 Spring Boot 的源码中寻找一下。

搜一下输出日志，会看到下面一段输出：

```
2018-01-31 11:07:45.333  INFO 27251 --- [           main] b.a.s.Authentication-
    Manager-Configuration :

Using default security password: 3cb4ffd9-f567-49a7-9170-8a7cdf1d60f8
```

这段日志是 AuthenticationManagerConfiguration 类中输出的。通过这个 Authentication ManagerConfiguration 类的关键如下代码：

```
@Configuration
@ConditionalOnBean(ObjectPostProcessor.class)
@ConditionalOnMissingBean({ AuthenticationManager.class, AuthenticationProvider.
    class,
    UserDetailsService.class })
@Order(0)
public class AuthenticationManagerConfiguration {
  …
  @Bean
  public InMemoryUserDetailsManager inMemoryUserDetailsManager(
      ObjectProvider<PasswordEncoder> passwordEncoder) throws Exception {
    String password = UUID.randomUUID().toString();
    logger.info(String.format("%n%nUsing default security password: %s%n",-password));
    String encodedPassword = passwordEncoder
        .getIfAvailable(PasswordEncoderFactories::createDelegatingPasswordE-ncoder)
        .encode(password);
    return new InMemoryUserDetailsManager(
        User.withUsername("user").password(encodedPassword).roles().build());
  }

}
```

可以看出，是 InMemoryUserDetailsManager 这个内存认证 Bean 中管理了用户名和密码。其中，该行代码：中设置了默认的用户名是 user，密码是 passwordEncoder 加密

```
User.withUsername("user").password(encodedPassword).roles().build()
```

UUID.randomUUID().toString() 这个字符串之后的值。其中的 User 类实现了 UserDetails 接口和 CredentialsContainer 接口。这个 User 类中的属性如下：

```
public class User implements UserDetails, CredentialsContainer {
    private String password;
    private final String username;
    private final Set<GrantedAuthority> authorities;
    private final boolean accountNonExpired;
    private final boolean accountNonLocked;
    private final boolean credentialsNonExpired;
    private final boolean enabled;
    ...
}
```

可见，security 默认的用户名是 user，默认密码是应用启动的时候，通过 UUID 算法随机生成；如果提供了 passwordEncoder，使用 passwordEncoder 进行加密。

输入用户名、密码，点击登录（Login），浏览器返回信息后我们可以看到系统已经登录成功，只不过一个测试页面都没有，直接进入了一个默认的 Error Page。我们可以写一个获取当前登录用户认证信息的 HTTP 接口来测试一下。代码如下：

```
@RestController
class CurrentAuthentication() {
    @GetMapping(value = ["", "/"])
    fun auth(): Authentication? {
          val request = (RequestContextHolder.getRequestAttributes() as Servlet-
            RequestAttributes).request
        // 从 Spring Security 当前 session 中获取 SPRING_SECURITY_CONTEXT
        val authentication = (request.session.getAttribute("SPRING_SECURITY-
            _CONTEXT") as SecurityContext).authentication
        return authentication
    }
}
```

重启应用，从控制台中找到密码，输入用户名、密码登录。登录成功后，可以发现输出结果如下：

```
{
  "authorities": [

  ],
  "details": {
    "remoteAddress": "127.0.0.1",
    "sessionId": "67CDDBBA47AFE55C532E6897A00C6FF0"
  },
```

```
  "authenticated": true,
  "principal": {
    "password": null,
    "username": "user",
    "authorities": [

    ],
    "accountNonExpired": true,
    "accountNonLocked": true,
    "credentialsNonExpired": true,
    "enabled": true,
    "name": "user"
  },
  "credentials": null,
  "name": "user"
}
```

另外，在控制台中我们可以看到 DefaultSecurityFilterChain 如下：

```
org.springframework.security.web.context.request.async.WebAsyncManagerIntegration-
    Filter
org.springframework.security.web.context.SecurityContextPersistenceFilter
org.springframework.security.web.header.HeaderWriterFilter
org.springframework.security.web.authentication.logout.LogoutFilter
org.springframework.security.web.authentication.UsernamePasswordAuthenticationFilter
org.springframework.security.web.authentication.ui.DefaultLoginPageGeneratingFilter
org.springframework.security.web.authentication.www.BasicAuthentica-tionFilter
org.springframework.security.web.savedrequest.RequestCacheAwareFilter
org.springframework.security.web.servletapi.SecurityContextHolderAwareRequestFilter
org.springframework.security.web.authentication.AnonymousAuthen-ticationFilter
org.springframework.security.web.session.SessionManagementFilter
org.springframework.security.web.access.ExceptionTranslationFilter
org.springframework.security.web.access.intercept.FilterSecuri-tyInterceptor
```

这些过滤器是 Spring Security 实现安全权限的核心。

当然这只是一个初级的配置，更复杂的配置可以分不同角色，在控制范围上能够拦截到方法级别的权限控制。

10.3.2 中阶 Security：内存用户名密码认证

在上节中，我们什么都没做，就添加了 spring-boot-starter-security 依赖，整个应用就有了默认的认证安全机制。下面，我们在内存中定制用户名密码。

首先，写一个继承 WebSecurityConfigurerAdapter 的配置类 WebSecurityConfig。重写其中的配置方法：

```
fun configure(auth: AuthenticationManagerBuilder)
```

完整的 WebSecurityConfig 配置代码参考实例工程源代码。关键代码如下：

```
@Configuration
```

```
@EnableWebSecurity                                                   //1）
@EnableGlobalMethodSecurity(prePostEnabled = true, securedEnabled = true,jsr250-
    Enabled = true)                                                  //2）
open class WebSecurityConfig : WebSecurityConfigurerAdapter() {      //3）

    @Throws(Exception::class)
    override fun configure(auth: AuthenticationManagerBuilder) {     //4）
        val passwordEncoder = passwordEncoder()                      //5）
        auth.inMemoryAuthentication()                                //6）
            .passwordEncoder(passwordEncoder)                        //7）
            .withUser("root")                                        //8）
            .password(passwordEncoder.encode("root"))                //9）
            .roles("USER")                                           //10）
    }

    /**
     * 密码加密算法器
     *
     * @return
     */
    @Bean
    open fun passwordEncoder(): BCryptPasswordEncoder {
        return BCryptPasswordEncoder();                              //11）
    }

}
```

代码说明：

1）通过 @EnableWebSecurity 注解开启 Spring Security 的功能。

2）使用 @EnableGlobalMethodSecurity(prePostEnabled = true) 这个注解，可以开启 security 的注解，我们可以在需要控制权限的方法上面使用 @PreAuthorize、@PreFilter 这些注解。

3）WebSecurityConfig 继承 WebSecurityConfigurerAdapter 类，并重写它的方法来设置 Web 安全的细节。

4）覆盖重写 configure(auth: AuthenticationManagerBuilder) 方法，在内存中创建一个用户，该用户的名称为 root，密码为 root，用户角色为 USER。

5）获取密码加密算法器。

6）表明使用内存用户名密码认证。

7）设置密码加密器。

8）用户名是 user。

9）存放的密码需要通过 passwordEncoder 对象的 encode() 方法加密。

10）用户角色是 ROLE_USER(其中，Security 会默认加上 ROLE_ 前缀)。

11）密码加密器对象为 BCryptPasswordEncoder。

再次启动应用，访问 http://localhost:8080，页面自动跳转到：http://localhost:8080/login。输入用户名 root、密码 root，点击登录，如图 10-5 所示，将得到信息如图 10-6 所示。

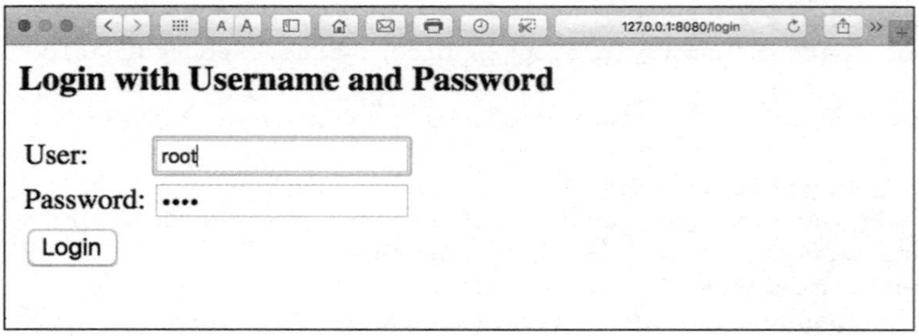

127.0.0.1:8080/login

Login with Username and Password

User: root

Password: ••••

Login

图 10-5 登录页面

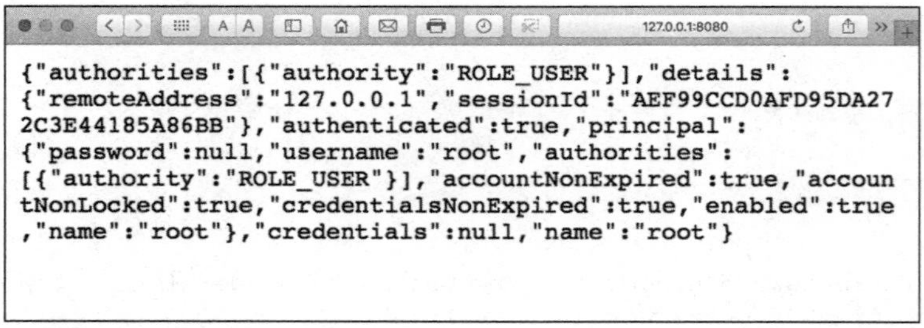

127.0.0.1:8080

{"authorities":[{"authority":"ROLE_USER"}],"details":{"remoteAddress":"127.0.0.1","sessionId":"AEF99CCD0AFD95DA272C3E44185A86BB"},"authenticated":true,"principal":{"password":null,"username":"root","authorities":[{"authority":"ROLE_USER"}],"accountNonExpired":true,"accountNonLocked":true,"credentialsNonExpired":true,"enabled":true,"name":"root"},"credentials":null,"name":"root"}

图 10-6 使用用户名 root、密码 root 登录成功

成功访问了 HTTP 接口，并输出了正确的响应结果。如果我们要配置多个用户、多个角色，例如，再添加一个 admin 用户，角色为 ADMIN。配置如下：

```
@Throws(Exception::class)
override fun configure(auth: AuthenticationManagerBuilder) {
    val passwordEncoder = passwordEncoder()
    auth.inMemoryAuthentication()
        .passwordEncoder(passwordEncoder)
        .withUser("user")
        .password(passwordEncoder.encode("user")) // 存放的密码需要通过 encode 加密
        .roles("USER")
        .and()
        .withUser("admin")
        .password(passwordEncoder.encode("admin"))
        .roles("ADMIN")
}
```

重启应用，使用 admin 用户登录，如图 10-7 所示，将得到信息如图 10-8 所示。

10.3.3 角色权限控制

当系统功能发展到一定程度时，会需求不同用户、不同角色使用系统。这样就要求我们的系统做到，把不同的系统功能模块，开放给对应的拥有其访问权限的用户使用。下面我们就介绍如何进行角色权限控制。

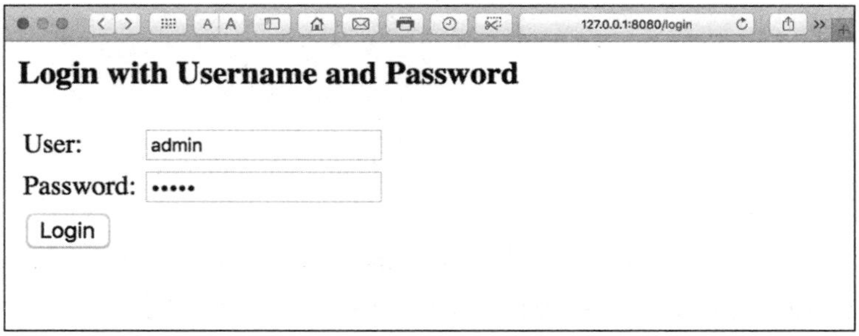

Login with Username and Password

User: admin

Password: •••••

Login

图 10-7 使用 admin 用户登录

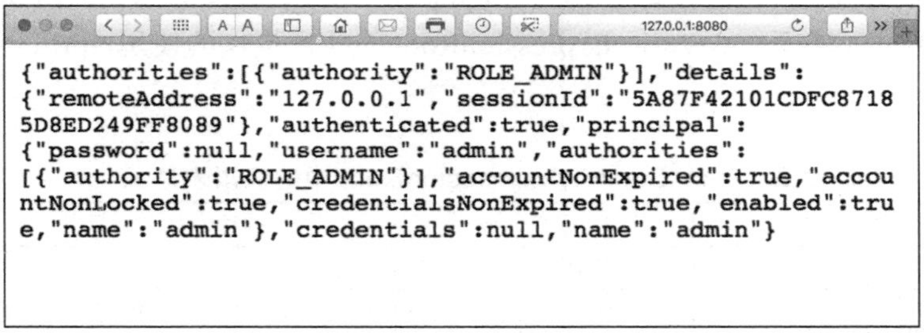

{"authorities":[{"authority":"ROLE_ADMIN"}],"details":{"remoteAddress":"127.0.0.1","sessionId":"5A87F42101CDFC87185D8ED249FF8089"},"authenticated":true,"principal":{"password":null,"username":"admin","authorities":[{"authority":"ROLE_ADMIN"}],"accountNonExpired":true,"accountNonLocked":true,"credentialsNonExpired":true,"enabled":true,"name":"admin"},"credentials":null,"name":"admin"}

图 10-8 使用 admin 用户登录结果

1. Spring Security 表达式

Spring Security 提供了 Spring EL 表达式，允许我们在定义 URL 路径访问 @RequestMapping 的方法上面添加注解，来控制访问权限。

在标注访问权限时，根据对应的表达式返回结果，控制访问权限。Spring Security 表达式操作对象的标准接口是 SecurityExpressionOperations。Security 提供的实现类是 SecurityExpressionRoot，例如，其中的代码片段如下：

```
public abstract class SecurityExpressionRoot implements SecurityExpression-
    Operations {     ...
    public final boolean hasAuthority(String authority) {
        return hasAnyAuthority(authority);
    }

    public final boolean hasAnyAuthority(String... authorities) {
        return hasAnyAuthorityName(null, authorities);
    }

    public final boolean hasRole(String role) {
        return hasAnyRole(role);
    }

    public final boolean hasAnyRole(String... roles) {
        return hasAnyAuthorityName(defaultRolePrefix, roles);
```

```
    }

    private boolean hasAnyAuthorityName(String prefix, String... roles) {
        ...
    }
    ...
}
```

通过源码，我们可以看出 hasRole 背后调用的是 hasAnyRole，hasAnyRole 调用了 hasAnyAuthorityName(defaultRolePrefix, roles)。而且 Role 的默认前缀是：

```
private String defaultRolePrefix = "ROLE_";
```

可见，我们在学习一个框架的时候，最好的方法就是阅读源码。通过源码，我们可以更深入地理解技术的本质。SecurityExpressionRoot 为我们提供的使用 Spring EL 表达式参见表 10-3。

表 10-3 SecurityExpressionRoot 中的权限表达式

表达式	描述
hasRole([role])	当前用户是否拥有指定角色
hasAnyRole([role1,role2])	多个角色通过逗号进行分隔。如果当前用户拥有指定角色中的任意一个则返回 true
hasAuthority([auth])	等同于 hasRole
hasAnyAuthority([auth1,auth2])	等同于 hasAnyRole
Principle	当前用户的 principle 对象
authentication	直接从 SecurityContext 获取的当前 Authentication 对象
permitAll	总是返回 true，表示允许所有请求
denyAll	总是返回 false，表示拒绝所有的
isAnonymous()	当前用户是否是一个匿名用户
isRememberMe()	当前用户是否是通过 Remember-Me 自动登录的
isAuthenticated()	当前用户是否已经登录认证成功了
isFullyAuthenticated()	如果当前用户既不是一个匿名用户，同时又不是通过 Remember-Me 自动登录的，则返回 true

2. 权限控制实例

下面我们来实现表 10-4 中的权限控制。

表 10-4 权限控制说明

URL	角色权限
/auth	hasAuthority('ROLE_ADMIN') 等价于 hasRole('ADMIN')
/admin	hasRole('ADMIN')
"", "/", "/home"	hasAnyRole('ADMIN','USER')
/user	hasRole('USER')or hasRole('ADMIN')

完整的实现代码在 CurrentAuthenticationController.kt 中。关键代码如下：

```
@RestController
open class CurrentAuthenticationController() {

    @GetMapping(value = ["/auth"])
    @PreAuthorize("hasAuthority('ROLE_ADMIN')") // (1)
    fun auth(): Authentication? {
        val request = (RequestContextHolder.getRequestAttributes() as ServletRequest-
            Attributes).request
            // 从 Spring Security 当前 session 中获取 SPRING_SECURITY_CONTEXT
            val authentication = (request.session.getAttribute("SPRING_SECURITY_
            CONTEXT") as SecurityContext).authentication
            return authentication
    }

    @GetMapping(value = ["/admin"])
    @PreAuthorize("hasRole('ADMIN')"
    fun admin(): String {
        return "ADMIN"
    }

    @GetMapping(value = ["", "/", "/home"])
    @PreAuthorize("hasAnyRole('ADMIN','USER')")
    fun index(): String {
        return "HOME"
    }

    @GetMapping(value = ["/user"])
    @PreAuthorize("hasRole('USER') or hasRole('ADMIN')")
    fun user(): String {
        return "USER"
    }
}
```

代码说明如下。（1）处在方法上添加 @PreAuthorize 这个注解，value="hasRole('ADMIN')") 是 Spring-EL expression。当表达式值为 true，标识这个方法可以被调用；如果表达式值是 false，标识此方法无权限访问。其中的 hasAuthority（'ROLE_ADMIN'）等价于 hasRole（'ADMIN'），表示该方法只有当用户拥有 ADMIN 角色的时候才允许访问。

为了使上面的安全认证注解生效，还需要在 WebSecurityConfig 类上面添加 @EnableWebSecurity 和 @EnableGlobalMethodSecurity(prePostEnabled = true) 注解：

```
@Conguration
@EnableWebSecurity(debug = true)
@EnableGlobalMethodSecurity(prePostEnabled = true, proxyTargetClass = true)
open class WebSecurityConfig : WebSecurityConfigurerAdapter() {
...
}
```

这里使用注解 @EnableGlobalMethodSecurity 开启 Spring Security 方法级安全，prePost-

Enabled = true 决定 Spring Security 的 @PreAuthorize,@PostAuthorize 等注解是否可用。proxy-TargetClass = true 表示使用基于子类代理（subclass-based CGLIB）替换基于接口（Java interface-based）的代理。默认值是 false。

对应的 WebSecurityConfig 配置代码如下：

```
@Configuration
@EnableWebSecurity(debug = true)
@EnableGlobalMethodSecurity(prePostEnabled = true, proxyTargetClass = true)
open class WebSecurityConfig : WebSecurityConfigurerAdapter() {

    @Throws(Exception::class)
    override fun configure(http: HttpSecurity) {
        http.csrf().disable()
            .authorizeRequests()
            .antMatchers("/login").permitAll()
            .anyRequest().authenticated()
            .and().formLogin()
            .and().logout().logoutSuccessUrl("/login")
    }

    @Throws(Exception::class)
    override fun configure(auth: AuthenticationManagerBuilder) {
        val passwordEncoder = passwordEncoder()
        auth.inMemoryAuthentication()
            .passwordEncoder(passwordEncoder)
            .withUser("user").roles("USER")
            .password(passwordEncoder.encode("user"))     // 存放的密码需要通过 encode
                                                          // 加密
            .and()
            .withUser("admin").roles("ADMIN", "USER")
            .password(passwordEncoder.encode("admin"))
    }

    /**
     * 密码加密算法
     *
     * @return
     */
    @Bean
    open fun passwordEncoder(): BCryptPasswordEncoder {
        return BCryptPasswordEncoder();
    }

}
```

在开发环境，我们可以开启 WebSecurity 的 debug 日志：@EnableWebSecurity-(debug = true)，方便定位问题。从日志中我们可以看到更多关于 Spring Security 内部运行的信息。

3. 运行测试

使用 user 用户登录（没有 ADMIN 角色），访问：http://127.0.0.1:8080/auth，页面提示鉴权失败，禁止访问，如图 10-9 所示。

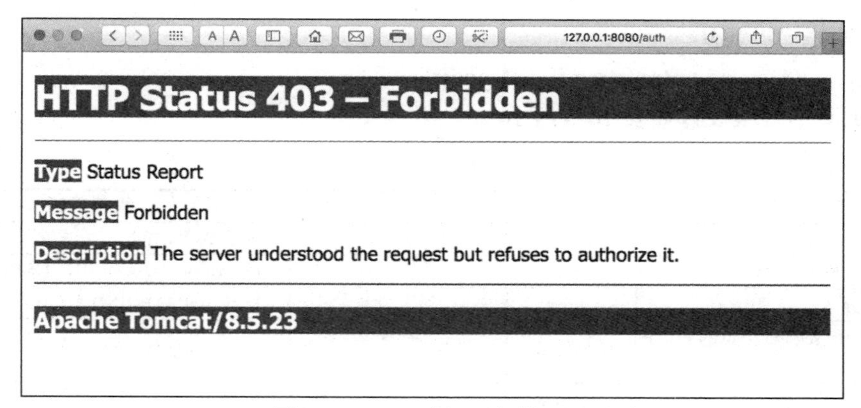

图 10-9　user 访问无权限页面

使用 admin 用户（拥有 ADMIN 角色），访问：http://127.0.0.1:8080/auth，页面成功输出接口返回数据，如图 10-10 所示。

127.0.0.1:8080/auth


```
{"authorities":[{"authority":"ROLE_ADMIN"},
{"authority":"ROLE_USER"}],"details":
{"remoteAddress":"127.0.0.1","sessionId":null},"authenticated"
:true,"principal":
{"password":null,"username":"admin","authorities":
[{"authority":"ROLE_ADMIN"},
{"authority":"ROLE_USER"}],"accountNonExpired":true,"accountNo
nLocked":true,"credentialsNonExpired":true,"enabled":true,"nam
e":"admin"},"credentials":null,"name":"admin"}
```


图 10-10　admin 访问有权限

提示　本小节完整的工程源代码网址：https://github.com/KotlinSpringBoot/demo_security_simple

10.3.4　进阶 Security：基于数据库的用户和角色权限

本节我们实现一个简单的 HTTP 接口测试平台 LightSword，系统使用数据库存储用户和角色，并实现系统页面的权限控制和安全认证。同时，系统还提供了 CSRF 的功能。

LightSword 系统开发的技术栈如下：

- 编程语言：Java、Kotlin
- 编程框架：Spring、Spring MVC、Spring Boot
- ORM 框架：Spring Data JPA
- 视图模板引擎：FreeMarker
- 安全框架：Spring Security

- 数据库：MySQL
- 前端：jQuery.js、Bootstrap.js、React.js

系统的页面权限设计如表 10-5 所示。

表 10-5 系统的页面权限设计

页面名称	URL	权限说明
登录页	/login	所有人可访问
普通用户权限页	/main	hasAnyRole('ADMIN','USER')
管理员页面	/admin	hasRole('ADMIN')
获取当前用户 session 信息的 HTTP 接口	/api/session	hasRole('ADMIN')
获取系统用户列表的 HTTP 接口	/api/users	hasRole('ADMIN')
获取 HTTP 接口测试记录	/httptestrecord/findAll?pageSize=10&pageNo=0	hasRole('ADMIN')
无权限提示页面	/403	当一个用户访问了其没有权限的页面，我们使用全局统一的异常处理页面提示。
HTTP 接口 Ajax 访问无权限提示信息		提示：{ "code"："403", "msg"：" 没有权限 " }

1. 数据库层设计

系统表结构设计如下：

```
▼ http_test_record
    id bigint(20) (auto increment)
    gmt_create datetime
    gmt_modify datetime
    is_deleted int(11)
    method varchar(50)
    post_data varchar(255)
    response_text longtext
    url varchar(255)
    author varchar(255)
    PRIMARY (id)
▼ role
    id bigint(20) (auto increment)
    role varchar(50)
    PRIMARY (id)
    UK_bjxn5ii7v7ygwx39et0wawu0q (role)
    UK_bjxn5ii7v7ygwx39et0wawu0q (role) UNIQUE
▼ user
    id bigint(20) (auto increment)
    password varchar(255)
    username varchar(50)
    PRIMARY (id)
    UK_sb8bbouer5wak8vyiiy4pf2bx (username)
    UK_sb8bbouer5wak8vyiiy4pf2bx (username) UNIQUE
▼ user_roles
    user_id bigint(20)
    roles_id bigint(20)
    PRIMARY (user_id, roles_id)
    FKj9553ass9uctjrmh0gkqsmv0d (roles_id)
```

对应的领域实体模型类如下：

```
▼ entity
  ▼ HttpTestRecord
      author: String
      gmtCreate
      gmtModify
      id: Long
      isDeleted
      method: String
      postData: String?
      responseText: String?
      url: String
  ▼ Role
      id: Long
      role: String
  ▼ User
      id: Long
      password: String
      roles: Set<Role>
      toString(): String
      username: String
```

主要代码分别如下。

用户模型类 User.kt：

```
@Entity
class User {
    @Id
    @GeneratedValue(strategy = GenerationType.IDENTITY)
    var id: Long = -1
    @Column(length = 50, unique = true)
    var username: String = ""
    var password: String = ""
    @ManyToMany(targetEntity = Role::class, fetch = FetchType.EAGER)
    lateinit var roles: Set<Role>

    override fun toString(): String {
        return "User(id=$id, username='$username', password='$password', roles=-$roles)"
    }
}
```

角色模型类 Role.kt：

```
@Entity
class Role {
    @Id
    @GeneratedValue(strategy = GenerationType.IDENTITY)
    var id: Long = -1
    @Column(length = 50, unique = true)
    var role: String = "ROLE_USER"
}
```

“接口测试记录”模型类 HttpTestRecord.kt：

```
@Entity
class HttpTestRecord {
    @Id
```

```
    @GeneratedValue(strategy = GenerationType.IDENTITY)
    var id: Long = -1
    @JsonFormat(pattern = "yyyy-MM-dd HH:mm:ss")
    var gmtCreate = Date()
    @JsonFormat(pattern = "yyyy-MM-dd HH:mm:ss")
    var gmtModify = Date()
    var isDeleted = 0

    var author: String = ""
    var url: String = ""
    @Column(length = 50)
    var method: String = ""
    var postData: String? = null
    @Lob
    var responseText: String? = null
}
```

开发调试阶段，我们在 application-dev.properties 中配置：

```
spring.jpa.hibernate.ddl-auto=update
```

为了方便测试，我们后面会在 Spring Boot 启动类中写一个用户测试数据的自动生成的 Bean，用来做测试数据的自动初始化工作。

2. 继承 WebSecurityConfigurerAdapter 自定义 Spring Security 配置

首先使用 Spring Security 帮我们做登录、登出的处理，以及当用户未登录时能访问“/login”页面，同时安全认证不拦截静态资源。

同样，我们要写一个继承 WebSecurityConfigurerAdapter 的配置类 WebSecurity-Config。代码如下，完整代码在 WebSecurityConfig.kt 中。关键代码如下：

```
@Configuration
@EnableWebSecurity

@EnableGlobalMethodSecurity(prePostEnabled = true, securedEnabled = true, jsr250-
    Enabled = true)                         // (1)
class WebSecurityConfig : WebSecurityConfigurerAdapter() {
    @Autowired lateinit var myAccessDeniedHandler: MyAccessDeniedHandler
    @Autowired lateinit var userDetailsService: MyUserDetailService

    @Throws(Exception::class)
    override fun configure(web: WebSecurity) {
        web.ignoring().antMatchers(
                "/css/**",
                "/fonts/**",
                "/js/**",
                "/plugins/**",
                "/images/**"            // 不拦截静态资源
        )
    }

    @Throws(Exception::class)
```

```
    override fun configure(http: HttpSecurity) {
        // http.csrf().disable()          // 使用 csrf 功能
        http.authorizeRequests()
                .antMatchers(
                        "/login/**"
                ).permitAll()
                .anyRequest().authenticated()
                .and()
                .formLogin()
                .loginPage("/login")    // 登录请求页
                .loginProcessingUrl("/login")
                                        // 登录 POST 请求路径
                .usernameParameter("username")
                                        // 登录用户名参数
                .passwordParameter("password")
                                        // 登录用户名密码
                .defaultSuccessUrl("/main")
                                        // 登录成功页面
                .and()
                .exceptionHandling()
                .accessDeniedHandler(myAccessDeniedHandler)
                                        //无权限处理器
                .and()
                .logout()               // 退出功能
                .logoutSuccessUrl("/login?logout")
                                        //退出成功 URL
    }

    @Throws(Exception::class)
    override fun configure(auth: AuthenticationManagerBuilder) {
        auth.userDetailsService(userDetailsService)
                                        // (2)
                .passwordEncoder(passwordEncoder())
                                        // (3)
    }

    /**
     * 密码加密算法
     *
     * @return
     */
    @Bean
    fun passwordEncoder(): BCryptPasswordEncoder {
        return BCryptPasswordEncoder();
    }
}
```

其中 AuthenticationManager 管理器使用我们自定义的 MyUserDetailService 来获取用户信息。具体的 MyUserDetailService 实现下面讲。用户密码使用 BCryptPasswordEncoder 加密算法。

3. 自定义实现 UserDetailService 接口

从数据库中获取用户信息的操作是必不可少的，首先来实现接口 UserDetailsService，这个接口需要我们实现一个方法：loadUserByUsername，即从数据库中取出用户名、密码以及权限相关的信息。最后返回一个 UserDetails 实现类。完整的代码参见本章源代码工程中的 MyUserDetailService.kt。关键代码如下：

```
@Service
class MyUserDetailService : UserDetailsService {
    val logger = LoggerFactory.getLogger(MyUserDetailService::class.java)
    @Autowired lateinit var userDao: UserDao

    override fun loadUserByUsername(username: String): UserDetails {
        // 去数据库 User 表，根据 username 查询用户是否存在
         val user = userDao.findByUsername(username) ?: throw  UsernameNotFound-
            Exception(username + " not found")
        logger.info("user = {}", user)
        val roles = user.roles  // 数据库中存的该用户的角色集合
        val authorities = mutableSetOf<SimpleGrantedAuthority>()
        roles.forEach {
            // 封装 org.springframework.security.core.userdetails.User 对象中的
            // Collection<? extends GrantedAuthority> authorities 参数
            authorities.add(SimpleGrantedAuthority(it.role))
        }
        return org.springframework.security.core.userdetails.User(
            // 此处为了区分我们本地系统中的 User 实体类，特意列出 Security 中的 User 类的全路径
                username,
                user.password,
                authorities
        )
    }
}
```

4. 实现 AccessDeniedHandler

一个系统既然有权限认证，必然需要进行无权限的提示交互处理。Spring Security 中提供了一个 AccessDeniedHandler 接口来处理无权限的请求。AccessDeniedHandler 接口代码如下：

```
public interface AccessDeniedHandler {

    void handle(HttpServletRequest request, HttpServletResponse response,
            AccessDeniedException accessDeniedException) throws IOException,
            ServletException;
}
```

我们自定义实现的访问无权限的处理器类代码在 My AccesssDenied Handler.kt 中，完整代码参考示例工程源代码。

5. 登录控制器

实现处理登录控制器代码如下：

```
@Controller
```

```
class LoginController {
    …
    @GetMapping(value = "/login")
    fun login(
            @RequestParam(value = "error", required = false) error: String?,
             @RequestParam(value = "logout", required = false) logout: String?): -
               ModelAndView {
        val model = ModelAndView()
        if (error != null) {
            model.addObject("error", " 不正确的用户名和密码 ")
        }
        if (logout != null) {
            model.addObject("msg", " 你已经成功退出 ")
        }
        model.viewName = "login" // login.ftl
        return model
    }

}
```

6. 登录页面表单与 CSRF

实现一个自定义的登录页面，如图 10-11 所示。

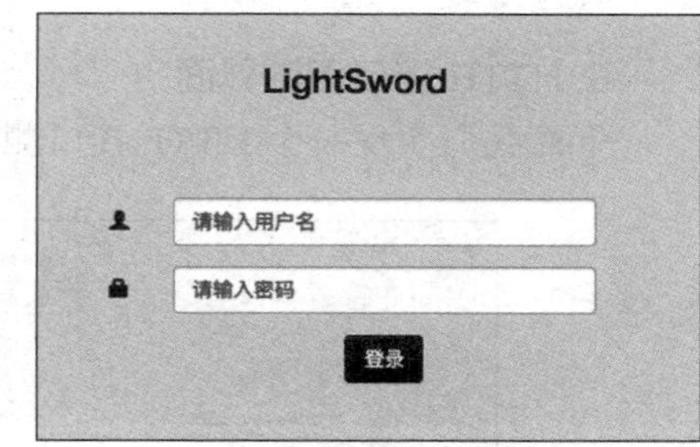

图 10-11　登录页面

前端登录界面的完整代码在 login.ftl 中。CSS 样式代码在 app.css 中。其中，登录表单中的 csrf 的代码如下：

```
<#include 'common/header.ftl'>
<div class="container text-info text-center">
    <div class="jumbotron">
        <h3 class=""> LightSword </h3>
        <form class="form-horizontal form-login"
              role="form"
              method="post"
              action="/login">
            <div class="form-group">
                <div class="col-sm-3">
                <${error!}
                ${msg!}
                </div>
            </div>
            <div class="form-group">
                    <input type="hidden" name="${_csrf.parameterName}" value="${_
                     csrf.token}"/>
                ……
        </form>
    </div>
</div>
<#include 'common/footer.ftl'>
```

使用一个 hidden 隐藏的输入框来存储 csrf 的值：

```
<input type="hidden" name="${_csrf.parameterName}" value="${_csrf.token}"/>
```

该表单域会在 POST 登录请求的时候被放到请求头中。

7. 用户退出

系统的退出界面如图 10-12 所示。

用户退出是一个表单，因为我们开启了 CSRF 功能，Spring Security 默认的退出 /logout 请求约定为 POST 请求。用户退出的按钮表单代码如下：

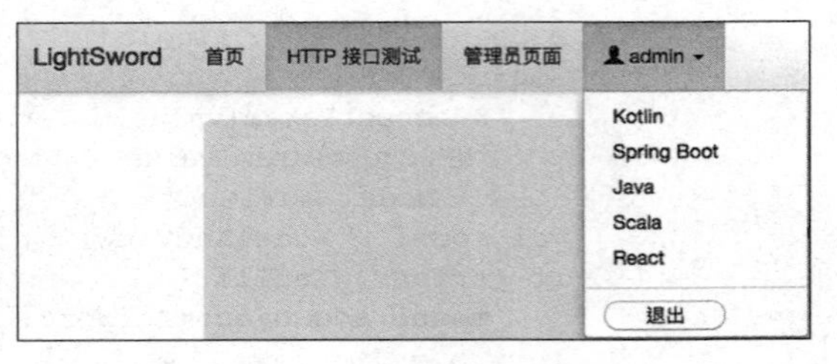

图 10-12 系统的退出界面

```
<form action="/logout" method="post">
    <input type="hidden" name="${_csrf.parameterName!}" value="${_csrf.token!}"/>
    <button type="submit" class="logout-btn">退出</button>
</form>
```

8. HTTP 接口测试界面

下面我们实现一个 HTTP 接口测试的页面，如图 10-13 所示。

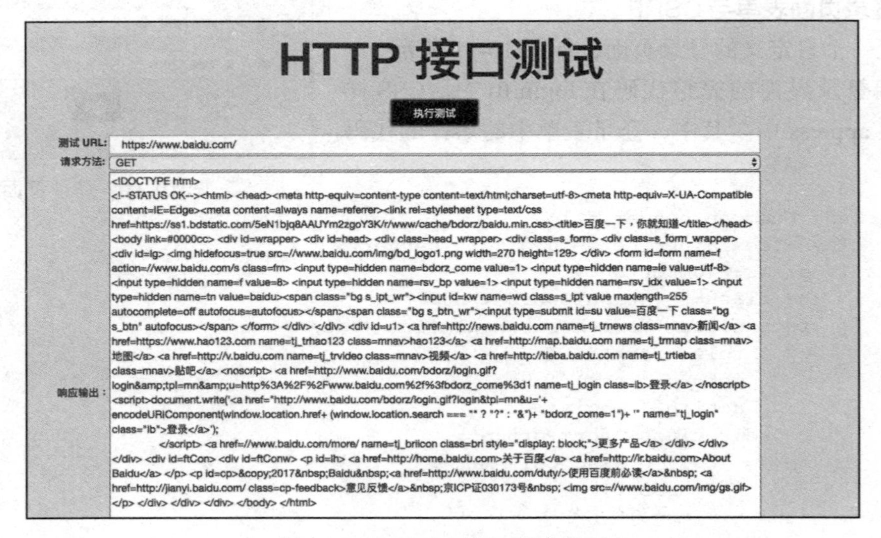

图 10-13 HTTP 接口测试界面

对应的前端代码我们使用 React 的 JSX 代码实现。相关代码示例如下：

```
var HttpTestAppPage = React.createClass({
                    …                   handleClick: function () {
    $.ajax({
        url: "/httpTest",
        type: 'POST',
        data: {url: this.state.url, method: this.state.method, postData: this.state.
            postData},
        dataType: 'json',
        beforeSend: function (xhr) {
```

```
                xhr.setRequestHeader('X-CSRF-TOKEN', token); // POST, header 'X-CSRF-
                TOKEN'
            },
            success: function (data) {
                this.setState({
                    result: data.result,
                    msg: data.msg,
                })
            }.bind(this),              // 修改 bind() 前的函数内部 this 变量指向。if 不 bind 的话,
                                       // 方法内部的 this 就是 $.ajax({ 这个对象 }), bind 传入的
                                       // this 应该是组件。 可以到 console 输出一下看看。
            error: function (msg) {
                console.log("error:" + msg);
            }.bind(this)
            })
        },

        handleUrlChange: function (e) {
            console.log('handleUrlChange')
            console.dir(e)
            this.setState({
                url: e.target.value
            })
        }
        ,

        handleMethodChange: function (e) {
            console.log('handleMethodChange')
            console.dir(e)
            this.setState({
                method: e.target.value
            })
        }
        ,

        render: function () {
        …
        }
        })
    ReactDOM.render(<HttpTestAppPage/>
            , document.getElementById("httpTestApp"))
```

完整的前端 JSX 代码参考工程源代码。

9. Ajax POST 请求与 CSRF

其中，执行测试的 Ajax 请求是 POST 请求 /httpTest，使用带 CSRF 请求需要在 Ajax 请求执行前在请求头中添加 CSRF 的信息：

```
beforeSend: function (xhr) {
    // POST, header 'X-CSRF-TOKEN'
    xhr.setRequestHeader('X-CSRF-TOKEN', token);
}
```

10.HTTP 接口测试执行代码

POST 请求 /httpTest 的后端代码在 HttpTestController.kt 中。相关代码如下：

```
@RestController
class HttpTestController {
    @Autowired lateinit var HttpTestRecordDao: HttpTestRecordDao

    @PostMapping("/httpTest")
    fun doTest(@Valid httpTestRequest: HttpTestRequest, bindingResult: BindingResult):
        HttpTestResult {
        val request = (RequestContextHolder.getRequestAttributes() as ServletRequest-
            Attributes).request
        // ${(Session["SPRING_SECURITY_CONTEXT"].authentication.principal.username)!}
        // 从 Spring Security 当前 session 中获取 SPRING_SECURITY_CONTEXT 对象
        val authentication = (request.session.getAttribute("SPRING_SECURITY_CONTEXT")
            as SecurityContext).authentication
        val username = (authentication.principal as User).username

        if (bindingResult.hasErrors()) {
            var msg = ""
            bindingResult.fieldErrors.forEach {
                msg += it.defaultMessage
            }
            return HttpTestResult(
                    success = false,
                    msg = msg,
                    result = "执行失败：提交数据错误"
            )
        }
        if ("GET" == httpTestRequest.method) {
            var result = HttpEngine.get(httpTestRequest.url)

            val HttpTestRecord = HttpTestRecord()
            HttpTestRecord.author = username
            HttpTestRecord.method = "GET"
            HttpTestRecord.url = httpTestRequest.url
            HttpTestRecord.responseText = result
            HttpTestRecordDao.save(HttpTestRecord)

            return HttpTestResult(
                    success = true,
                    msg = "执行成功",
                    result = result
            )
        } else if ("POST" == httpTestRequest.method) {
            var result = HttpEngine.post(httpTestRequest.url, httpTestRequest.postData)
            val HttpTestRecord = HttpTestRecord()
            HttpTestRecord.author = username
            HttpTestRecord.method = "POST"
            HttpTestRecord.postData = httpTestRequest.postData
            HttpTestRecord.url = httpTestRequest.url
            HttpTestRecord.responseText = result
```

```
            HttpTestRecordDao.save(HttpTestRecord)

            return HttpTestResult(
                    success = true,
                    msg = "执行成功",
                    result = result
            )
        } else {
            return HttpTestResult(
                    success = false,
                    msg = "",
                    result = "执行失败：提交数据错误"
            )
        }
    }

}

data class HttpTestRequest(@NotEmpty(message = "请求方法不能为空")
                           var method: String,
                           @NotEmpty(message = "请求 URL 不能为空")
                           var url: String,
                           var postData: String?)

data class HttpTestResult(var success: Boolean,
                          var msg: String,
                          var result: String?)
```

执行引擎的 HttpEngine.kt 的代码如下：

```
object HttpEngine {
    private val okhttpClient = OkHttpClient.Builder()
            .build()

    fun get(url: String): String? {
        val request = Request.Builder()
                .url(url)
                .get()
                .build()
        val response = okhttpClient.newCall(request).execute()
        return response.body()?.string()
    }

    fun post(url: String, data: String?): String? {
        //设置媒体类型 application/json ，表示传递的是一个 json 格式的对象
        val mediaType = MediaType.parse("application/json")
        val requestBody = RequestBody.create(mediaType, data);
        val request = Request.Builder()
                .url(url)
                .post(requestBody)
                .build()
        val response = okhttpClient.newCall(request).execute()
```

```
        return response.body()?.string()

    }
}
```

11. HTTP 接口测试记录列表

下面实现一个 HTTP 接口测试记录的列表页面，权限控制是 ADMIN 角色才有权限查看，如图 10-14 所示。

管理员页面

HTTP 接口测试记录：

搜索

ID	Author	URL	METHOD	CreateTime
1	-	http://127.0.0.1:8004/api/users	GET	2018-01-30 08:57:34
2	-	https://www.baidu.com/	GET	2018-01-30 08:58:14
3	admin	http://www.weather.com.cn/data/sk/101110101.html	GET	2018-01-30 09:58:16
4	admin	http://www.weather.com.cn/data/sk/110100.html	GET	2018-01-30 09:59:07
5	admin	http://www.weather.com.cn/data/sk/101110102.html	GET	2018-01-30 09:59:30
6	admin	http://www.weather.com.cn/data/sk/101110201.html	GET	2018-01-30 09:59:54
7	admin	http://www.weather.com.cn/data/sk/101120201.html	GET	2018-01-30 10:00:07
8	user	http://www.weather.com.cn/data/sk/101130201.html	GET	2018-01-30 10:00:36
9	user	http://www.weather.com.cn/data/sk/101140201.html	GET	2018-01-30 10:00:43
10	user	http://www.weather.com.cn/data/sk/101150201.html	GET	2018-01-30 10:00:48

显示第 1 到第 10 条记录，总共 23 条记录 每页显示 10 条记录　上一页 1 2 3 下一页

图 10-14　HTTP 接口测试记录的列表页面

这个列表页面，我们使用 jquery-3.1.1.js、bootstrap.js、bootstrap-table.js 来实现。静态资源文件目录如下：

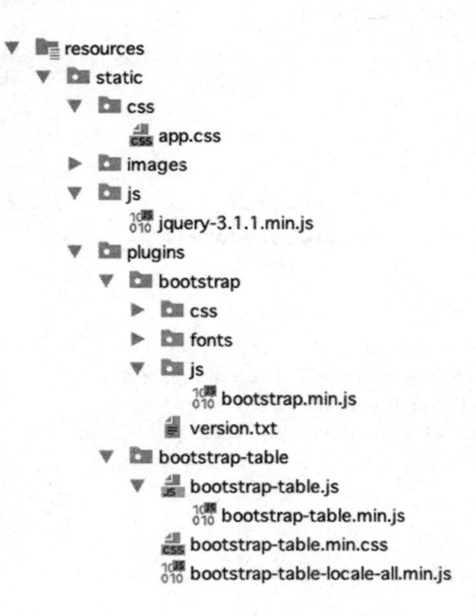

12. Ajax GET 请求与 CSRF

Ajax GET 请求，我们在 bootstrapTable 函数中配置 ajaxOptions 如下：

```
$('#http-test-record').bootstrapTable({
ajaxOptions: {
    beforeSend: function (xhr) {
        xhr.setRequestHeader(header, token); // GET 参数 '_csrf'
    }
}
```

Ajax 的 GET 请求参数是 '_csrf'。

后端 Controller 接口的代码是：

```
@GetMapping(value = ["/findAll"])
@PreAuthorize("hasRole('ADMIN')")
fun page(@RequestParam(value = "pageNo", defaultValue = "0") pageNo: Int,
         @RequestParam(value = "pageSize", defaultValue = "10") pageSize: Int,
         @RequestParam(value = "searchText", defaultValue = "") searchText: String
): Page<HttpTestRecord> {
    return HttpTestRecordDao.findAll(searchText, PageRequest.of(pageNo, pageSize))
}
```

HttpTestRecordDao.findAll 方法实现是：

```
interface HttpTestRecordDao : JpaRepository<HttpTestRecord, Long> {
    @Query(value = """
        select a from #{#entityName} a
        where concat(a.url, '|', a.method)
        like %:searchText%
        """)
    fun findAll(@Param("searchText") searchText: String, pageable: Pageable):Page<
       HttpTestRecord>
}
```

13. 测试运行

为了方便测试用户权限功能，我们给数据库初始化 User，Role 测试数据进去：

```
fun main(args: Array<String>) {
    SpringApplicationBuilder().initializers(
        beans {
            bean {
                ApplicationRunner {
                    try {
                        val roleDao = ref<RoleDao>()
                        // 普通用户角色
                        val roleUser = Role()
                        roleUser.role = "ROLE_USER"
                        val r1 = roleDao.save(roleUser)

                        // 超级管理员角色
                        val roleAdmin = Role()
                        roleAdmin.role = "ROLE_ADMIN"
```

```
                        val r2 = roleDao.save(roleAdmin)

                        val userDao = ref<UserDao>()
                        // 普通用户
                        val user = User()
                        user.username = "user"
                        user.password = BCryptPasswordEncoder().encode("user")
                        val userRoles = setOf(r1)
                        user.roles = userRoles
                        userDao.save(user)

                        // 超级管理员用户
                        val admin = User()
                        admin.username = "admin"
                        admin.password = BCryptPasswordEncoder().encode("admin")
                        val adminRoles = setOf(r1, r2)
                        admin.roles = adminRoles
                        userDao.save(admin)

                    } catch (e: Exception) {
                    }
                }
            }
        }
    ).sources(KsbWithSecurityApplication::class.java).run(*args)
}
```

启动应用，访问 http://127.0.0.1:8004/，系统跳转到登录页面。输入用户名 user 密码 user，可以查看首页如图 10-15 所示。

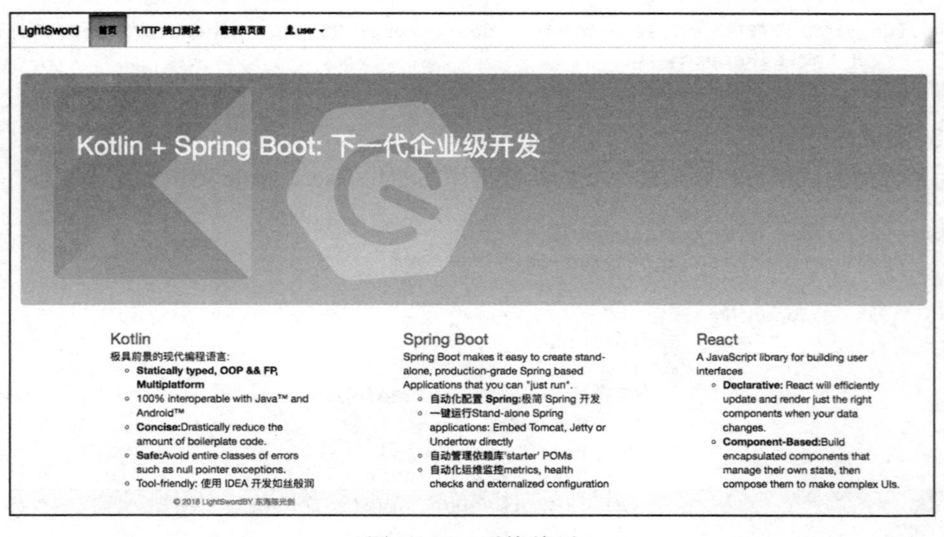

图 10-15 系统首页

HTTP 接口测试页面如图 10-16 所示。

访问管理员页面，会出现 403 信息，提示无权限。点击退出，系统重定向到 http://127.0.0.1:8004/login?logout 页面，然后，输入 ADMIN 角色的用户名 admin 密码 admin，我们可以看到系统可以访问管理员页面，如图 10-17 所示。

图 10-16　HTTP 接口测试页面

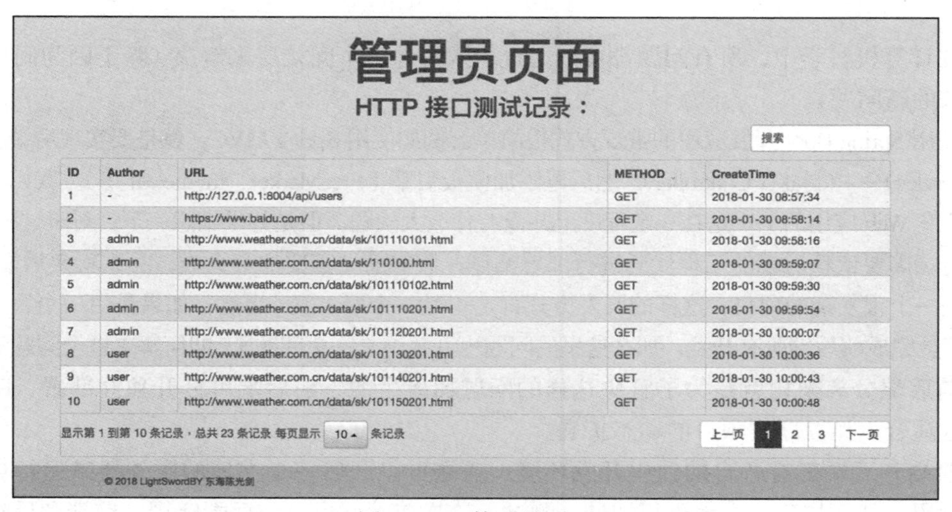

图 10-17　管理员页面

提示　本节工程实例源代码地址为 https://github.com/KotlinSpringBoot/ksb_with_security。

10.4　本章小结

通过本章的学习，我们可以发现 Spring Boot 集成 Security 开发非常方便。Spring Security 主要做两件事：认证，授权。Spring Security 是一个强大的和高度可定制的身份验证和访问控制框架。Spring Security 为基于 Java EE 的企业应用程序提供一个全面的解决方案。

Chapter 11 第 11 章

Spring Boot 集成 React.js 开发前后端分离项目

在计算机科学中，所有问题都可以通过引入另外一个间接层来解决（除了因为间接层太多带来的新问题）。

使用 Spring Boot 开发应用的常规方式很简单，例如使用 Spring MVC，就是要实现后端业务逻辑（Service）+ 控制器（Controller），然后再添加模板引擎（FreeMarker、thymeleaf 等）。我们通常是这样开发 Web 应用的——这在一定程度上并没有什么大问题。但是如果细想一下：对于这种方式，首先前端需要依赖后端的工程代码进行部署测试，假如前端需要加个页面，那么就要后端实现，如果是一个很复杂的项目，这样的开发方式肯定不是一个好方案（当然，如果是由一个很牛的全栈程序员完成前后端所有开发，那么这里的问题也许并不是一个问题），如何解决这个问题呢？

前后端分离架构就是为了解决这样的问题而诞生的。前后端可分开独立部署、跨域访问，这适合大型互联网项目的动态扩展。

前端和后端都有各自的独立开发环境。前端可以开启一个本地的服务器来运行自己的前端代码，从而快速方便地进行调试、测试。这样前端就可以与后端解耦。前端和后端没有强依赖关系，前端只是利用后端的数据，后端只是给前端提供数据。这样，前端可以使用数据接口的 Mock 技术来与后端实现“并行开发”，到了项目后期再进行集成联调。

Spring Boot 和 React.js 是后端开发和前端开发的代表框架。本章就来介绍基于 React.js + Spring Boot 的前后端分离项目的实战开发。

11.1 Web 前端技术简史

本节介绍前端简史和前后端分离架构的背景和优势。Web 前端的三大核心技术是：

HTML、JavaScript、CSS。

1. Web 简史

Web 简史大致如下：

- 1990 年，Tim Berners-Lee 和 Robert Cailliau 共同发明了 Web。
- 1994 年，Web 真正走出实验室。1994 年可以看作前端历史的起点，这一年 10 月 13 日网景推出了第一版 Navigator；这一年，Tim Berners-Lee 创建了 W3C；哈坤 · 利（Hakon Lie）等提出了 CSS。还是这一年，为动态 Web 网页设计的服务端脚本 PHP 诞生。
- 1995 年网景推出了 JavaScript，实现了客户端的计算任务（如表单验证）。
- 1996 年微软推出了 iframe 标签，实现了异步的局部加载。
- 1999 年 W3C 发布 HTML 4 标准。同年，微软推出用于异步数据传输的 ActiveX。随即各大浏览器厂商模仿实现了 XMLHttpRequest。这标识着 Ajax 的诞生，但是 Ajax 这个词是在六年后问世的，特别是在谷歌使用 Ajax 技术打造了 Gmail 和谷歌地图之后，Ajax 获得了巨大的关注。Ajax 是 Web 网页迈向 Web 应用的关键技术，它标识着 Web 2.0 时代的到来。2006 年，XMLHttpRequest 被 W3C 正式纳入标准。
- 2014 年，经过 8 年的艰苦努力，W3C 发布了 HTML 5 标准，其设计目的是在移动设备上支持多媒体。

2. ECMAScript 标准

ECMAScript 是由 ECMA（European Computer Manufactures Association，欧洲计算机制造联合会）制定的脚本程序设计语言标准，例如 JavaScript 是 ECMA-262 标准的实现和扩展。ECMAScript 标准制定的成员包括 Microsoft、Mozilla、Google 等公司。

ECMAScript 的发展简史大致如下：

- 1997 年 ECMAScript 1.0 诞生。
- 1998 年 6 月，ECMAScript 2.0 版发布。
- 1999 年 12 月，ECMAScript 3.0 版发布，成为 JavaScript 的通行标准，得到了广泛支持。
- 2007 年 10 月，ECMAScript 4.0 版草案发布，由于 4.0 版的目标过于激进，各方对于这个标准发生了严重分歧。ECMA 开会决定，中止 ECMAScript 4.0 的开发，改善它的小部分现有功能，发布为 ECMAScript 3.1，而将其他激进的设想扩大范围，放入以后的版本。会后不久，ECMAScript 3.1 就改名为 ECMAScript 5。
- 2009 年 12 月，ECMAScript 5.0 版正式发布。2011 年 6 月，ECMAscript 5.1 版发布，并且成为 ISO 国际标准（ISO/IEC 16262:2011）。
- 2015 年 6 月 17 日，ECMAScript 6（ES6）发布正式版本，即 ECMAScript 2015，其目标是使 JavaScript 可用于编写大型的应用程序，成为企业级的开发语言。

3. 从后端走向前端

早期的网页开发是由后端主导的，前端能做的也就是操作一下 DOM。

2006 年 John Resig 发布了 jQuery，jQuery 主要用于操作 DOM，其优雅的语法、符合直觉的事件驱动型的编程思维使其极易上手，因此很快风靡全球，大量基于 jQuery 的插件构成了一个庞大的生态系统。

2008 年问世的谷歌 V8 引擎改变了这一局面。现代浏览器的崛起终结了微软的垄断时代，前端的计算能力一下子变得过剩了。标准组织也非常配合地在 2009 年发布了 ECMAScript 5，前端的装备得到了整体性的提高，前端技术繁荣昌盛起来，进入了一个令人目不暇接的新时代。

2009 年 Ryan Dahl 发布了 Node，Node 是一个基于 V8 引擎的服务端 JavaScript 运行环境，类似于一个虚拟机，也就是说 JS 在服务端语言中有了一席之地。如果说 Ajax 是前端的第一次飞跃，那么 Node 可算作前端的第二次飞跃。它意味着 JavaScript 走出了浏览器的藩篱，迈出了全端化的第一步。

2009 年 AngularJS 诞生，随后被谷歌收购。

2010 年 Backbone.js 诞生。

2011 年 React 和 Ember 诞生。

2014 年 Vue.js 诞生。

2014 年第五代 HTML 标准发布。H5 是由浏览器厂商主导，与 W3C 合作制定的一整套 Web 应用规范，至今仍在不断补充新的草案。这一系列规范的勃勃雄心是“占领所有屏幕”。

11.2 前后端分离架构

在实际项目架构设计中，前后端分离的方式很多。具体采用何种方案进行分离，在什么层面 / 维度进行分离，这些是要根据实践中的具体情况来进行取舍的，例如需要考虑业务复杂度、人机交互、系统性能、Server 与 Client 的计算能力等，系统架构最终的本质还是“权衡的艺术”，在权衡诸多影响因素后寻求出一条最优路径。

比较好的架构设计是要让团队中的每个人发挥最大的潜能并能更好地协作，同时能够支持系统模块间的松耦合。

前后端分离的系统架构一般如图 11-1 所示。

其中，React.js 框架（类似的有 Vue.js、Angular 等）的客户端 MVC 是目前 Web 前端架构设计的主流。

以 Spring 框架生态为主的服务端（编程语言 Java + Kotlin）只负责提供数据接口，返回的数据格式通常为 JSON 格式、XML 格式或者特定的二进制流等。

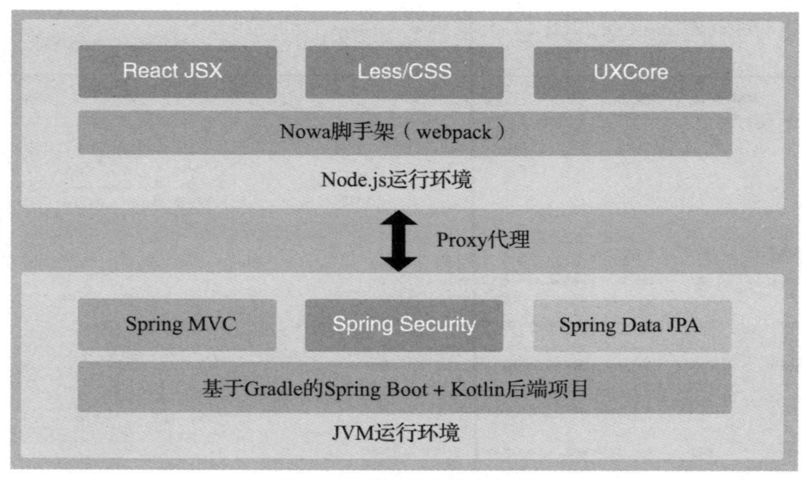

图 11-1　前后端分离架构

11.3　项目实战

本节通过一个简单的前后端分离项目，实现登录、展示 HTTP 接口测试列表页面。后端技术栈如下：

- ❑ 数据库层：MySQL、Spring Data JPA
- ❑ Web 层：Spring、Spring MVC、Spring Security
- ❑ 编程语言：Java、Kotlin
- ❑ 构建工具：Gradle
- ❑ 运行环境：JVM

前端技术栈如下：

- ❑ 前端框架：React.js(https://reactjs.org/)
- ❑ 样式语言：CSS、Less(http://lesscss.org/)
- ❑ 前端 UI 组件库：UXCore(http://uxco.re/)
- ❑ 项目脚手架：Nowa(https://nowa-webpack.github.io/)
- ❑ 运行环境：Node.js(https://nodejs.org/)

关于上面的前端技术在此不做介绍，感兴趣的朋友可以参考相应的文档学习。

11.3.1　系统功能介绍

系统主要有登录、HTTP 接口测试列表页（后端分页表格、支持搜索）两个页面。其中，登录界面如图 11-2 所示。

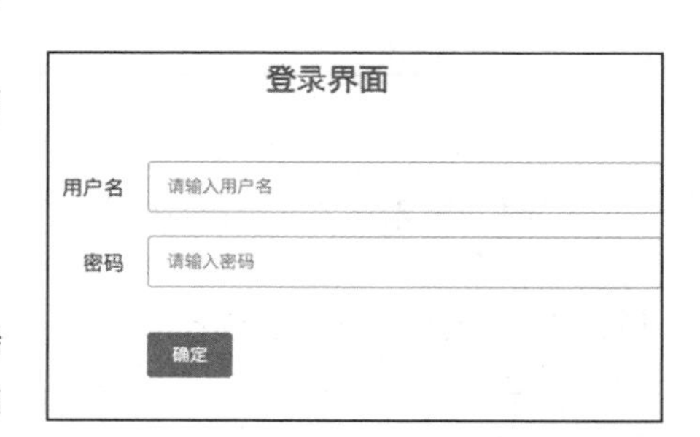

图 11-2　登录界面

HTTP 接口测试列表页面如图 11-3 所示。

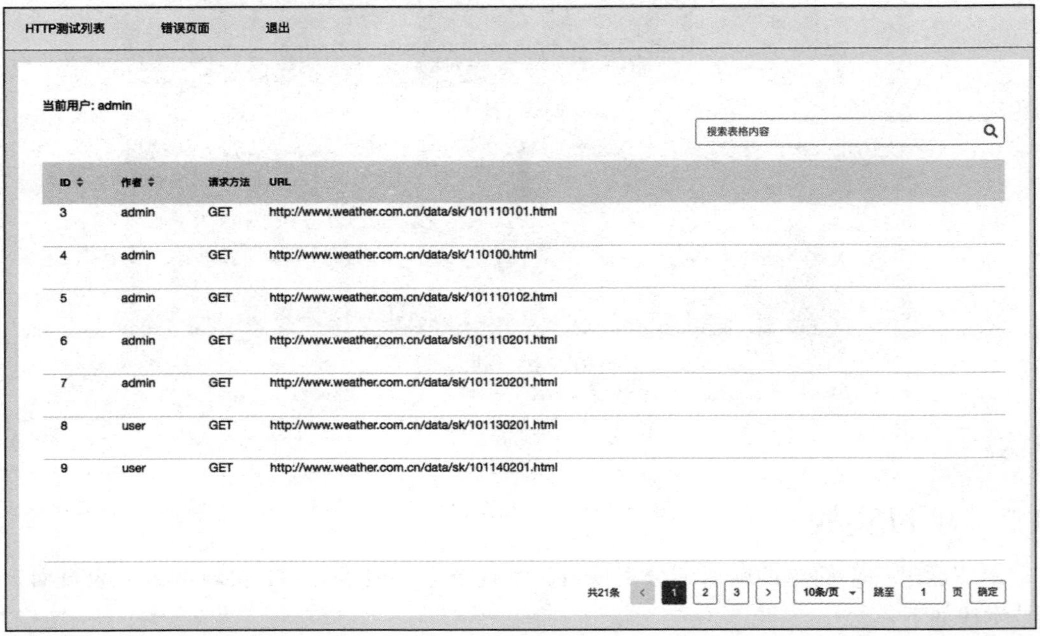

HTTP测试列表 错误页面 退出

当前用户: admin

搜索表格内容

ID	作者	请求方法	URL
3	admin	GET	http://www.weather.com.cn/data/sk/101110101.html
4	admin	GET	http://www.weather.com.cn/data/sk/110100.html
5	admin	GET	http://www.weather.com.cn/data/sk/101110102.html
6	admin	GET	http://www.weather.com.cn/data/sk/101110201.html
7	admin	GET	http://www.weather.com.cn/data/sk/101120201.html
8	user	GET	http://www.weather.com.cn/data/sk/101130201.html
9	user	GET	http://www.weather.com.cn/data/sk/101140201.html

共21条 < 1 2 3 > 10条/页 跳至 1 页 确定

图 11-3 列表页面

表格支持分页、搜索功能，如图 11-4 所示。

当前用户: admin

baidu

ID	作者	请求方法	URL
21	admin	GET	https://www.baidu.com/
22	admin	GET	https://www.baidu.com/
23	admin	GET	https://www.baidu.com/

图 11-4 搜索 baidu 关键字

下面分别来实现上面的功能。

11.3.2 实现登录后端接口

后端采用 Spring Security 实现用户登录和认证。其中，登录相关的安全配置在 WebSecurity-Config.kt 类中，完整的代码参考实例工程源代码。

前后端分离架构中，前端后端交互通常需要跨域支持。如果后端不处理，会报跨域禁止访问的错误

```
▶XHR finished loading: GET "http://127.0.0.1:3000/login".                    jquery.js?27d9:9600
XHR failed loading: OPTIONS "http://127.0.0.1:8008/login/success".
▶{values: {…}, pass: true}                                                   LoginForm.jsx:93
OPTIONS http://127.0.0.1:8008/login/success 403 ()                           :8008/login/success:1
OPTIONS http://127.0.0.1:8008/login/success 403 ()                           :8008/login/success:1
Failed to load http://127.0.0.1:8008/login/success: Response to preflight :3000/#/home?_k=hclz17:1
request doesn't pass access control check: No 'Access-Control-Allow-Origin' header is present on
the requested resource. Origin 'http://127.0.0.1:3000' is therefore not allowed access. The
response had HTTP status code 403.
```

这个时候，只需要在 /login/success 的 Controller 接口代码中添加 @CrossOrigin 注解即可：

```
@RestController
class LoginController {

    @CrossOrigin(origins = ["http://127.0.0.1:3000"],
            maxAge = 3600,
            methods = [RequestMethod.GET, RequestMethod.POST])
    @GetMapping("/login/success")
    fun success(): Result<String> {
        return Result("1", "login success")
    }

    @CrossOrigin(origins = ["http://127.0.0.1:3000"],
            maxAge = 3600,
            methods = [RequestMethod.GET, RequestMethod.POST])
    @GetMapping("/login/failure")
    fun failure(): Result<String> {
        return Result("0", "login failure: username or password incorrect!")
    }

}
```

再次登录执行 POST 请求，才成功。

11.3.3　实现登录前端页面

首先使用 Nowa 创建前端 React 项目（关于 Nowa 命令行的使用，可参考 https://nowa-webpack.github.io/nowa/），得到的项目目录如下：

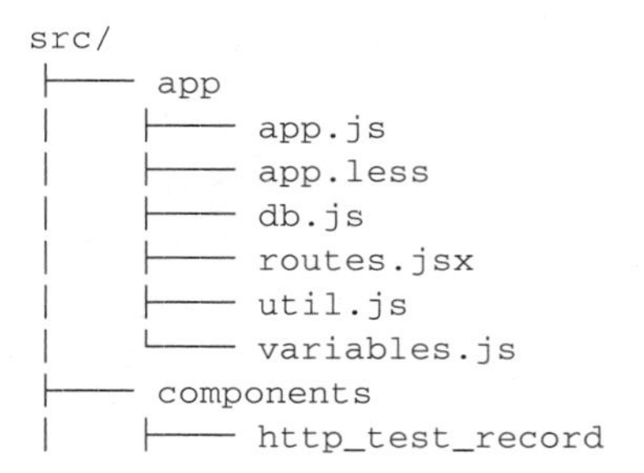

```
src/
├── app
│   ├── app.js
│   ├── app.less
│   ├── db.js
│   ├── routes.jsx
│   ├── util.js
│   └── variables.js
├── components
│   ├── http_test_record
```

```
│   │   ├── HttpTestRecord.css
│   │   ├── HttpTestRecord.jsx
│   │   ├── HttpTestRecord.less
│   │   └── index.js
│   └── login-form
│       ├── LoginForm.css
│       ├── LoginForm.jsx
│       ├── LoginForm.less
│       └── index.js
├── i18n
│   ├── en.js
│   ├── index.js
│   └── zh-cn.js
├── images
│   └── README.md
└── pages
    ├── error
    │   ├── PageError.jsx
    │   ├── PageError.less
    │   └── index.js
    ├── home
    │   ├── PageHome.jsx
    │   ├── PageHome.less
    │   ├── index.js
    │   └── logic.js
    └── http_test_record
        ├── PageHttpTestRecord.css
        ├── PageHttpTestRecord.jsx
        ├── PageHttpTestRecord.less
        ├── index.js
        └── logic.js
```

其中登录 React 组件的代码在 LoginForm.jsx 中。前端部分的代码请参考示例工程源代码。

11.3.4 实现列表展示后端接口

本节使用 UXCore 中的表格组件来实现后端分页接口：http://127.0.0.1:3000/httptestrecord/findAllUxCore?__api=nattyFetch&__stamp=1517590240370&pageSize=7¤tPage=1&searchTxt=baidu。其中，后端分页的表格分页的数据格式的约定 (http://uxco.re/components/table/) 如下：

```
{
    "content":{
        "data":[
            {
                "id":'1'
                "grade":"grade1",
                "email":"email1",
                "firstName":"firstName1",
                "lastName":"lastName1",
                "birthDate":"birthDate1",
                "country":"country1",
```

```
                "city":"city1"
            }
            ...

        ],
        "currentPage":1,
        "totalCount":30
    }
}
```

对应的后端的 DTO 数据类是 PageDtoUxCore。后端接口实现在 HttpTestRecordController.kt 中。完整代码请参考实例工程。

前端表格的代码在 HttpTestRecord.jsx 中。前端在 abc.json 配置文件中添加前端请求转发代理到后端 8008 端口的命令，配置如下：

```
{
    ...
    "proxy": "http://127.0.0.1:8008", // 前端请求转发代理到后端 8008 端口
    ...
  }
}
```

11.3.5　前后端联调测试

后端项目使用 Gradle BootRun 启动，监听 8008 端口。前端使用 Nowa Server 启动，默认在 3000 端口。浏览器打开：http://127.0.0.1:3000/，显示登录界面，使用用户名 admin、密码 admin 登录，如图 11-5 所示。

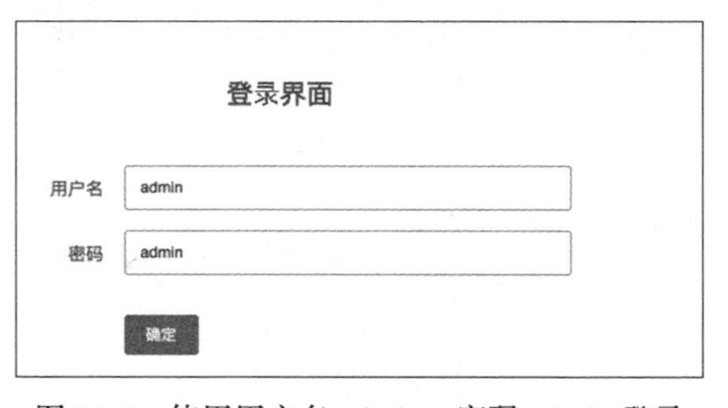

图 11-5　使用用户名 admin、密码 admin 登录

登录成功，跳转到 HTTP 测试列表页面 http://127.0.0.1:3000/#/http_test_record，如图 11-6 所示。

该列表页面支持前后翻页、模糊搜索功能。点击退出，重新访问 http://127.0.0.1:3000/，使用用户名 user、密码 user 登录系统，跳转 HTTP 测试列表页面，但是没有数据。因为我们配置了：

```
@GetMapping(value = ["/findAllUxCore"])
@PreAuthorize("hasRole('ADMIN')")
```

只有 admin 角色的用户才能访问该接口数据。打开浏览器 Console，可以看到该接口的返回是：

```
{
    "code":"403",
    "msg": " 没有权限 "
}
```

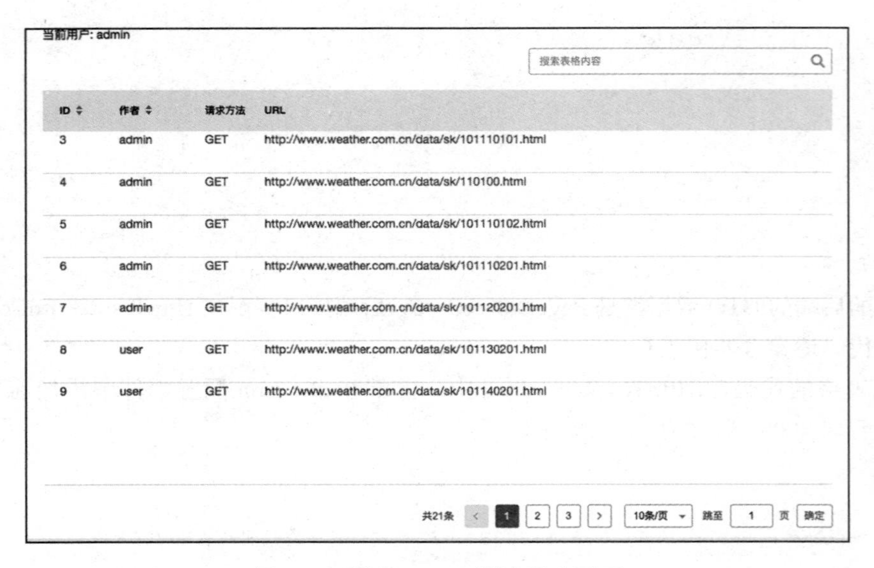

图 11-6　跳转 HTTP 测试列表页面

界面如图 11-7 所示。

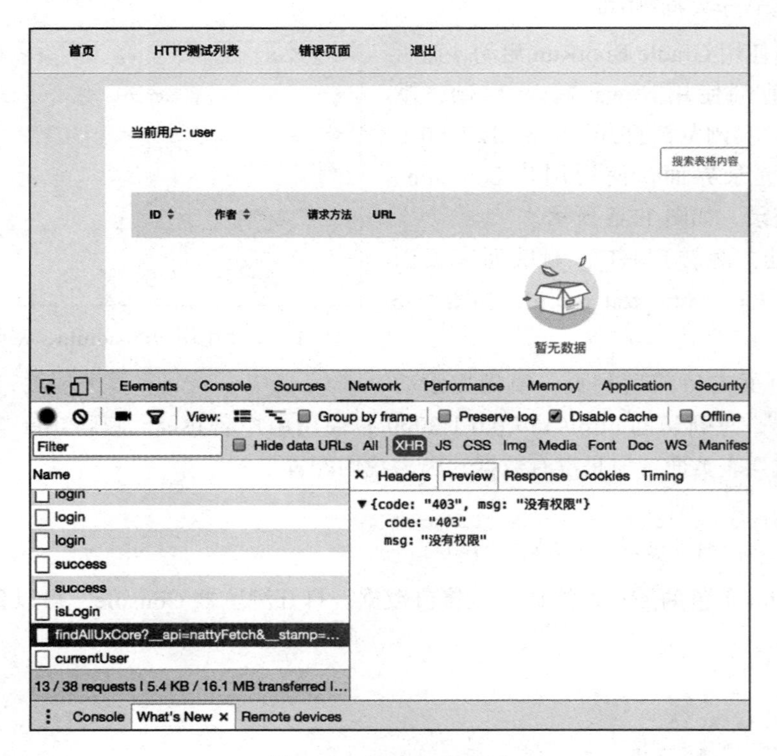

图 11-7　没有权限提示

这个 /findAllUxCore 请求无权限的处理类在 MyAccessDeniedHandler 中，完整代码参考后端项目源代码。

11.4　本章小结

前后端分离可谓大势所趋。让专业的前端去做前端的事情，后端就专注于后端业务逻辑的实现。前后端分离和微服务一样，渐渐地影响了新的大型系统的架构。微服务和前后端分离架构的理念是相通的：解耦——解耦复杂的业务逻辑，解耦架构。前后端分离意味着，前后端之间使用 JSON 数据接口来通信，前端可以独立于后端项目的进度依赖。后端选用的技术栈不影响前端。前端技术栈可以充分发挥 React、Vue、Node.js、ES6/7 等带来的开发福利。

提示　本章前端项目源代码地址为 https://github.com/KotlinSpringBoot/demo_spring_mvc_front。

本章后端项目源代码地址为 https://github.com/KotlinSpringBoot/ksb_with_security/tree/front_back_end_2018.2.2。

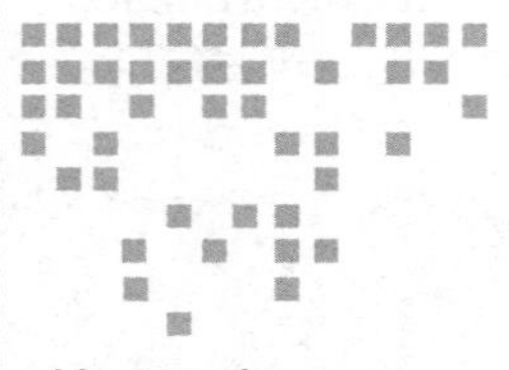

Chapter 12 第 12 章

任务调度与邮件服务开发

定时任务是后端开发过程中一项十分常见的需求，例如：数据报表、异步的后台业务逻辑的处理、日志分析处理、垃圾数据清理、定时更新缓存等场景。Spring Boot 集成了一整套的定时任务工具，让我们只需要专注地完成逻辑，剩下的基础调度工作将自动完成。

发送邮件应该是网站的必备功能之一，例如：注册验证，忘记密码或者给用户发送营销信息，又诸如系统本身的运行状态预警监控通知等。最早期的时候我们会使用 JavaMail 相关 API 二方包来写发送邮件的相关代码，后来 Spring 推出了 JavaMailSender 更加简化了邮件发送的过程。Spring Boot 对 Spring 的 JavaMailSender 进行了更加简单的封装，我们只需要引入起步依赖 spring-boot-starter-mail，然后在属性配置文件中配置邮件属性即可。

本章将详细介绍使用 Spring Boot 如何开发定时任务，以及如何开发任务调度服务和邮件服务。

12.1 定时任务

本节通过项目实例来具体介绍如何在 Spring Boot 项目中开发定时任务。主要内容包括两个方面：实现简单的静态定时任务，以及基于数据库中存储 cron 表达式实现动态配置定时任务。

12.1.1 通用实现方法

在企业级应用系统里，“定时任务”是一个重要的功能，很多地方需要定时执行一项任务。比如，定时消息的通知、业务数据的定时结算、缓存数据的定时更新等。定时任务实现方式通常有如下 3 种方式：

- 使用 Java 自带的 java.util.Timer 类。Timer 提供了一个 java.util.TimerTask 任务支持任务调度。使用这种方式可以让程序按照某一个频度执行，但不能在指定时间运行。因为功能过于单一，使用较少。
- 使用 Quartz：Quartz 是一个功能比较强大的调度器，当然使用起来也相对麻烦。Quartz 支持程序在指定时间执行，也可以按照某一个频度执行。
- 使用 Spring 框架自带的 Schedule 模块：可以将 Schedule 看成一个轻量级的 Quartz，在 Spring Boot 中使用 Schedule 模块比 Quartz 简单。下面我们主要介绍在 Spring Boot 中使用 Schedule 实现定时任务的开发。

另外，定时任务执行方式可分为：单线程（串行）和多线程（并行）。

在 Spring Boot 下开发定时任务需要在启动类上增加一个 @EnableScheduling 注解来开启定时任务功能。

下面我们先来实现一个简单的静态定时任务。

12.1.2 静态定时任务

在 Spring Boot 中实现一个静态定时任务相当极简，只需要两步：

1) 使用 @Component 注解标注一个类，在其中需要定时执行的方法上添加 @Scheduled 注解。代码示例如下：

```
@Component
class DemoScheduleJob {

    @Scheduled(cron = "0/10 * * * * ?")        // 每隔 1 分钟执行 1 次
    fun job1() {
        println(" 执行任务 job1: ${Date()}")
    }

    @Scheduled(fixedDelay = 3000)              // 单位 ms
    fun job2() {
        println(" 执行任务 job2: ${Date()}")
    }

    @Scheduled(fixedRate = 5000)               // 单位 ms
    fun job3() {
        println(" 执行任务 job3: ${Date()}")
    }

}
```

其中，@Scheduled 中的参数说明如下：

- @Scheduled（fixedRate=5000）：上一次开始执行时间点后 5 秒再次执行。
- @Scheduled（fixedDelay=3000）：上一次执行完毕时间点后 3 秒再次执行。

- @Scheduled（cron="0/10 * * * * ?"）：按照 cron 表达式规则执行。关于 cron 表达式我们在下面小节中介绍。

2) 在 Spring Boot 启动类上添加 @EnableScheduling 注解，启用 Spring Schedule 定时调度功能。代码如下：

```
@SpringBootApplication
@EnableScheduling
class CmsApplication

fun main(args: Array<String>) {
    runApplication<CmsApplication>(*args)
}
```

启动应用，我们可以看到控制台打印任务执行记录：

```
执行任务job1: Sun Feb 04 21:44:40 CST 2018
执行任务job3: Sun Feb 04 21:44:41 CST 2018
…
```

可以看出，job1 每隔 10s 执行一次，job2 每隔 3s 执行一次，job3 每隔 5s 执行一次。

12.1.3 Cron 简介

本节简单介绍 Spring 定时任务中的 Cron 表达式。

cron 的表达式用于配置 CronTrigger 实例。cron 的表达式是字符串，实际上是由 7 个子表达式，分别如下：

```
1 Seconds
2 Minutes
3 Hours
4 Day-of-Month
5 Month
6 Day-of-Week
7 Year（可选字段，Spring 中不实用该子域）
例如
0 0 16 ? * FRI
```

表示在每星期五下午 16：00 执行。

Spring Schedule 中的 cron 表达式的值支持 6 个域的表达式，也就是不能设定年，如果超过六个则会报错。框架中相关的源码如下：

```
private void parse(String expression) throws IllegalArgumentException {
    String[] fields = StringUtils.tokenizeToStringArray(expression, " ");
    if (!areValidCronFields(fields)) {
        throw new IllegalArgumentException(String.format(
                "Cron expression must consist of 6 fields (found %d in \"%s\")",
                    fields.length, expression));
    }
```

```
        setNumberHits(this.seconds, fields[0], 0, 60);
        setNumberHits(this.minutes, fields[1], 0, 60);
        setNumberHits(this.hours, fields[2], 0, 24);
        setDaysOfMonth(this.daysOfMonth, fields[3]);
        setMonths(this.months, fields[4]);
        setDays(this.daysOfWeek, replaceOrdinals(fields[5], "SUN,MON,TUE,WED,
            THU,FRI,SAT"), 8);
        if (this.daysOfWeek.get(7)) {
            // Sunday can be represented as 0 or 7
            this.daysOfWeek.set(0);
            this.daysOfWeek.clear(7);
        }
    }

    private static boolean areValidCronFields(String[] fields) {
        return (fields != null && fields.length == 6);
    }
```

Cron 表达式中的每个字段都有对应的有效值范围，如表 12-1 所示。

表 12-1 Cron 表达式字段取值范围

字段	范　围
Seconds（秒）	0 ～ 59
Minutes（分）	0 ～ 59
Hours（时）	0 ～ 23
Day-of-Month（天）	可以用数字 1 ～ 31 中的任意一个值，但要注意一些特别的月份
Month（月）	可以用 0 ～ 11 或用字符串“JAN，FEB，MAR，APR，MAY，JUN，JUL，AUG，SEP，OCT，NOV，DEC”表示
Day-of-Week（每周）	可以用数字 1 ～ 7 表示（1 = 星期日）或用字符口串“SUN，MON，TUE，WED，THU，FRI，SAT”表示

Cron 表达式中的特殊字符的意义见表 12-2。

表 12-2 Cron 表达式中特殊字符的意义

特殊字符	说　明
/	表示为“每隔”一段时间。例如 0 0/15 * * * ? 其中的 0/15 表示从 0 分钟开始，每隔 15 分钟执行一次
?	表示每月的某一天，或周别中的某一天。例如 0 0 0/1 ? * ? * 从 0 点开始，每小时执行 1 次
W	表示最近工作日。例如 0 0 0 15W ? ? * 表示每月 15 号最近的那个工作日
L	用于每月，或每周，表示为每月的最后一天。例如 0 0 0 L ? ? * 表示本月最后一天

（续）

特殊字符	说　明
#	A#B 用来指定每月的第 A 周的星期 B。例如 0 0 0 0 0 1#3 * 表示每月第 1 周的星期 3

提示 上面这些语法不大好记，可以使用一个在线生成 cron 表达式的工具：http://cron.qqe2.com/。

这里给出常用的 Cron 表达式实例，如表 12-3 所示。

表 12-3 常用的 Cron 表达式实例

Cron 表达式	说　明
0 15 10 * * ?	每天 10 点 15 分触发
0 * 14 * * ?	每天下午的 2 点到 2 点 59 分每分钟触发
0 0/5 14 * * ?	每天下午的 2 点到 2 点 59 分，整点开始每隔 5 分触发
0 0/5 14，18 * * ?	每天下午的 2 点到 2 点 59 分、18 点到 18 点 59 分：整点开始，每隔 5 分触发
0 0-5 14 * * ?	每天下午的 2 点到 2 点 05 分每分触发
0 15 10 ? * 6L	每月最后一周的星期五的 10 点 15 分触发
0 15 10 ? * 6#3	每月的第三周的星期五触发
0 0 10，14，16 * * ?	每天 10 点，14 点，16 点
0 0/30 9-17 * * ?	每天朝九晚五时间内每半小时
0 0 12 ? * WED	表示每个星期三中午 12 点
0 0 12 * * ?	每天中午 12 点触发
0 10，44 14 ? 3 WED	每年三月的星期三的下午 2：10 和 2：44 触发
0 15 10 ? * MON-FRI	周一至周五的上午 10：15 触发
0 15 10 15 * ?	每月 15 日上午 10：15 触发
0 15 10 L * ?	每月最后一日的上午 10：15 触发
0 15 10 ? * 6L	每月的最后一个星期五上午 10：15 触发
0 15 10 ? * 6#3	每月的第三个星期五上午 10：15 触发
0 15 10 ? * *	每天上午 10：15 触发

12.1.4 动态定时任务

本节介绍基于数据库中存储 cron 表达式实现动态配置定时任务的执行。我们只需要实现 SchedulingConfigurer 接口中的 configureTasks() 方法来定制任务执行的调度逻辑即可。

1. SchedulingConfigurer 接口

SchedulingConfigurer 接口是任务调度器 Schedule 的配置接口，实现其中的 configureTasks 方法可以实现任务的配置。SchedulingConfigurer 接口定义如下：

```
@FunctionalInterface
public interface SchedulingConfigurer {
    void configureTasks(ScheduledTaskRegistrar var1);
}
```

配合数据库动态执行定时任务的完整实例代码在 CustomScheduleConfig .kt 中。其中的关键代码如下（代码行的说明见注释）：

```
@Configuration
class CustomScheduleConfig : SchedulingConfigurer {
    …
    override fun configureTasks(taskRegistrar: ScheduledTaskRegistrar) {

        taskRegistrar.addTriggerTask(
            // 1）添加任务执行 Runnable
            {
                println("${df.format(Date())} 执行任务 1)，线程：${Thread.current
                    Thread().name}")
            },
            // 2）设置执行周期 (Trigger)
            { triggerContext ->
                // 3）从数据库获取执行周期
                val cron = cronTriggerDao.findByTaskId(CRAW_JIANSHU_TECH_ARTICLE
                    _TASK_ID)
                // 4）合法性校验
                var cronExpression: String? = DEFAULT_CRON
                if (!StringUtils.isEmpty(cron?.cron)) {
                    cronExpression = cron?.cron
                }
                // 5）返回执行周期 (Date)
                CronTrigger(cronExpression).nextExecutionTime(triggerContext)
            }
        )
    }
}
```

2. 从数据库中读取 cron 表达式值

在上节的 Schedule 配置类中，我们通过数据库查询对应任务的 cron 表达式的值。其中，CronTriggerDao 的代码如下：

```
interface CronTriggerDao : JpaRepository<CronTrigger, Long> {

    @Query("""
        select a from #{#entityName} a where a.taskId = :taskId
    """)
    fun findByTaskId(@Param("taskId") taskId: Int): CronTrigger?
}
```

CronTrigger 实体类数据结构如下：

```
@Entity
class CronTrigger {
```

```
    @Id
    @GeneratedValue(strategy = GenerationType.IDENTITY)
    var id: Long = -1
    var gmtCreate = Date()
    var gmtModify = Date()
    var isDeleted = 0

    @Column(unique = true)
    var taskId: Int = -1
    var cron: String = ""
}
```

3. 数据库初始化数据

在运行上面的代码之前，我们需要在数据中初始化 2 条任务执行周期的数据：

```
INSERT INTO `cms_spider`.`cron_trigger` (`cron`, `gmt_modify`, `gmt_create`,`is_
    deleted`, `task_id`) VALUES ('0 */1 * * * ?',now(), now(), 0,'1');
INSERT INTO `cms_spider`.`cron_trigger` (`cron`, `gmt_create`, `gmt_modify`,`is
    _deleted`, `task_id`) VALUES ('0/7 * * * * ?', '2018-02-04 03:38:54','2018-
    02-04 03:38:54', '0', '2');
```

cron_trigger 表初始化的数据如下：

	id	cron	gmt_create	gmt_modify	is_deleted	task_id
1	1	0/10 * * * * ?	2018-02-04 03:38:54	2018-02-04 03:38:54	0	1
2	2	0/7 * * * * ?	2018-02-04 03:38:54	2018-02-04 03:38:54	0	2

启动应用，观察控制台日志，我们可以看出当前的任务 1 的执行周期是每隔 10s 执行 1 次：

```
2018-02-04 22:35:00 执行任务 1，线程：pool-1-thread-1
2018-02-04 22:35:10 执行任务 1，线程：pool-1-thread-1
2018-02-04 22:35:20 执行任务 1，线程：pool-1-thread-1
2018-02-04 22:35:30 执行任务 1，线程：pool-1-thread-1
…
```

下面，我们手动把数据库中的 task_id=1 的记录中的 cron 的值更新为：

```
0/5 * * * * ?
```

如下所示：

	id	cron	gmt_create	gmt_modify	is_deleted	task_id
1	1	0/5 * * * * ?	2018-02-04 03:38:54	2018-02-04 03:38:54	0	1
2	2	0/7 * * * * ?	2018-02-04 03:38:54	2018-02-04 03:38:54	0	2

再次观察控制台日志，可以发现任务 1 的执行周期动态改变为每隔 5s 执行 1 次：

```
2018-02-04 22:37:00 执行任务 1，线程：pool-1-thread-1
2018-02-04 22:37:10 执行任务 1，线程：pool-1-thread-1
2018-02-04 22:37:20 执行任务 1，线程：pool-1-thread-1
2018-02-04 22:37:25 执行任务 1，线程：pool-1-thread-1
……
```

12.1.5　多线程执行任务

从上节的执行日志中可以看出任务的执行都是单线程的。我们可以设置多线程执行定时任务，只需要在 SchedulingConfigurer 接口的 configureTasks 方法实现中添加线程池即可。代码如下：

```
@Configuration
class CustomScheduleConfig : SchedulingConfigurer {
    ...
    override fun configureTasks(taskRegistrar: ScheduledTaskRegistrar) {

        //设定一个长度 10 的定时任务线程池
        taskRegistrar.setScheduler(Executors.newScheduledThreadPool(10))

        taskRegistrar.addTriggerTask(
            ...
        )
    }
}
```

重启应用，观察控制台日志，可以看出多线程执行任务的输出：

```
2018-02-04 22:41:30 执行任务 1，线程：pool-1-thread-1
2018-02-04 22:41:35 执行任务 1，线程：pool-1-thread-3
2018-02-04 22:41:40 执行任务 1，线程：pool-1-thread-5
2018-02-04 22:41:45 执行任务 1，线程：pool-1-thread-5
```

另外，我们可以在 configureTasks（taskRegistrar ： ScheduledTaskRegistrar）方法中添加多个任务，代码如下：

```
@Configuration
class CustomScheduleConfig : SchedulingConfigurer {
    val DEFAULT_CRON = "0 0/1 * * * ? "
    val CRAW_JIANSHU_TECH_ARTICLE_TASK_ID = 1
    val GANK_IMAGE_CRAW_TASK_ID = 2

    val logger = LoggerFactory.getLogger(CustomScheduleConfig::class.java)

    @Autowired lateinit var cronTriggerDao: CronTriggerDao

    @Autowired lateinit var crawTechArticleService: CrawTechArticleService
    @Autowired lateinit var crawImagesService: CrawImageService

    val df = SimpleDateFormat("yyyy-MM-dd HH:mm:ss")

    override fun configureTasks(taskRegistrar: ScheduledTaskRegistrar) {

        //设定一个长度 10 的定时任务线程池
        taskRegistrar.setScheduler(Executors.newScheduledThreadPool(10))

        taskRegistrar.addTriggerTask(
            //1）添加任务执行 Runnable
```

```
            {
                println("${df.format(Date())} 执行任务 1）线程：${Thread.current
                    Thread().name}")
            },
            //2）设置执行周期（Trigger）
            { triggerContext ->
                //3）从数据库获取执行周期
                val cron = cronTriggerDao.findByTaskId(CRAW_JIANSHU_TECH_ARTICLE
                    _TASK_ID)
                //4）合法性校验
                var cronExpression: String? = DEFAULT_CRON
                if (!StringUtils.isEmpty(cron?.cron)) {
                    cronExpression = cron?.cron
                }
                //5）返回执行周期（Date）
                CronTrigger(cronExpression).nextExecutionTime(triggerContext)
            }
        )

        taskRegistrar.addTriggerTask(// 添加任务 2）
            {
                println("${df.format(Date())} 执行任务 2）线程：${Thread.current
                    Thread().name}")
            },
            { triggerContext ->
                val cron = cronTriggerDao.findByTaskId(GANK_IMAGE_CRAW_TASK_ID)
                var cronExpression: String? = DEFAULT_CRON
                if (!StringUtils.isEmpty(cron?.cron)) {
                    cronExpression = cron?.cron
                }
                CronTrigger(cronExpression).nextExecutionTime(triggerContext)
            }
        )

    }
}
```

重新启动应用，我们可以发现 Spring Schedule 使用线程池成功调度了任务 1）和任务 2）的执行：

```
2018-02-04 21:13:53.580  INFO 58967 --- [           main] com.ak47.cms.cms.Cms
    ApplicationKt        : Started CmsApplicationKt in 29.738 seconds(JVM running
    for 30.837)
2018-02-04 21:13:56 执行任务 2，线程：pool-1-thread-1
2018-02-04 21:14:00 执行任务 2，线程：pool-1-thread-3
2018-02-04 21:14:00 执行任务 1，线程：pool-1-thread-2
…
```

在 Spring Boot 下开发定时任务，只需要简单的注解即可实现静态定时任务，基于数据库的动态定时任务也只需要实现 SchedulingConfigurer 接口中的 configureTasks 完成任务的注册和执行周期的动态修改即可。

本节实例工程源代码地址为 https://github.com/AK-47-D/cms-spider/tree/dynamic_schedule_2018.2.4。

12.2 开发任务调度服务

本节介绍 Spring Boot 中怎样实现任务的异步执行。

12.2.1 同步与异步

"异步"（asynchronous）与"同步"（synchronous）相对，异步不用阻塞当前线程来等待处理完成，而是允许后续操作，直至其他线程将处理完成，并回调通知此线程。也就是说，异步永远是非阻塞的（non-blocking）。

同步操作的程序，会按照代码的顺序依次执行，每一行程序都必须等待上一个程序执行完成之后才能执行。哪些情况建议使用同步交互呢？例如，银行的转账系统，对数据库的保存操作等等，都会使用同步交互操作。

异步操作的程序，在代码执行时，不等待异步调用的语句返回结果就执行后面的程序。当任务间没有先后顺序依赖逻辑的时候，可以使用异步。异步编程的主要困难在于，构建程序的执行逻辑时是非线性的，这需要将任务流分解成很多小的步骤，再通过异步回调函数的形式组合起来。

12.2.2 同步任务执行

下面通过一个简单示例来直观的理解什么是同步任务执行。

编写一个 SyncTask 类，创建三个处理函数：doTaskA()、doTaskB()、doTaskC() 来分别模拟三个任务执行的操作，操作消耗时间分别设置为：1000ms、2000ms、3000ms。代码如下：

```
package com.easy.springboot.demo_async_task.task

import org.springframework.stereotype.Component

@Component
class SyncTask {

    fun doTaskA() {
        println("开始任务A")
        val start = System.currentTimeMillis()
        Thread.sleep(1000)
        val end = System.currentTimeMillis()
        println("结束任务A，耗时：" + (end - start) + "ms")
    }

    fun doTaskB() {
        println("开始任务B")
        val start = System.currentTimeMillis()
```

```
        Thread.sleep(2000)
        val end = System.currentTimeMillis()
        println("结束任务B，耗时：" + (end - start) + "ms")
    }

    fun doTaskC() {
        println("开始任务C")
        val start = System.currentTimeMillis()
        Thread.sleep(3000)
        val end = System.currentTimeMillis()
        println("结束任务C，耗时：" + (end - start) + "ms")
    }
}
```

下面我们来写一个单元测试，在测试用例中顺序执行 doTaskA()、doTaskB()、doTaskC() 三个函数：

```
package com.easy.springboot.demo_async_task

import com.easy.springboot.demo_async_task.task.AsyncTask
import com.easy.springboot.demo_async_task.task.SyncTask
import org.junit.Test
import org.junit.runner.RunWith
import org.springframework.beans.factory.annotation.Autowired
import org.springframework.boot.test.context.SpringBootTest
import org.springframework.test.context.junit4.SpringRunner

@RunWith(SpringRunner::class)
@SpringBootTest
class DemoAsyncTaskApplicationTests {
    @Autowired lateinit var syncTask: SyncTask
    @Autowired lateinit var asyncTask: AsyncTask

    @Test
    fun testSyncTask() {
        println("开始测试SyncTask")
        val start = System.currentTimeMillis()
        syncTask.doTaskA()
        syncTask.doTaskB()
        syncTask.doTaskC()
        val end = System.currentTimeMillis()
        println("结束测试SyncTask，耗时：" + (end - start) + "ms")
    }

}
```

执行上面的单元测试，可以在控制台看到类似如下输出：

```
开始测试SyncTask
开始任务A
结束任务A，耗时：1004ms
开始任务B
结束任务B，耗时：2005ms
开始任务C
```

```
结束任务 C，耗时：3002ms
结束测试 SyncTask，耗时：6012ms
```

任务 A、任务 B、任务 C 依次按照其先后顺序执行完毕，总共耗时：6012ms。

12.2.3　异步任务执行

上面的同步任务的执行，虽然顺利地执行完了三个任务，但我们可以看到执行时间比较长，是这三个任务时间的累加。若这三个任务本身之间不存在依赖关系，可以并发执行的话，同步顺序执行在执行效率上就比较差了。这个时候，我们可以考虑通过异步调用的方式来实现“异步并发”地执行。下面介绍如何设计异步任务执行。

1. 编写 AsyncTask 类

在 Spring Boot 中，我们只需要通过使用 @Async 注解就能简单地将原来的同步函数变为异步函数，编写 AsyncTask 类，代码编写如下：

```
package com.easy.springboot.demo_async_task.task

import org.springframework.scheduling.annotation.Async
import org.springframework.scheduling.annotation.AsyncResult
import org.springframework.stereotype.Component
import java.util.*
import java.util.concurrent.Future

@Component
open class AsyncTask {

    @Async("asyncTaskExecutor")
    open fun doTaskA(): Future<String> {
        println(" 开始任务 A")
        val start = System.currentTimeMillis()
        Thread.sleep(1000)
        val end = System.currentTimeMillis()
        println(" 结束任务 A，耗时：" + (end - start) + "ms")
        return AsyncResult("TaskA DONE")
    }

    @Async("asyncTaskExecutor")
    open fun doTaskB(): Future<String> {
        println(" 开始任务 B")
        val start = System.currentTimeMillis()
        Thread.sleep(2000)
        val end = System.currentTimeMillis()
        println(" 结束任务 B，耗时：" + (end - start) + "ms")
        return AsyncResult("TaskB DONE")
    }

    @Async("asyncTaskExecutor")
    open fun doTaskC(): Future<String> {
        println(" 开始任务 C")
        val start = System.currentTimeMillis()
        Thread.sleep(3000)
```

```
        val end = System.currentTimeMillis()
        println("结束任务C，耗时：" + (end - start) + "ms")
        return AsyncResult("TaskC DONE")
    }
}
```

上面的异步执行的任务，都返回一个 Future<String> 类型的结果对象 AsyncResult。这个对象中保存了任务的执行状态。我们可以通过轮询这个结果来等待任务执行完毕，这样我们可以在上面三个任务都执行完毕后，再继续做一些事情。

2. 自定义线程池

上面代码中，asyncTaskExecutor 是我们自定义的线程池。代码如下：

```
@Configuration
open class TaskExecutorPoolConfig {
    @Bean("asyncTaskExecutor")
    open fun taskExecutor(): Executor {
        val executor = ThreadPoolTaskExecutor()
        executor.corePoolSize = 10 //线程池维护线程的最少数量
        executor.maxPoolSize = 20  //线程池维护线程的最大数量
        executor.setQueueCapacity(100)
        executor.keepAliveSeconds = 30  //线程池维护线程所允许的空闲时间，TimeUnit.SECONDS
        executor.threadNamePrefix = "asyncTaskExecutor-"
        //线程池对拒绝任务的处理策略：CallerRunsPolicy 策略，当线程池没有处理能力的时候，
        //该策略会直接在 execute 方法的调用线程中运行被拒绝的任务；如果执行程序已关闭，则会
        //丢弃该任务 executor.setRejectedExecutionHandler(ThreadPoolExecutor.Caller
        //RunsPolicy())
        return executor
    }
}
```

我们通过使用 ThreadPoolTaskExecutor 创建了一个线程池，同时设置了以下这些参数，简单说明如下：

- 核心线程数 10：线程池创建时候初始化的线程数。
- 最大线程数 20：线程池最大的线程数，只有在缓冲队列满了之后才会申请超过核心线程数的线程。
- 缓冲队列 100：用来缓冲执行任务的队列。
- 允许线程的空闲时间 30 秒：当超过了核心线程出之外的线程在空闲时间到达之后会被销毁。
- 线程池名的前缀：设置好了之后可以方便我们定位处理任务所在的线程池。
- 线程池对拒绝任务的处理策略：这里采用了 CallerRunsPolicy 策略，当线程池没有处理能力的时候，该策略会直接在 execute 方法的调用线程中运行被拒绝的任务；如果执行程序已关闭，则会丢弃该任务。

3. 启用 @EnableAsync

为了让 @Async 注解能够生效，还需要在 Spring Boot 的入口类上配置 @EnableAsync，

代码如下：

```
@SpringBootApplication
@EnableAsync
open class DemoAsyncTaskApplication

fun main(args: Array<String>) {
    runApplication<DemoAsyncTaskApplication>(*args)
}
```

4. 单元测试

同样，我们来编写一个单元测试用例来测试一下异步执行这三个任务所花费的时间，代码如下：

```
package com.easy.springboot.demo_async_task

import com.easy.springboot.demo_async_task.task.AsyncTask
import com.easy.springboot.demo_async_task.task.SyncTask
import org.junit.Test
import org.junit.runner.RunWith
import org.springframework.beans.factory.annotation.Autowired
import org.springframework.boot.test.context.SpringBootTest
import org.springframework.test.context.junit4.SpringRunner

@RunWith(SpringRunner::class)
@SpringBootTest
class DemoAsyncTaskApplicationTests {
    @Autowired lateinit var syncTask: SyncTask
    @Autowired lateinit var asyncTask: AsyncTask

    @Test
    fun testAsyncTask() {
        println("开始测试AsyncTask")
        val start = System.currentTimeMillis()

        val r1 = asyncTask.doTaskA()
        val r2 = asyncTask.doTaskB()
        val r3 = asyncTask.doTaskC()

        while (true) {
            //三个任务都调用完成，退出循环等待
            if (r1.isDone && r2.isDone && r3.isDone) {
                break
            }
            Thread.sleep(100)
        }

        val end = System.currentTimeMillis()
        println("结束测试AsyncTask，耗时：" + (end - start) + "ms")
    }

}
```

我们使用了一个死循环来等待三个任务都调用完成，当满足条件 r1.isDone && r2.isDone && r3.isDone 就退出循环等待。

执行上面的测试代码，可以在控制台看到类似如下输出：

```
开始测试 AsyncTask
开始任务 A
开始任务 B
开始任务 C
结束任务 A，耗时：1002ms
结束任务 B，耗时：2004ms
结束任务 C，耗时：3004ms
结束测试 AsyncTask，耗时：3125ms
```

我们可以看到，通过异步调用，任务 A、任务 B、任务 C 异步执行完毕总共耗时 3125ms。相比于同步执行，无疑大大地减少了程序的总运行时间。

提示 本节实例工程源代码地址为 https://github.com/EasySpringBoot/demo_async_task。

12.3 开发邮件服务

邮件服务在互联网早期就已经出现，如今已成为人们互联网生活中必不可少的一项服务。Spring Boot 实现发送邮件服务相当简单，只需要简单的三步即可完成。

第 1 步 添加起步依赖

在 build.gradle 中添加 spring-boot-starter-mail 如下：

```
buildscript {
    ext {
        kotlinVersion = '1.2.20'
        springBootVersion = '2.0.1.RELEASE'
    }
    repositories {
        mavenCentral()
    }
    dependencies {
        classpath("org.springframework.boot:spring-boot-gradle-plugin:${spring
            BootVersion}")
        classpath("org.jetbrains.kotlin:kotlin-gradle-plugin:${kotlinVersion}")
        classpath("org.jetbrains.kotlin:kotlin-allopen:${kotlinVersion}")
    }
}

...

dependencies {
    compile('org.springframework.boot:spring-boot-starter-mail') //发邮件的依赖
    compile('org.springframework.boot:spring-boot-starter-web')
```

```
    compile('com.fasterxml.jackson.module:jackson-module-kotlin')
    compile("org.jetbrains.kotlin:kotlin-stdlib-jdk8")
    compile("org.jetbrains.kotlin:kotlin-reflect")
    testCompile('org.springframework.boot:spring-boot-starter-test')
}
```

第 2 步　配置 spring.mail.* 属性

在 application.properties 配置文件中配置邮件相关的属性如下：

```
spring.mail.host=smtp.163.com
spring.mail.username=15868187925@163.com
spring.mail.password=************
spring.mail.properties.mail.smtp.auth=true
spring.mail.properties.mail.smtp.starttls.enable=true
spring.mail.properties.mail.smtp.starttls.required=true
```

其中，spring.mail.password 是发件人邮箱的密码。spring.mail.host=smtp.163.com 是 163 邮箱的 SMTP 服务器地址。

第 3 步　实现发送邮件的代码

实现发送邮件的实例完整代码如下：

```
package com.easy.springboot.demo_send_mail

import org.springframework.beans.factory.annotation.Autowired
import org.springframework.mail.SimpleMailMessage
import org.springframework.mail.javamail.JavaMailSender
import org.springframework.stereotype.Service

@Service
class MailTool {

    @Autowired lateinit var mailSender: JavaMailSender

    fun send(): Boolean {
        try {
            val message = SimpleMailMessage()
            message.setFrom("15868187925@163.com")          // 发送者
            message.setTo("15868187925@163.com")            // 接收者
            message.setSubject(" 邮件主题 ")                 // 邮件主题
            message.setText(" 这是邮件内容 ")                // 邮件内容
            mailSender.send(message)                        // 发送邮件
            println(" 发送成功 ")
            return true
        } catch (e: Exception) {
            e.printStackTrace()
            return false
        }
    }
}
```

下面我们来写一个单元测试的代码来测试一下邮件发送的效果。测试代码如下：

```
package com.easy.springboot.demo_send_mail

import org.springframework.beans.factory.annotation.Autowired
import org.springframework.mail.SimpleMailMessage
import org.springframework.mail.javamail.JavaMailSender
import org.springframework.stereotype.Service

@Service
class MailTool {

    @Autowired lateinit var mailSender: JavaMailSender

    fun send(): Boolean {
        try {
            val message = SimpleMailMessage()
            message.setFrom("15868187925@163.com")              //发送者
            message.setTo("15868187925@163.com")                //接收者
            message.setSubject("邮件主题")                      //邮件主题
            message.setText("这是邮件内容")                      //邮件内容
            mailSender.send(message)                            //发送邮件
            println("发送成功")
            return true
        } catch (e: Exception) {
            e.printStackTrace()
            return false
        }
    }
}
```

运行上面的测试代码登入邮件中，可以发现已经收到了我们的测试邮件，如图 12-1 所示。

图 12-1 测试邮件发送效果

12.3.1 发送富文本邮件

上面我们介绍的是发送纯文本的邮件内容。很多情况下，我们的邮件内容需要富文本展示。我们只需要使用 JavaMail 的 MimeMessage，就可以支持更加复杂的邮件格式和内容。代码如下：

```
fun sendHtml(): Boolean {
    //使用 JavaMail 的 MimeMessage，支持更加复杂的邮件格式和内容
    val mimeMessage = mailSender.createMimeMessage()
    //创建 MimeMessageHelper 对象，处理 MimeMessage 的辅助类
    val helper = MimeMessageHelper(mimeMessage, true)
    return try {
        //使用辅助类 MimeMessage 设定参数
        helper.setFrom("15868187925@163.com")              //发送者
        helper.setTo("15868187925@163.com")                //接收者
        helper.setSubject("富文本邮件主题")                  //邮件主题
        helper.setText("<h1>这是富文本邮件内容</h1><div style='color:red'>你好，Spring
            Boot！我是 Kotlin。</div>", true)//邮件内容，第 2 个参数设置为 true 表示是 html 文本
        mailSender.send(mimeMessage)                       //发送邮件
```

```
            println(" 发送成功 ")
            true
        } catch (e: Exception) {
            e.printStackTrace()
            false
        }
    }
```

其中，helper.setText() 中的 setText（String text，boolean html）方法第 2 个参数设置为 true，表示这段文本是 HTML 文本。写一段测试代码如下：

```
package com.easy.springboot.demo_send_mail

import org.junit.Test
import org.junit.runner.RunWith
import org.springframework.beans.factory.annotation.Autowired
import org.springframework.boot.test.context.SpringBootTest
import org.springframework.test.context.junit4.SpringRunner

@RunWith(SpringRunner::class)
@SpringBootTest
class DemoSendMailApplicationTests {

    @Autowired lateinit var MailTool: MailTool

    @Test
    fun testSendHtml() {
        MailTool.sendHtml()
    }

}
```

运行上面的测试代码，可以发现收到了富文本邮件，如图 12-2 所示。

图 12-2　富文本邮件

12.3.2　发送带附件的富文本邮件

邮件中发送附件也是非常常见的需求，这个实现起来也比较简单，代码如下：

```
fun sendHtmlWithAttach(): Boolean {
    val mimeMessage = mailSender.createMimeMessage()
    val helper = MimeMessageHelper(mimeMessage, true)
    return try {
        helper.setFrom("15868187925@163.com")                  // 发送者
        helper.setTo("15868187925@163.com")                    // 接收者
        helper.setSubject(" 富文本带附件的邮件主题 ")           // 邮件主题
        helper.setText("<h1> 这是富文本带附件的邮件内容 </h1><div style='color:red'>
            你好，Spring Boot！ 我是 Kotlin。</div>", true)     // 邮件内容
        // 加载文件资源，作为附件
        val file = ClassPathResource("kotlin.png")
        // 加入附件
```

```
            helper.addAttachment("attachment.jpg", file)
            mailSender.send(mimeMessage)                                   // 发送邮件
            println(" 发送成功 ")
            true
        } catch (e: Exception) {
            e.printStackTrace()
            false
        }
    }
```

其中，使用 helper.addAttachment（"attachment.jpg"，file）这一句来给邮件添加附件。写一段测试代码：

```
package com.easy.springboot.demo_send_mail

import org.junit.Test
import org.junit.runner.RunWith
import org.springframework.beans.factory.anno-tation.Autowired
import org.springframework.boot.test.context.
   SpringBootTest
import org.springframework.test.context.junit4.
   SpringRunner

@RunWith(SpringRunner::class)
@SpringBootTest
class DemoSendMailApplicationTests {

    @Autowired lateinit var MailTool: MailTool

    @Test
    fun testSendHtmlWithAttach() {
        MailTool.sendHtmlWithAttach()
    }

}
```

执行测试，可以收到如图 12-3 所示的带附件的富文本邮件。

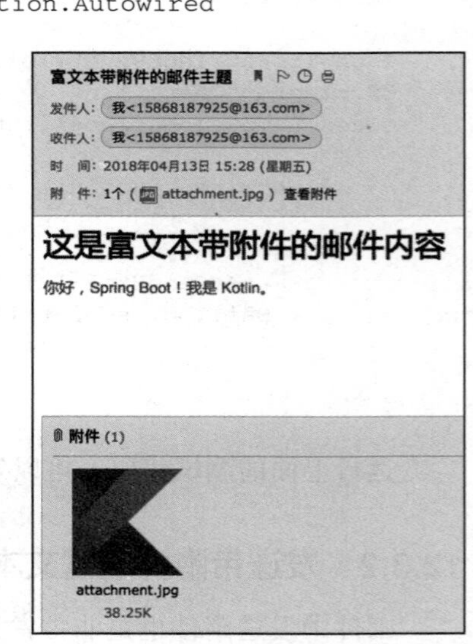

图 12-3 带附件的富文本邮件

提示 本节实例工程源代码地址为 https://github.com/EasySpringBoot/demo_send_mail。

12.4 本章小结

通过本章的学习，我们已经知道如何使用 Spring Boot 来开发动态定时任务、实现任务的异步执行，以及开发邮件服务。

在系统功能中，我们经常会发送一些邮件通知给用户，可以看到使用 Spring Boot 开发邮件服务也非常简单。

第 13 章　Chapter 13

Spring Boot 集成 WebFlux 开发响应式 Web 应用

IBM 的研究称，整个人类文明所获得的全部数据中，有 90% 是过去两年内产生的。在此背景下，包括 NoSQL，Hadoop，Spark，Storm，Kylin 在内的大批新技术应运而生。其中以 RxJava 和 Reactor 为代表的响应式（Reactive）编程技术针对的就是经典的大数据 4V（Volume，Variety，Velocity，Value）中的 Velocity，即高并发问题，而在 Spring 5 中，引入了对响应式编程的支持。

本章介绍 Spring Boot 如何集成 Spring 5 中的 WebFlux 开发响应式 Web 应用。

13.1　响应式宣言及架构

响应式宣言和敏捷宣言一样，包含了 4 组关键词：

- Responsive（可响应的）。要求系统尽可能做到在任何时候都能及时响应。
- Resilient（可恢复的）。要求系统即使出错了，也能保持可响应性。
- Elastic（可伸缩的）。要求系统在各种负载下都能保持可响应性。
- Message Driven（消息驱动的）。要求系统通过异步消息连接各个组件。

可以看到，对于任何一个响应式系统，首先要保证的就是可响应性，否则就称不上是响应式系统。

引用一张来自 Spring 5 框架官方文档中的图（参见图 13-1）。

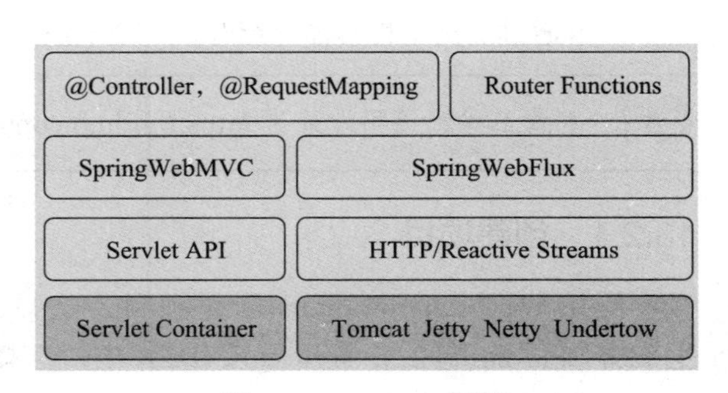

图 13-1　Spring 5 框架

左侧是传统的基于 Servlet 的 Spring Web MVC 框架。右侧是 Spring 5.0 新引入的基于 Reactive Streams 的 Spring WebFlux 框架。从上到下依次是如下三个新组件：

- Router Functions
- WebFlux
- Reactive Streams

下面分别作简要介绍。

1. Router Functions

对标 @Controller、@RequestMapping 等标准的 Spring MVC 注解，提供一套函数式风格的 API，用于创建 Router、Handler 和 Filter。

2. WebFlux: 核心组件

协调上下游各个组件提供响应式编程支持。

3. Reactive Streams

一种支持背压（Backpressure）的异步数据流处理标准，主流实现有 RxJava 和 Reactor，Spring WebFlux 默认集成的是 Reactor。

在 Web 容器的选择上，Spring WebFlux 既支持像 Tomcat、Jetty 这样的传统容器（前提是支持 Servlet 3.1 Non-Blocking IO API），又支持像 Netty、Undertow 那样的异步容器。不管是何种容器，Spring WebFlux 都会将其输入 / 输出流适配成 Flux<DataBuffer> 格式，以便进行统一处理。

值得一提的是，除了新的 Router Functions 接口，Spring WebFlux 同时支持使用老的 Spring MVC 注解声明 Reactive Controller。和传统的 MVC Controller 不同，Reactive Controller 操作的是非阻塞的 ServerHttpRequest 和 ServerHttpResponse，而不再是 Spring MVC 里的 HttpServletRequest 和 HttpServletResponse。

13.2 项目实战

本节通过实例工程具体介绍开发一个响应式 Web 应用程序的过程。

提示 本章工程源代码地址为 https://github.com/EasyKotlin/kotlin-with-webflux。

13.2.1 创建项目

使用 http://start.spring.io/ 创建项目，选择 Reactive Web 起步依赖，如图 13-2 所示。生成好项目 zip 包后，解压导入 IDEA 中，选择 Gradle 构建项目，如图 13-3 所示。配置 Gradle 本地环境，如图 13-4 所示。

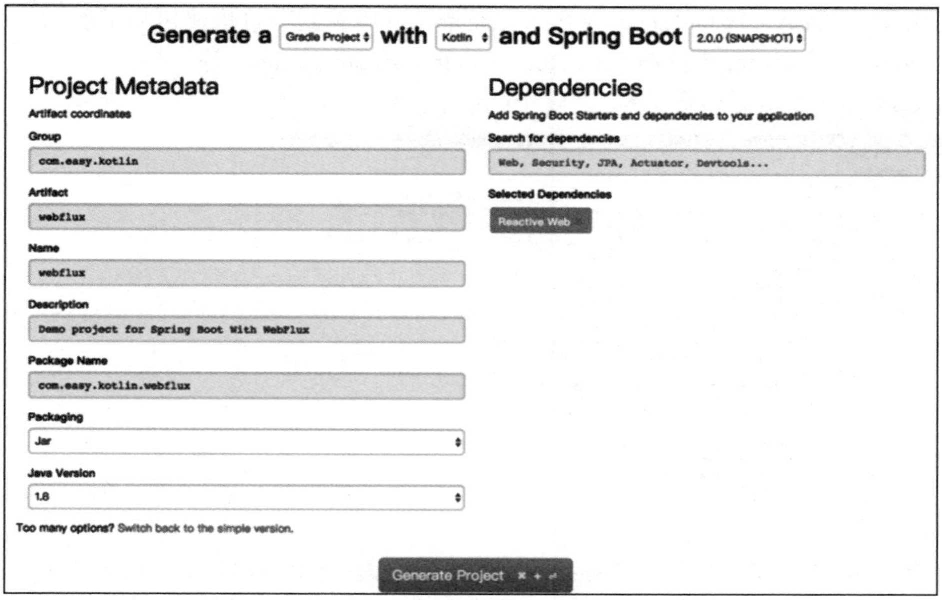

Generate a Gradle Project with Kotlin and Spring Boot 2.0.0 (SNAPSHOT)

Project Metadata

Artifact coordinates

Group

com.easy.kotlin

Artifact

webflux

Name

webflux

Description

Demo project for Spring Boot With WebFlux

Package Name

com.easy.kotlin.webflux

Packaging

Jar

Java Version

1.8

Too many options? Switch back to the simple version.

Dependencies

Add Spring Boot Starters and dependencies to your application

Search for dependencies

Web, Security, JPA, Actuator, Devtools...

Selected Dependencies

Reactive Web

Generate Project

图 13-2　选择 Reactive Web 起步依赖

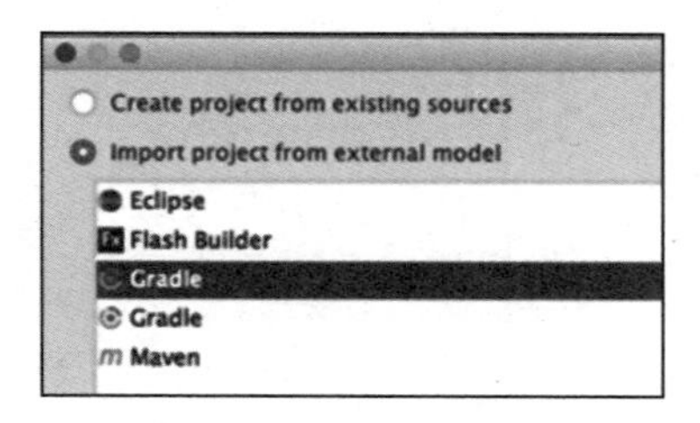

图 13-3　选择 Gradle 构建

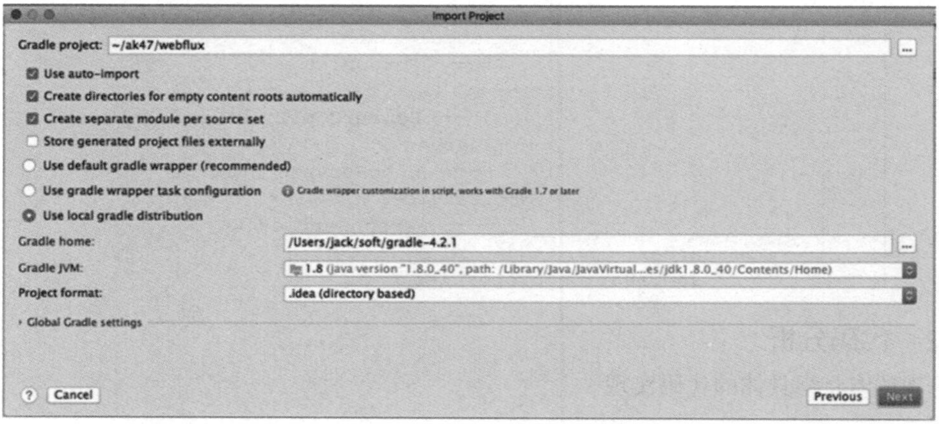

Import Project

Gradle project: ~/ak47/webflux

Use auto-import

Create directories for empty content roots automatically

Create separate module per source set

Store generated project files externally

Use default gradle wrapper (recommended)

Use gradle wrapper task configuration　Gradle wrapper customization in script, works with Gradle 1.7 or later

Use local gradle distribution

Gradle home: /Users/jack/soft/gradle-4.2.1

Gradle JVM: 1.8 (java version "1.8.0_40", path: /Library/Java/JavaVirtual...es/jdk1.8.0_40/Contents/Home)

Project format: .idea (directory based)

Global Gradle settings

Cancel　Previous　Next

图 13-4　配置 Gradle 本地环境

完成导入 IDEA，等待项目构建初始化完毕，可以看到项目依赖树如下所示。可以看到，在 webflux 的 starter 中依赖了 reactor、reactive-streams、netty 等。

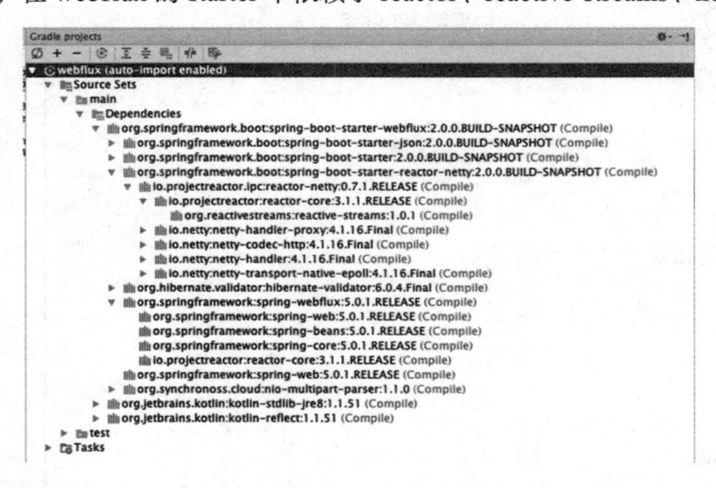

Spring Initializr 将会帮我们自动生成一个样板工程。下面我们分别来加入 dao、handler、model、router、service 层等模块的代码。完整的项目的代码目录结构设计如下：

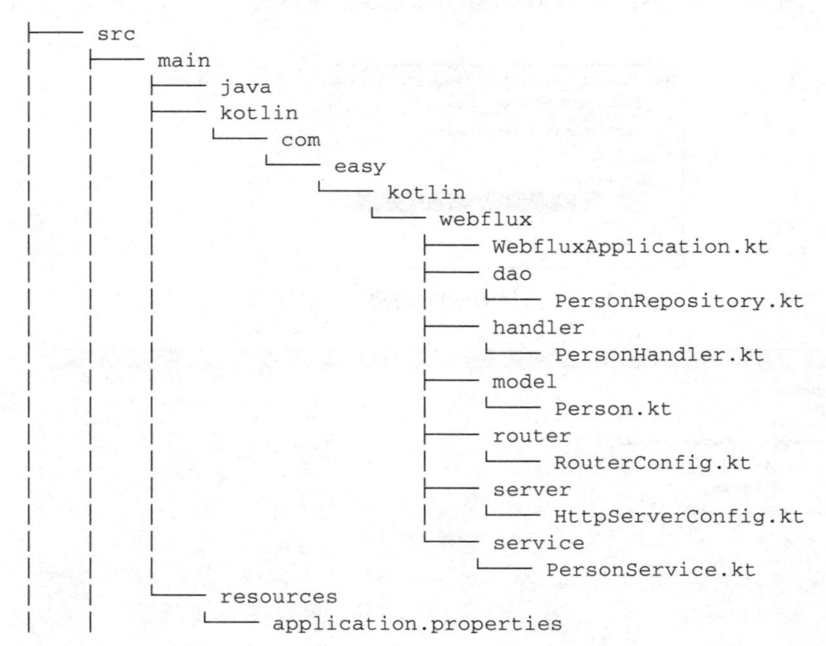

```
├── src
│   ├── main
│   │   ├── java
│   │   ├── kotlin
│   │   │   └── com
│   │   │       └── easy
│   │   │           └── kotlin
│   │   │               └── webflux
│   │   │                   ├── WebfluxApplication.kt
│   │   │                   ├── dao
│   │   │                   │   └── PersonRepository.kt
│   │   │                   ├── handler
│   │   │                   │   └── PersonHandler.kt
│   │   │                   ├── model
│   │   │                   │   └── Person.kt
│   │   │                   ├── router
│   │   │                   │   └── RouterConfig.kt
│   │   │                   ├── server
│   │   │                   │   └── HttpServerConfig.kt
│   │   │                   └── service
│   │   │                       └── PersonService.kt
│   │   └── resources
│   │       └── application.properties
```

13.2.2 代码分析

本节具体介绍具体的代码实现。

1. model 层

Person 对象模型代码是：

```
class Person(@JsonProperty("name") val name: String, @JsonProperty("age") valage:
    Int) {

    override fun toString(): String {
        return "Person{" +
            "name='" + name + '\'' +
            ", age=" + age +
            '}'
    }
}
```

2. serice 层

接口 PersonRepository.kt 的代码如下：

```
interface PersonRepository {

    fun getPerson(id: Int): Mono<Person>

    fun allPeople(): Flux<Person>

    fun savePerson(person: Mono<Person>): Mono<Void>
}
```

服务层的实现类 PersonService.kt 代码如下：

```
@Service
class PersonService : PersonRepository {
    var persons: MutableMap<Int, Person> = hashMapOf()

    constructor() {
        this.persons[1] = Person("Jack", 20)
        this.persons[2] = Person("Rose", 16)
    }

    // 根据 id 获取 Mono 对象包装的 Person 数据
    override fun getPerson(id: Int): Mono<Person> {
        return Mono.justOrEmpty(this.persons[id])
    }
    // 返回所有 Person 数据，包装在 Flux 对象中
    override fun allPeople(): Flux<Person> {
        return Flux.fromIterable(this.persons.values)
    }

    override fun savePerson(person: Mono<Person>): Mono<Void> {
        return person.doOnNext {
            val id = this.persons.size + 1
            persons.put(id, it)
            println("Saved ${person} with ${id}")
        }.thenEmpty(Mono.empty())

    }
}
```

其中，Mono 和 Flux 是由 Reactor 提供的两个 Reactor 的类型：Flux<T> 和 Mono<T>。

Flux 单词本身的意思是“流”。Flux 类似 RaxJava 的 Observable，它可以触发零个或者多个事件，并根据实际情况结束处理或触发错误。

Mono 这个单词本身的意思是“单子”的意思。Mono 最多只触发一个事件，它跟 RxJava 的 Single 和 Maybe 类似，可以把 Mono<Void> 用于在异步任务完成时发出通知。

Spring 同时支持其他响应式流实现，如 RxJava。

控制器层 PersonHandler.kt 代码如下：

```
@Service
class PersonHandler {

    @Autowired lateinit var repository: PersonRepository

    fun getPerson(request: ServerRequest): Mono<ServerResponse> {
        val personId = Integer.valueOf(request.pathVariable("id"))!!
        val notFound = ServerResponse.notFound().build()
        val personMono = this.repository.getPerson(personId)
        return personMono
            .flatMap { person -> ServerResponse.ok().contentType(APPLICATION_
                JSON).body(fromObject(person)) }
            .switchIfEmpty(notFound)
    }

    fun createPerson(request: ServerRequest): Mono<ServerResponse> {
        val person = request.bodyToMono(Person::class.java)
        return ServerResponse.ok().build(this.repository.savePerson(person))
    }

    fun listPeople(request: ServerRequest): Mono<ServerResponse> {
        val people = this.repository.allPeople()
        return ServerResponse.ok().contentType(APPLICATION_JSON).body(people,
            Person::class.java)
    }

}
```

这里我们没有真实地连接数据库进行操作，只是在内存中模拟了数据的返回。

3. handler 请求路由处理层

RouterConfig.kt 配置请求路由，把请求映射到相应的 Handler 处理方法。代码如下：

```
@Configuration
class RouterConfig {

    @Autowired lateinit var personHandler: PersonHandler

    @Bean
    fun routerFunction(): RouterFunction<*> {
        return route(GET("/api/person").and(accept(APPLICATION_JSON)),
```

```
                HandlerFunction { personHandler.listPeople(it) })
            .and(route(GET("/api/person/{id}").and(accept(APPLICATION_JSON)),
                    HandlerFunction { personHandler.getPerson(it) }))
    }

}
```

这里我们将 /api/person 的 GET 请求映射到 personHandler.listPeople() 方法处理；/ 将 api/person/{id} 的 GET 请求映射到 personHandler.getPerson() 方法来处理。

Reactive Web 服务器配置类于 HttpServerConfig.kt 配置基于 Netty 的 Reactive Web Server。我们配置端口号为 application.properties 文件中 server.port 的值。代码如下：

```
@Configuration
class HttpServerConfig {
    @Autowired
    lateinit var environment: Environment

    @Bean
    fun httpServer(routerFunction: RouterFunction<*>): HttpServer {
        val httpHandler = RouterFunctions.toHttpHandler(routerFunction)
        val adapter = ReactorHttpHandlerAdapter(httpHandler)
        val server = HttpServer.create("localhost", environment.getProperty
            ("server.port").toInt())
        server.newHandler(adapter)
        return server
    }

}
```

这个项目入口类是 WebfluxApplication，代码如下：

```
@SpringBootApplication
class WebfluxApplication

fun main(args: Array<String>) {
    runApplication<WebfluxApplication>(*args)
}
```

4. 运行测试

直接在 IDEA 中启动运行应用，在控制台启动日志中，可以看到路由映射的信息：

```
Mapped ((GET && /api/person) && Accept: [application/json]) -> com.easy.kotlin.
    webflux.router.RouterConfig$routerFunction$1@46292372
((GET && /api/person/{id}) && Accept: [application/json]) -> com.easy.kotlin.
    webflux.router.RouterConfig$routerFunction$2@126be319
……
2017-11-04 00:39:50.459  INFO 2884 --- [ctor-http-nio-1] r.ipc.netty.tcp.BlockingNetty
    Context      : Started HttpServer on /0:0:0:0:0:0:0:0:9000
2017-11-04 00:39:50.459  INFO 2884 --- [           main] o.s.b.web.embedded.netty.
    NettyWebServer  : Netty started on port(s): 9000
2017-11-04 00:39:50.466  INFO 2884 --- [           main] c.e.kotlin.webflux.Webflux
```

```
ApplicationKt  : Started WebfluxApplicationKt in 5.047 seconds (JVM running
for 6.276)
```

直接在命令行执行 curl 请求相应的 url，可以看到对应的输出：

```
$ curl http://127.0.0.1:9000/api/person
  [{"name":"Jack","age":20},{"name":"Rose","age":16}]

$ curl http://127.0.0.1:9000/api/person/1
{"name":"Jack","age":20}

$ curl http://127.0.0.1:9000/api/person/2
{"name":"Rose","age":16}
```

13.3 本章小结

Spring Web MVC 是一个命令式的编程框架，可以很方便地进行开发和调试。在很多情况下，命令式的编程风格就可以满足，但当我们的应用需要高可伸缩性，那么 Reactive 非堵塞方式是最适合的。所以，需要根据实际情况去决定采用 Spring 5 Reactive 或者是 Spring Web MVC 命令式框架。

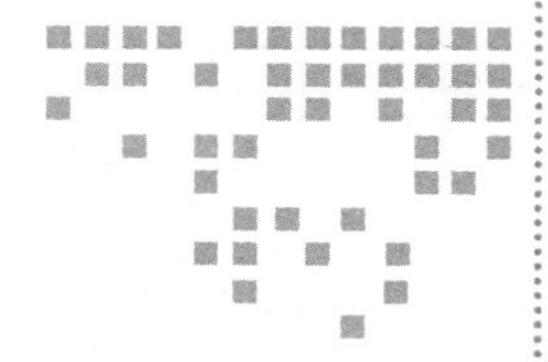

第 14 章 Chapter 14

Spring Boot 缓存

我们知道一个系统的瓶颈通常发生在与数据库交互的过程中。内存的速度远远快于硬盘速度。所以，当我们需要重复地获取相同的数据时，我们一次又一次地请求数据库或者远程服务，这无疑是性能上的浪费（这会导致大量的时间被耗费在数据库查询或者远程方法调用上致使程序性能恶化），于是有了“缓存”。

本章介绍在 Spring Boot 项目开发中怎样来使用 Spring Cache 实现数据的缓存。

14.1 Spring Cache 简介

在 Spring 3.1 中，引入了对 Cache 的支持。在 spring-context 包中定义了 org.springframework.cache.CacheManager 和 org.springframework.cache.Cache 接口用来统一不同的缓存的技术。其中，CacheManager 是 Spring 提供的各种缓存技术抽象接口，Cache 接口包含缓存的常用操作：增加、删除、读取等。

针对不同的缓存技术，需要实现不同的 CacheManager，Spring 支持的常用 CacheManager 如表 14-1 所示。

表 14-1 Spring 支持的常用 CacheManager

CacheManager	描 述
SimpleCacheManager	使用简单的 Collection 来存储缓存
ConcurrentMapCacheManager	使用 java.util.concurrent.ConcurrentHashMap 实现的 Cache
NoOpCacheManager	仅测试用，不会实际存储缓存

（续）

CacheManager	描　述
EhCacheCacheManager	集成使用 EhCache 缓存技术。EhCache 是一个纯 Java 的进程内缓存框架，具有快速、精干等特点，是 Hibernate 中默认的 CacheProvider，也是 Java 领域应用最为广泛的缓存
JCacheCacheManager	支持 JCache（JSR-107）标准的实现作为缓存技术，如 Apache Commons JCS
CaffeineCacheManager	使用 Caffeine 作为缓存技术。Caffeine 是使用 Java 8 对 Guava 缓存的重写版本，在 Spring Boot 2.0 中将取代 Guava。如果出现 Caffeine，CaffeineCacheManager 将会自动配置。使用 spring.cache.cache-names 属性可以在启动时创建缓存
CompositeCacheManager	用于组合 CacheManager，即可以从多个 CacheManager 中轮询得到相应的缓存

Spring Cache 的使用方法和原理类似于 Spring 对事务管理的支持，都是 AOP 的方式。其核心思想是：当我们在调用一个缓存方法时会把该方法参数和返回结果作为一个键值对存放在缓存中，等到下次利用同样的参数来调用该方法时将不再执行该方法，而是直接从缓存中获取结果进行返回。

Spring Cache 提供了 @Cacheable、@CachePut、@CacheEvict 等注解，在方法上使用。通过注解 Cache 可以实现类似于事务一样、缓存逻辑透明的应用到我们的业务代码上，且只需要更少的代码就可以完成。

14.2 Cache 注解

Spring 中提供了 4 个注解来声明缓存规则，如下所示：

- @Cacheable——主要针对方法配置，根据方法的请求参数对其结果进行缓存。
- @CachePut——主要针对方法配置，根据方法的请求参数对其结果进行缓存，和 @Cacheable 不同的是，每次都会触发真实方法的调用。
- @CacheEvict——主要针对方法配置，能够根据一定的条件对缓存进行清空。
- @Caching——用来组合使用其他注解，可以同时应用多个 Cache 注解。

下面我们分别来简单介绍。

1. @Cacheable

该注解中的属性值如下：

- value：缓存名，必填。
- key：可选属性，可以使用 SPEL 标签自定义缓存的 key。
- condition：属性指定发生的条件。

代码示例如下：

```
@Cacheable("userList")                    // 标识读缓存操作
```

```
override fun findAll(): List<User> {
    return userDao.findAll()
}

@Cacheable(cacheNames = ["user"], key = "#id")          //如果缓存存在，直接读取缓存值；如果不存
                                                        //在调用目标方法，并将方法返回结果放入缓存
override fun findOne(id: Long): User {
    return userDao.getOne(id)
}
```

2. @CachePut

使用该注解标识的方法，每次都会执行目标逻辑代码，并将结果存入指定的缓存中。之后另一个方法就可以直接从相应的缓存中取出缓存数据，而不需要再去查询数据库。@CachePut 注解的属性说明如下：

- value：缓存名，必填。
- key：可选属性，可以使用 SPEL 标签自定义缓存的 key。

代码示例如下：

```
@Transactional
@CachePut(cacheNames = ["user"], key = "#user.id")      //写入缓存，key 为 user.id ；
                                                        //一般可以标注在 save 方法上面
override fun saveUser(user: User): User {
    return userDao.save(user)


}
```

3. @CacheEvict

标记要清空缓存的方法，当这个方法被调用后，即会清空缓存。@CacheEvict 注解属性说明如下：

- value：必填。
- key：可选（默认是所有参数的组合）。
- condition：缓存的条件。
- allEntries：是否清空所有缓存内容，默认为 false。如果指定为 true，则方法调用后将立即清空所有缓存。
- beforeInvocation：是否在方法执行前就清空，默认为 false。如果指定为 true，则在方法还没有执行的时候就清空缓存。默认情况下，如果方法执行抛出异常，则不会清空缓存。

代码示例如下：

```
@Transactional
@CacheEvict(cacheNames = ["user"], key = "#id") //根据 key（值为 id）来清除缓存 ；一般标
                                                //注在 delete,update 方法上面
override fun updatePassword(id: Long, password: String): Int {
```

```
    return userDao.updatePassword(id, password)
}
```

4. @Caching

@Caching 注解的源码如下，从中可以看到同时使用了 Cacheable、CachePut、CacheEvict 方法：

```
public @interface Caching {
    Cacheable[] cacheable() default {};

    CachePut[] put() default {};

    CacheEvict[] evict() default {};
}
```

使用 @Caching 注解可以实现在同一个方法上同时使用多种注解，例如：

```
@Caching(evict={@CacheEvict(“u1”),@CacheEvict(“u2”,allEntries=true)})
```

14.3 项目实战

本节将通过完整的项目案例来讲解 Spring Cache 的具体使用方法。

1. 创建项目

首先使用 Spring Initializr 创建基于 Gradle、Kotlin 的 Spring Boot 项目。使用的 Kotlin 版本和 Spring Boot 版本如下：

```
kotlinVersion = '1.2.20'
springBootVersion = '2.0.1.RELEASE'
```

2. 添加依赖

添加 spring-boot-starter-cache 项目依赖如下：

```
dependencies {
    compile('org.springframework.boot:spring-boot-starter-cache')
}
```

3. 数据库配置

本项目需要连接真实的数据库，我们使用 MySQL，同时 ORM 框架选用 JPA。所以我们在项目依赖中添加如下依赖：

```
runtime('mysql:mysql-connector-java')
compile('org.springframework.boot:spring-boot-starter-data-jpa')
compile('org.springframework.boot:spring-boot-starter-web')
```

在本地测试数据库中创建 schema 如下：

```
CREATE SCHEMA `demo_cache` DEFAULT CHARACTER SET utf8 ;
```

在 application.properties 中配置数据库连接信息如下：

```
spring.datasource.url=jdbc:mysql: //localhost:3306/demo_cache?useUnicode=true&
    characterEncoding=UTF8&useSSL=false
spring.datasource.username=root
spring.datasource.password=root
spring.datasource.driverClassName=com.mysql.jdbc.Driver
spring.jpa.database=MYSQL
spring.jpa.show-sql=true
spring.jpa.hibernate.ddl-auto=create-drop
spring.jpa.properties.hibernate.dialect=org.hibernate.dialect.MySQL5Dialect
spring.jpa.hibernate.naming.physical-strategy=org.springframework.boot.orm.jpa.
    hibernate.SpringPhysicalNamingStrategy
```

4. 实体类

为了简单起见，我们设计一个用户实体，包含 3 个字段：id、username、password。具体的代码如下：

```
package com.easy.springboot.demo_cache
import javax.persistence.*

@Entity
class User {

    @Id
    @GeneratedValue(strategy = GenerationType.IDENTITY)
    var id: Long = 0

    @Column(unique = true, length = 100)
    var username: String = ""
    @Column(length = 100)
    var password: String = ""

}
```

5. 数据访问层

使用 JPA 写 Dao 层代码是一件相当快乐的事情，不需要我们去写那么多样板化的 CRUD 方法。代码如下：

```
package com.easy.springboot.demo_cache

import org.springframework.data.jpa.repository.JpaRepository
import org.springframework.data.jpa.repository.Modifying
import org.springframework.data.jpa.repository.Query
import org.springframework.data.repository.query.Param

interface UserDao : JpaRepository<User, Long> {
    @Query("update #{#entityName} a set a.password = :password where a.id=:id")
    @Modifying
    fun updatePassword(@Param("id") id: Long, @Param("password") password:String):
        Int
}
```

其中，需要注意的是这里的 updatePassword() 函数需要添加 @Modifying 注解，否则会报如下错误：

```
org.hibernate.hql.internal.QueryExecutionRequestException: Not supported for
    DML operations [update com.easy.springboot.demo_cache.User a set a.password
    =:password where id=:id]
at org.hibernate.hql.internal.ast.QueryTranslatorImpl.errorIfDML(QueryTranslatorImpl.
    java:311)
```

6. 业务层代码

我们通常是在业务逻辑层来使用缓存服务。我们的接口定义如下：

```
interface UserService {
    fun findAll(): List<User>
    fun saveUser(u: User): User
    fun updatePassword(id:Long, password: String): Int
    fun findOne(id: Long): User
}
```

对应的实现类代码是：

```
package com.easy.springboot.demo_cache

import org.springframework.beans.factory.annotation.Autowired
import org.springframework.cache.annotation.CacheEvict
import org.springframework.cache.annotation.CachePut
import org.springframework.cache.annotation.Cacheable
import org.springframework.stereotype.Service
import org.springframework.transaction.annotation.Transactional

@Service
open class UserServiceImpl : UserService {


    @Autowired lateinit var userDao: UserDao

    @Cacheable("userList")                                  // 标识读缓存操作
    override fun findAll(): List<User> {
        return userDao.findAll()
    }

    @Transactional
    @CachePut(cacheNames = ["user"], key = "#user.id")      // 写入缓存，key 为 user.id ；
                                                            // 一般可以标注在 save 方法上面
    override fun saveUser(user: User): User {
        return userDao.save(user)
    }

    @Transactional
    @CacheEvict(cacheNames = ["user"], key = "#id")   // 根据 key （值为 id）来清除缓存 ；一般
                                                      // 标注在 delete,update 方法上面
    override fun updatePassword(id: Long, password: String): Int {
```

```
        return userDao.updatePassword(id, password)
    }

    @Cacheable(cacheNames = ["user"], key = "#id") // 如果缓存存在，直接读取缓
        存值；如果不存在调用目标方法，并将方法返回结果放入缓存
    override fun findOne(id: Long): User {
        return userDao.getOne(id)
    }

}
```

7. 测试 Controller

为了看到缓存的效果，我们编写 UserController 代码来进行测试缓存的效果。代码如下：

```
package com.easy.springboot.demo_cache

import org.springframework.beans.factory.annotation.Autowired
import org.springframework.web.bind.annotation.GetMapping
import org.springframework.web.bind.annotation.PathVariable
import org.springframework.web.bind.annotation.RestController

@RestController
class UserController {
    @Autowired lateinit var userService: UserService

    @GetMapping("/user/list")
    fun findAll(): List<User> {
        return userService.findAll()
    }

    @GetMapping("/user/save")
    fun save(user: User): User {
        return userService.saveUser(user)
    }

    @GetMapping("/user/updatePassword")
    fun updatePassword(id: Long, password: String): Int {
        return userService.updatePassword(id, password)
    }

    @GetMapping("/user/{id}")
    fun findOne(@PathVariable("id") id: Long): User {
        return userService.findOne(id)
    }

}
```

8. 启用 Cache 功能

在 Spring Boot 项目中启用 Spring Cache 注解的功能非常简单。只需要在启动类上添加 @EnableCaching 注解即可。实例代码如下：

```
@SpringBootApplication
```

```
@EnableCaching
open class DemoCacheApplication

fun main(args: Array<String>) {
    ...
}
```

9. 数据库初始化测试数据

为了方便测试，我们在数据库中初始化三条用户数据进行测试。初始化代码如下：

```
fun main(args: Array<String>) {
    SpringApplicationBuilder().initializers(
            beans {
                bean {
                    ApplicationRunner {
                        initUser()
                    }
                }
            }
    ).sources(DemoCacheApplication::class.java).run(*args)
}

private fun BeanDefinitionDsl.BeanDefinitionContext.initUser() {
    val userDao = ref<UserDao>()
    try {
        val user = User()
        user.username = "user"
        user.password = "user"
        userDao.save(user)

        val jack = User()
        jack.username = "jack"
        jack.password = "123456"
        userDao.save(jack)

        val admin = User()
        admin.username = "admin"
        admin.password = "admin"
        userDao.save(admin)
    } catch (e: Exception) {
        e.printStackTrace()
    }
}
```

其中，BeanDefinitionDsl 是 Spring 5 中提供的基于 Kotlin 的函数式风格的 Bean 注册 DSL（Functional bean definition Kotlin DSL）。

10. 运行测试

启动项目，访问 http://localhost:8080/user/list，返回如下信息：

```
[
    {
```

```
        "id": 1,
        "username": "user",
        "password": "user"
    },
    {
        "id": 2,
        "username": "jack",
        "password": "123456"
    },
    {
        "id": 3,
        "username": "admin",
        "password": "admin"
    }
]
```

通过调用接口 http://localhost:8080/user/save?username=who&password=xxx，向数据库中新增一条记录。我们去数据库中查看，可以发现数据新增成功。但是在此访问 http://localhost:8080/user/list，依然返回上面的三条数据。这表明在下面的代码中：

```
@Cacheable("userList")                          // 标识读缓存操作
override fun findAll(): List<User>
```

这里 findAll() 函数的执行确实走了缓存，而没有去查询数据库。

我们再来测试一下 @CacheEvict 与 @Cacheable 注解的功能。对应的是下面的这段代码：

```
@Transactional
@CacheEvict(cacheNames = ["user"], key = "#id") // 根据 key（值为 id）来清除缓存；一般标注
                                                // 在 delete,update 方法上面
override fun updatePassword(id: Long, password: String): Int {
    return userDao.updatePassword(id, password)
}

@Cacheable(cacheNames = ["user"], key = "#id")  // 如果缓存存在，直接读取缓存值；如果不存在调用目
                                                // 标方法，并将方法返回结果放入缓存
override fun findOne(id: Long): User {
    return userDao.getOne(id)
}
```

首先，访问 http://localhost:8080/user/1 得到的结果是：

```
{
    "id": 1,
    "username": "user",
    "password": "user"
}
```

此时，我们调用被 @CacheEvict 标注的 updatePassword() 函数，该注解会清空 id=1 的缓存。访问接口 http://localhost:8080/user/updatePassword?id=1&password=ppp，返回值为 1，表明成功更新 1 条数据。此时，我们再次访问 http://localhost:8080/user/1 得到的结果是：

```
{
    "id": 1,
    "username": "user",
    "password": "ppp"
}
```

这表明缓存被成功更新了。最后，我们手工去数据库修改 id=1 的用户数据：

```
UPDATE `demo_cache`.`user` SET `password`='mmm' WHERE `id`='1';
```

更改完成后，我们再次访问 http://localhost:8080/user/1 得到的结果依然是：

```
{
    "id": 1,
    "username": "user",
    "password": "ppp"
}
```

这表明，此时 id=1 的 User 数据依然是从缓存中读取的，并没有去查询数据库。

14.4 本章小结

通常情况下，使用内置的 Spring Cache 只适用于单体应用，因为这些缓存的对象是存储在内存中的。在大型分布式的系统中，缓存对象往往会非常大，这个时候我们就会有专门的缓存服务器（集群）来存储这些数据了，例如 Redis。

我们可以把一些经常查询的数据放到 Redis 中缓存起来，不用每次都查询数据库。这样也不用直接占用大量内存了。关于 Redis 的使用我们将在下一章中介绍。

Spring Cache 对这些缓存实现都做了非常好的集成适配，所以使用起来可以说是“相当平滑”。另外，我们通常会设计一级缓存、二级缓存，本书限于篇幅就不详细介绍了。

本章示例工程源代码地址为：https://github.com/EasySpringBoot/demo_cache。

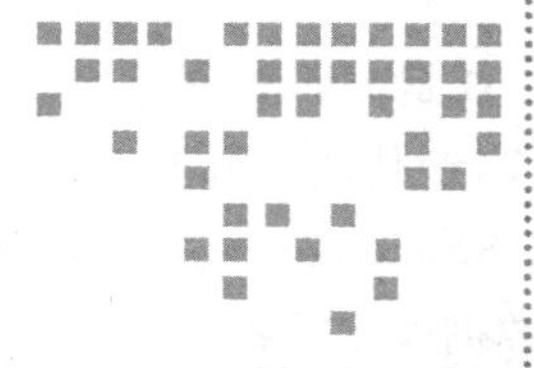

第 15 章 Chapter 15

使用 Spring Session 集成 Redis 实现 Session 共享

通常在 Web 开发中，Session 会话管理是很重要的一部分，用于存储与用户相关的一些数据。在 Java Web 系统中的 Session 一般由 Tomcat 容器来管理。使用特定的容器虽然可以很好地实现会话管理，但是基于 Tomcat 的会话插件太依赖于容器，并且对于 Tomcat 各个版本的支持不是特别的好。重写 Tomcat 的 session 管理，代码耦合度高，不利于维护。而使用开源的 Spring Session 框架，既不需要修改 Tomcat 配置，又无须重写代码，只需要配置相应的参数即可完成分布式系统中的 Session 共享管理。

本章我们来介绍在 Spring Boot 应用中如何使用 Spring Session 集成 Redis 实现分布式系统中的 Session 共享，从而实现 Spring Boot 应用的水平扩展。我们通常优先采用水平扩展架构来提升系统的可用性和系统性能，但是更多的应用导致管理更加复杂。

15.1 Spring Session 简介

对于 Spring Boot 应用，会话管理是一个难点。Spring Boot 应用水平扩展通常有如下两个问题需要解决：

1）负载均衡。将用户请求平均派发到水平部署的任意一台 Spring Boot 应用服务器上。可以用一个反向代理服务器来实现，例如使用 Nginx 作为反向代理服务器。在 Spring Cloud 中，我们使用 Zuul（智能路由）集成 Eureka（服务发现）、Hystrix（断路器）和 Ribbon（客户端负载均衡）来实现。

2）共享 Session。单个 Spring Boot 应用的 Session 由 Tomcat 来管理。如果部署多个 Spring Boot 应用，对于同一个用户请求，实现在这些应用之间共享 Session 通常有如下两种方式：

- Session 复制：Web 服务器通常都支持 Session 复制，一台应用的 Session 信息改变将立刻复制到其他集群的 Web 服务器上。
- 集中式 Session 共享：所有 Web 服务器都共享同一个 Session，Session 通常存放在 Redis 数据库服务器上。

Session 复制的缺点是效率较低，性能差。所以 Spring Boot 应用采用集中式 Session 共享。架构如图 15-1 所示。

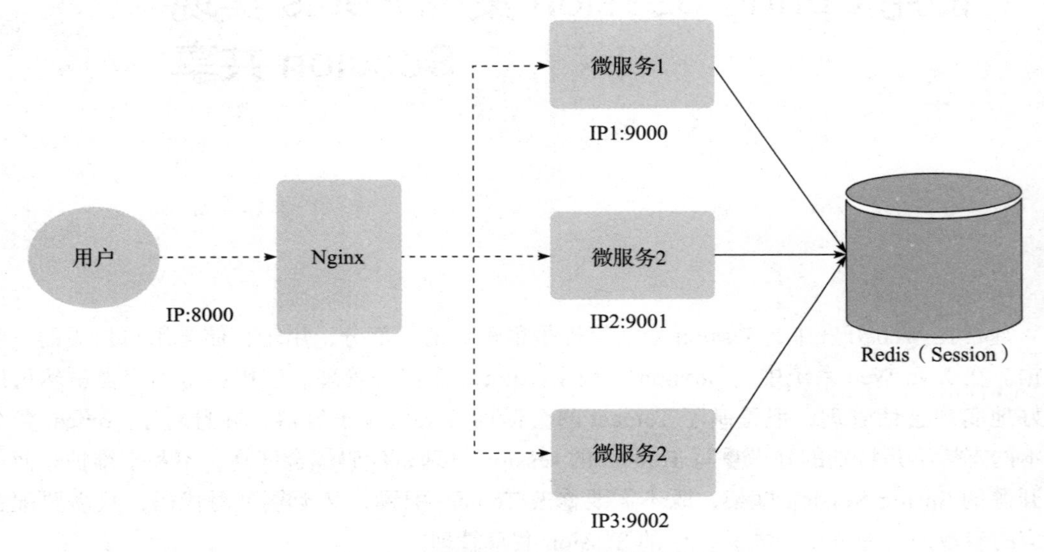

图 15-1 集中式 Session 共享架构图

这是一个通用的分布式系统架构，包含了三个独立运行的微服务应用。微服务 1 部署在一台 Tomcat 服务器上（IP1 ：9000），微服务 2 部署在两台 Tomcat 服务器（IP2：9001、IP3 ：9002）上采用水平扩展。架构采用 Nginx 作为反向代理，Nginx 提供统一的入口。Spring Boot 应用微服务 1 和微服务 2，都采用 Spring Session 实现各个子系统共享同一个 Session，该 Session 统一存放在 Redis 中。微服务 1 和微服务 2 独立部署的，支持水平扩展，最终整合成一个大的分布式系统。

Session 一直都是我们做分布式系统架构时需要解决的一个难题，过去我们可以从 Serlvet 容器上解决，比如开源 servlet 容器 -tomcat 提供的 tomcat-redis-session-manager、memcached-session-manager。或者通过 Nginx 之类的负载均衡做 ip_hash，路由到特定的服务器上。而使用 Spring Session 来管理分布式 session，则完全实现了与具体的容器无关。Spring Session 是 Spring 的项目之一，GitHub 地址：https://github.com/spring-projects/spring-session。

Spring Session 提供了一套创建和管理 Servlet HttpSession 的方案。Spring Session 提供

了集群 Session（Clustered Sessions）功能，默认采用外置的 Redis 来存储 Session 数据，以此来解决 Session 共享的问题。

使用 Spring Session 可以非常简易地把 Session 存储到第三方存储容器，框架提供了 Redis、JVM 的 map、mongo、gemfire、hazelcast、jdbc 等多种存储 Session 的容器的方式。

15.2 Redis 简介

Redis 是目前使用的非常广泛的内存数据库，相比 Memcached，它支持更加丰富的数据类型。

15.2.1 Redis 是什么

Redis 是完全开源免费的，遵守 BSD 协议，是一个高性能的 key-value 数据库。

Redis 支持数据的持久化，可以将内存中的数据保存在磁盘中，重启的时候可以再次加载进行使用。Redis 不仅仅支持简单的 key-value 类型的数据，同时还提供 list，set，zset，hash 等数据结构的存储。Redis 支持数据的备份，即 master-slave 模式的数据备份。

Redis 优势如下：

- 性能极高。Redis 能读的速度是 110000 次 /s，写的速度是 81000 次 /s。
- 丰富的数据类型。Redis 支持二进制案例的 Strings，Lists，Hashes，Sets 及 Ordered Sets 数据类型操作。
- 原子性。Redis 的所有操作都是原子性的，意思就是要么成功执行要么失败完全不执行。单个操作是原子性的。多个操作也支持事务，即原子性，通过 MULTI 和 EXEC 指令包起来。
- 丰富的特性。Redis 还支持 publish/subscribe，通知，key 过期等特性。

Redis 运行在内存中但是可以持久化到磁盘，所以在对不同数据集进行高速读写时需要权衡内存，因为数据不能大于硬件内存。在内存数据库方面的另一个优点是，相比在磁盘上相同的复杂数据结构，在内存中操作起来非常简单，这样 Redis 可以做很多内部复杂性很强的事情。同时，在磁盘格式方面是紧凑的、以追加的方式产生的，因为并不需要进行随机访问。

15.2.2 安装 Redis

使用下面的命令下载安装 redis：

```
$ wget http://download.redis.io/releases/redis-4.0.9.tar.gz
$ tar xzf redis-4.0.9.tar.gz
$ cd redis-4.0.9
$ make
```

启动 redis server 进程命令如下：

```
$ src/redis-server
```

打开 redis client 命令：

```
$ src/redis-cli
redis> set foo bar
OK
redis> get foo
"bar"
```

这样我们就简单完成了 redis 的环境配置。

如果需要在远程 redis 服务上执行命令，同样我们使用的也是 redis-cli 命令。语法格式如下：

```
$ redis-cli -h host -p port -a password
```

代码实例：

```
$redis-cli -h 127.0.0.1 -p 6379 -a "123456"
```

连接到主机为 127.0.0.1，端口为 6379，密码为 123456 的 redis 服务上。

使用 * 号获取所有配置项命令：

```
redis 127.0.0.1:6379> config get *
    1) "dbfilename"
    2) "dump.rdb"
    3) "requirepass"
    4) "123456"
    5) "masterauth"
    ...
```

15.2.3 设置 Redis 密码

通常我们会设置 redis 密码，命令如下：

```
127.0.0.1:6379> config set requirepass 123456
OK
```

测试密码：

```
127.0.0.1:6379> info
NOAUTH Authentication required.
127.0.0.1:6379> set x 0
(error) NOAUTH Authentication required.
```

提示无权限。使用密码授权登录：

```
127.0.0.1:6379> auth 123456
OK
```

```
127.0.0.1:6379> set x 0
OK
127.0.0.1:6379> get x
"0"
```

15.2.4　Redis 数据类型

Redis 支持五种数据类型：string（字符串），hash（哈希），list（列表），set（集合）及 zset（有序集合）。下面分别介绍。

1. string

string 是 Redis 最基本的类型，你可以理解成与 Memcached 一样的类型，一个 key 对应一个 value。string 类型是二进制安全的。意思是 Redis 的 string 可以包含任何数据。比如 jpg 图片或者序列化的对象。一个键最大能存储 512MB。

代码实例如下：

```
redis 127.0.0.1:6379> set name "Spring Boot Plus Kotlin"
OK
redis 127.0.0.1:6379> get name
"Spring Boot Plus Kotlin"
```

在以上实例中我们使用了 Redis 的 set 和 get 命令。键为 name，对应的值为 "Spring Boot Plus Kotlin"。

2. hash

Redis 中的 hash 是一个键值（key=>value）对集合。Redis hash 是一个 string 类型的 field 和 value 的映射表，hash 适用于存储对象。

代码实例如下：

```
redis> HMSET myhash field1 "Hello" field2 "World"
"OK"
redis> HGET myhash field1
"Hello"
redis> HGET myhash field2
"World"
```

以上实例中 hash 数据类型存储了包含用户脚本信息的用户对象。实例中我们使用了 Redis HMSET，HGETALL 命令，user : 1 为键值。每个 hash 可以存储的键值对为 $2^{32}-1$。

3. List

Redis 中的 list 是简单的字符串列表，按照插入顺序排序。你可以添加一个元素到列表的头部（左边）或者尾部（右边）。

代码实例如下：

```
127.0.0.1:6379> lpush mylist redis
```

```
(integer) 1
127.0.0.1:6379> lpush mylist springboot
(integer) 2
127.0.0.1:6379> lpush mylist kotlin
(integer) 3
127.0.0.1:6379> lpush mylist kotlin
(integer) 4
127.0.0.1:6379> lrange mylist 0 10
1) "kotlin"
2) "kotlin"
3) "springboot"
4) "redis"
```

列表最多可存储的元素为 $2^{32}-1$。

4. set

Redis 的 set 是 string 类型的无序集合。集合是通过哈希表实现的，所以添加，删除，查找的复杂度都是 $O(1)$。

使用 sadd 命令添加一个 string 元素到 key 对应的 set 集合中，成功返回 1，如果元素已经在集合中返回 0，如果 key 对应的 set 不存在则返回错误。

向集合添加一个或多个成员命令：

```
SADD key member1 [member2]
```

代码示例：

```
127.0.0.1:6379> sadd myset redis
(integer) 1
127.0.0.1:6379> sadd myset springboot
(integer) 1
127.0.0.1:6379> sadd myset kotlin
(integer) 1
127.0.0.1:6379> sadd myset kotlin
(integer) 0
```

获取集合的成员数：

```
SCARD key
```

代码示例：

```
127.0.0.1:6379> scard myset
(integer) 3
```

返回集合中的所有成员：

```
SMEMBERS key
```

代码示例：

```
127.0.0.1:6379> smembers myset
```

```
1) "kotlin"
2) "redis"
3) "springboot"
```

注意：以上实例中 kotlin 添加了两次，但根据集合内元素的唯一性，第二次插入的元素将被忽略。集合中最大的成员数为 $2^{32}-1$。

5. zset

Redis zset 和 set 一样，也是 string 类型元素的集合，且不允许重复的成员。不同的是每个元素都会关联一个 double 类型的分数。redis 正是通过分数来为集合中的成员进行从小到大的排序。zset 的成员是唯一的，但分数（score）却可以重复。集合是通过哈希表实现的，所以添加、删除、查找的复杂度都是 $O(1)$。集合中最大的成员数为 $2^{32}-1$。

代码实例如下：

```
127.0.0.1:6379> ZADD mysortedset 1 redis
(integer) 1
127.0.0.1:6379> ZADD mysortedset 2 mongodb
(integer) 1
127.0.0.1:6379> ZADD mysortedset 3 mysql
(integer) 1
127.0.0.1:6379> ZADD mysortedset 3 mysql
(integer) 0
127.0.0.1:6379> ZADD mysortedset 4 mysql
(integer) 0
127.0.0.1:6379> ZRANGE mysortedset 0 10 WITHSCORES
1) "redis"
2) "1"
3) "mongodb"
4) "2"
5) "mysql"
6) "4"
```

在以上实例中我们通过命令 ZADD 向 redis 的有序集合中添加了三个值并关联上分数。我们重复添加了 MySQL，分数以最后添加的元素为准。

15.2.5　Spring Boot 集成 Redis

在项目中添加 spring-boot-starter-data-redis 依赖，然后在 application.properties 中配置 spring.redis.* 属性即可使用 StringRedisTemplate 模板类来操作 Redis 了。Spring Data Redis 是对访问 redis 客户端的一个包装适配，支持 Jedis、JRedis、SRP、Lettuce 四种开源的 Redis 客户端。RedisTemplate 是对 Redis 的 CRUD 的高级封装，而 RedisConnection 提供了简单封装。

一个简单的代码示例如下：

```
@RestController
class RedisTemplateController {
```

```
    @Autowired lateinit var stringRedisTemplate: StringRedisTemplate

    @RequestMapping(value = ["/redis/{key}/{value}"], method = [RequestMethod.GET])
    fun redisSave(@PathVariable key: String, @PathVariable value: String): String {

        val redisValue = stringRedisTemplate.opsForValue().get(key)

        if (StringUtils.isEmpty(redisValue)) {
            stringRedisTemplate.opsForValue().set(key, value)
            return String.format("设置[key=%s,value=%s]成功! ", key, value)
        }

        if (redisValue != value) {
            stringRedisTemplate.opsForValue().set(key, value)
            return String.format("更新[key=%s,value=%s]成功! ", key, value)
        }

        return String.format("redis中已存在[key=%s,value=%s]的数据! ", key, value)
    }

    @RequestMapping(value = ["/redis/{key}"], method = [RequestMethod.GET])
    fun redisGet(@PathVariable key: String): String? {
        return stringRedisTemplate.opsForValue().get(key) //String 类型的 value
    }

    @RequestMapping(value = ["/redisHash/{key}/{field}"], method = [RequestMethod.
        GET])
    fun redisHashGet(@PathVariable key: String, @PathVariable field: String): String? {
        return stringRedisTemplate.opsForHash<String, String>().get(key, field) //Hash
            类型的 value
    }
}
```

StringRedisTemplate继承了RedisTemplate。RedisTemplate是一个泛型类，而String-Redis-Template则不是，它只能对key=String，value=String的键值对进行操作，Redis-Template可以对任何类型的key-value键值对操作。

StringRedisTemplate封装了对Redis的一些常用的操作。StringRedisTemplate使用的是StringRedisSerializer。RedisTemplate使用的序列类在操作数据的时候，比如说存入数据，会将数据先序列化成字节数组，然后再存入Redis数据库，这个时候打开Redis查看的时候，你会看到你的数据不是以可读的形式展现的。在使用StringRedisSerializer操作redis数据类型的时候必须要set相对应的序列化。从StringRedisTemplate类的构造函数代码可以看出：

```
public class StringRedisTemplate extends RedisTemplate<String, String> {
    public StringRedisTemplate() {
        RedisSerializer<String> stringSerializer = new StringRedisSerializer();
        setKeySerializer(stringSerializer);
        setValueSerializer(stringSerializer);
        setHashKeySerializer(stringSerializer);
```

```
            setHashValueSerializer(stringSerializer);
        }
        ...
    }
```

StringRedisTemplate 和 RedisTemplate 各自序列化的方式不同，但最终都得到了一个字节数组，殊途同归，StringRedisTemplate 使用的是 StringRedisSerializer 类；RedisTemplate 使用的是 JdkSerializationRedisSerializer 类。反序列化的，则是一个得到 String，一个得到 Object。

测试 redis 操作

请求 http://127.0.0.1:9000/redis/x/1，输出："更新 [key=x，value=1] 成功！"。

再次请求 http://127.0.0.1:9000/redis/x/1，输出："redis 中已存在 [key=x，value=1] 的数据！"。

请求 http://127.0.0.1:9000/redis/x，输出：1。

请求 http://127.0.0.1:9000/redisHash/spring:session:sessions:06830c1b-8157-46fc-b84a-a086aa8c8d45/lastAccessedTime，输出一段不可读的对象数据："...java.lang.Long；...java.lang.Number..."。

提示　更多关于 Redis 的介绍参考：

https://redis.io/download

http://try.redis.io/

15.3　项目实战

本节通过完整的项目实例介绍在 Spring Boot 应用中如何使用 Redis 实现共享 Session。在分布式系统中，Sessiong 共享有很多解决方案，其中使用 Redis 缓存是最常用的方案之一。

1. 创建项目

创建两个 Spring Boot 应用 demo_microservice_api_book、demo_microservice_api_user，它们的 Session 都使用同一个 Redis 数据库存储。

2. 添加依赖

在 build.gradle 中添加 spring-session-data-redis 就可以使用 Redis 来存储 Session。

3. 配置 Redis

为了简单起见，我们在这里就使用单点 Redis 模式。在实际生产中，为了保障高可用性，通常是一个 Redis 集群。在 application.properties 中配置 Redis 信息如下：

```
spring.application.name=demo_microservice_api_user
```

```
server.port=9001
################# Redis 基础配置 #################
spring.redis.host=127.0.0.1
spring.redis.password=123456
spring.redis.port=6379
# 连接超时时间 单位 ms（毫秒）
spring.redis.timeout=3000ms
################# Redis 线程池设置 #################
# 连接池中的最大空闲连接，默认值是 8。
spring.redis.jedis.pool.max-idle=10
# 连接池中的最小空闲连接，默认值是 0。
spring.redis.jedis.pool.min-idle=20
# 连接池最大活跃数。默认值 8。如果赋值为 -1，则表示不限制；如果 pool 已经分配了 maxActive
    个 jedis 实例，则此时 pool 的状态为 exhausted（耗尽）。
spring.redis.jedis.pool.max-active=10
# 等待可用连接的最大时间，单位毫秒，默认值为 -1ms，表示永不超时。如果超过等待时间，则直
    接抛出 JedisConnectionException
spring.redis.jedis.pool.max-wait=3000ms
```

4. 配置 Session 存储类型

在 application.properties 中配置存储 Session 的类型为 Redis：

```
################# 使用 Redis 存储 Session 设置 #################
# Redis|JDBC|Hazelcast|none
spring.session.store-type=Redis
```

spring-boot-autoconfigure 的源代码中使用 RedisAutoConfiguration 来加载 Redis 的配置类 RedisProperties。其中 RedisAutoConfiguration 会加载 application.properties 文件的前缀为“spring.redis”的属性。其中“spring.redis.sentinel”是哨兵模式的配置，“spring.redis.cluster”是集群模式的配置。

当我们添加 spring.session.store-type=Redis 这行配置，指定 Session 的存储方式为 Redis，可以看到控制台输出的日志为：

```
c.e.s.d.SessionController :
org.springframework.session.web.http.SessionRepositoryFilter.SessionRepository
    RequestWrapper.HttpSessionWrapper
```

我们可以看到，Session 已经使用了 HttpSessionWrapper 这个包装类实现，HttpSessionWrapper 背后真正负责 Session 管理的适配器类是 HttpSessionAdapter。RedisOperationsSessionRepository 是采用 Redis 存储 Session 的核心业务逻辑实现。其中的变量 DEFAULT_NAMESPACE = "spring:session" 定义了 Spring Session 存储在 Redis 中的默认命名空间。其中的 getSessionKey() 方法如下：

```
String getSessionKey(String sessionId) {
    return this.namespace + "sessions:" + sessionId;
}
```

通过方法源码，我们可以知道 session id 存储的 Key 是 spring:session:sessions:{session

Id}。我们稍后去 Redis 中查看验证。

按照上面的步骤在另一个项目中再次配置一次，启动后，该项目也会自动进行了 session 共享。

5. 测试 Session 数据

分别在两个 Spring Boot 应用中编写获取 Session 数据的 Controller 类 SessionController，代码相同，如下所示：

```
@RestController
class SessionController {
    val log = LoggerFactory.getLogger(SessionController::class.java)
    @RequestMapping(value = "/session")
    fun getSession(request: HttpServletRequest): SessionInfo {
        val session = request.session
        log.info(session.javaClass.canonicalName)
        log.info(session.id)

        val SessionInfo = SessionInfo()
        SessionInfo.id = session.id
        SessionInfo.creationTime = session.creationTime
        SessionInfo.lastAccessedTime = session.lastAccessedTime
        SessionInfo.maxInactiveInterval = session.maxInactiveInterval
        SessionInfo.isNew = session.isNew
        return SessionInfo
    }

    class SessionInfo {
        var id = ""
        var creationTime = 0L
        var lastAccessedTime = 0L
        var maxInactiveInterval = 0
        var isNew = false
    }
}
```

在本机部署 demo_microservice_api_book，端口号为 9000。部署 demo_microservice_api_user 两个运行实例，端口号分别为 9001、9002。即使用 gradle bootJar 打可执行 jar 包，然后在命令行分别执行如下代码：

```
$ java -jar demo_microservice_api_user-0.0.1-SNAPSHOT.jar  --server.port=9001
$ java -jar demo_microservice_api_user-0.0.1-SNAPSHOT.jar  --server.port=9002
```

访问 http://127.0.0.1:9000/session，得到输出如下：

```
{
    "id": "06830c1b-8157-46fc-b84a-a086aa8c8d45",
    "creationTime": 1523693635249,
    "lastAccessedTime": 1523697391616,
    "maxInactiveInterval": 1800,
    "new": false
}
```

访问 http://127.0.0.1:9001/session，得到输出如下：

```
{
    "id": "06830c1b-8157-46fc-b84a-a086aa8c8d45",
    "creationTime": 1523693635249,
    "lastAccessedTime": 1523697427153,
    "maxInactiveInterval": 1800,
    "new": false
}
```

访问 http://127.0.0.1:9002/session，得到输出如下：

```
{
    "id": "06830c1b-8157-46fc-b84a-a086aa8c8d45",
    "creationTime": 1523693635249,
    "lastAccessedTime": 1523697440377,
    "maxInactiveInterval": 1800,
    "new": false
}
```

我们可以看到，这三个独立运行的应用，都共享了同一个 Session Id。通过 Redis 客户端命令行 redis-cli 输入如下命令，查看所有"spring:session:"开头的 keys：

```
127.0.0.1:6379> keys spring:session:*
...
15) "spring:session:sessions:expires:06830c1b-8157-46fc-b84a-a086aa8c8d45"
16) "spring:session:sessions:06830c1b-8157-46fc-b84a-a086aa8c8d45"
17) "spring:session:sessions:expires:d2193501-1d0b-4f1a-9b50-cd01949ce998"
```

我们可以看到，spring:session:sessions 的值跟我们在浏览器中得到结果一样。正如我们看到的一样，session id 在 Redis 中存储的 Key 是 spring:session:sessions:{sessionId}。

通过 redis-cli 查看 Redis 存储的所有 key 命令如下：

```
127.0.0.1:6379> keys *
…
4) "spring:session:sessions:c3304842-d3a1-42f5-936c-fb73606beda7"
5) "mylist"
6) "spring:session:sessions:expires:c3304842-d3a1-42f5-936c-fb73606beda7"
7) "spring:session:expirations:1523691300000"
…
```

执行 type 命令可以获取一个 key 存储的数据类型，例如：

```
127.0.0.1:6379> type "spring:session:sessions:c3304842-d3a1-42f5-936c-fb73606beda7"
hash
```

其中 "spring:session:sessions:c3304842-d3a1-42f5-936c-fb73606beda7" 为其中的一个 key 值。表明出该 key 存储在现在 Redis 服务器中的类型为 hash。此时操作这个数据就必须使用 hset、hget 等操作方法。否则会报错：

```
127.0.0.1:6379> get "spring:session:sessions:c3304842-d3a1-42f5-936c-fb73606beda7"
```

```
(error) WRONGTYPE Operation against a key holding the wrong kind of value
```

例如，获取在哈希表中指定 key 为 "spring:session:sessions:c4a6db26-d86d-47db-b53c-a10d3b997e40" 的所有字段和值的命令如下：

```
127.0.0.1:6379> hgetall "spring:session:sessions:c4a6db26-d86d-47db-b53c-a10d3b997e40"
1) "maxInactiveInterval"
…
3) "lastAccessedTime"
…
```

单独获取 maxInactiveInterval、creationTime 值的命令如下：

```
127.0.0.1:6379> hget "spring:session:sessions:c4a6db26-d86d-47db-b53c-a10d3b997e40"
maxInactiveInterval
…
127.0.0.1:6379> hget "spring:session:sessions:c4a6db26-d86d-47db-b53c-a10d3b997e40"
    creationTime
…
```

提示　demo_microservice_api_book 工程源代码地址为
https://github.com/EasySpringBoot/demo_microservice_api_book/tree/demo_session。
demo_microservice_api_user 工程源代码地址为
https://github.com/EasySpringBoot/demo_microservice_api_user/tree/demo_session。

15.4　本章小结

我们在 Spring Session 的基础上完成了 Spring Boot 应用的水平扩展。通过 Spring Boot + Redis 来实现 Session 的共享非常简单，而且用处也极大，配合 nginx 进行负载均衡，便能实现分布式的应用了。

此处，我们没有对 Redis 进行主从、读写分离等配置。而且，nginx 的单点故障也是我们应用的障碍，可以使用 ZooKeeper 进行负载均衡。限于篇幅，本书暂不作详细介绍。

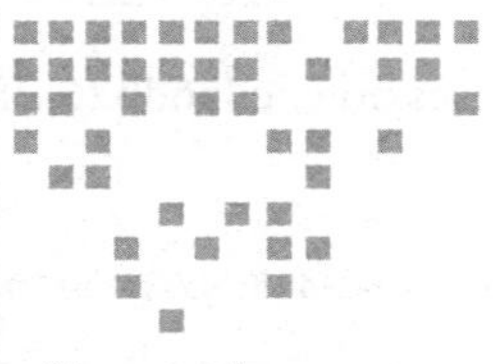

Chapter 16 第 16 章

使用 Zuul 开发 API Gateway

Spring Boot 是构建单个微服务应用的理想选择，但是我们还需要以某种方式将这些微服务互相联系起来，这就是 Spring Cloud Netflix 所要解决的问题。Netflix 提供了各种组件，比如：Eureka 服务发现与 Ribbon 客户端负载均衡的结合，为内部“微服务”提供通信支持。

本章介绍如何通过使用 Netflix Zuul 实现一个微服务 API Gateway，进而实现简单代理转发和过滤器功能。

16.1 API Gateway 简介

API Gateway 是随着微服务这个概念兴起的一种架构模式，它用于解决微服务过于分散，没有一个统一的出入口进行流量管理的问题。

不同的微服务一般有不同的网络域名（或 IP 地址），而通常情况下，在大规模分布式架构系统中，外部的客户端可能需要调用多个服务的接口才能完成一个业务逻辑。比如，在京东上下单购买一个商品的场景，通常会包含数据服务、订单服务、支付服务等。如果客户端直接单独和这些微服务进行通信，可能会存在如下问题：

- 客户端会多次请求不同微服务，增加客户端的复杂性。
- 存在跨域请求，在一定场景下处理相对复杂。
- 认证复杂，每一个服务都需要独立认证。

诸如上述问题，我们可以引入一个中间代理层——API Gateway 来解决。API Gateway 是介于客户端和服务器端之间的中间层，作为微服务网关，所有的外部请求都会先经过 API Gateway，其架构如图 16-1 所示。

这样客户端只需要和 API Gateway 交互，而无需单独去调用特定微服务的接口，而且方便监控，易于认证，减少客户端和各个微服务之间的交互次数。

16.2 Zuul 简介

API Gateway 常见的选型有：

- ❑ 基于 Openresty 的 Kong。
- ❑ 基于 Go 的 Tyk。
- ❑ 基于 Java 的 Zuul。

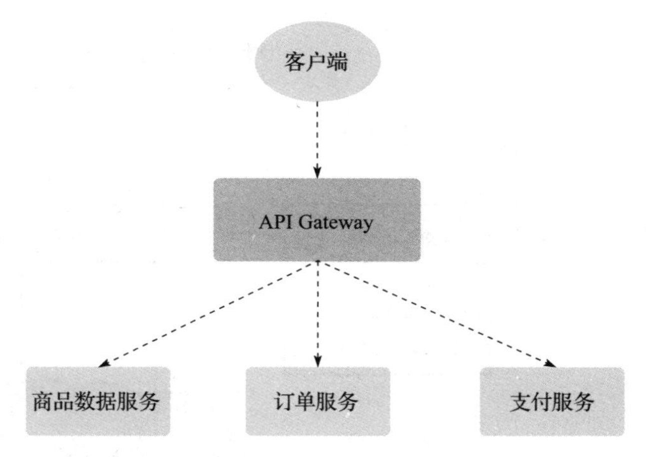

图 16-1 API Gateway 微服务网关架构

常规的选择我们会使用 Nginx 作为代理。但是 Netflix 带来了它自己的解决方案——智能路由 Zuul。它带有许多有趣的功能，可以用于身份验证、服务迁移、分级卸载以及各种动态路由选项。同时，它是使用 Java 编写的。

Zuul 是 Netflix 开源的微服务网关，可以和 Eureka、Ribbon、Hystrix 等组件配合使用。Zuul 本质上是一个 Web Servlet 应用。Zuul 在云平台上提供动态路由、监控、弹性、安全等边缘服务的框架。Zuul 相当于设备和 Web 网站后端所有请求的前门。

Netflix Zuul 提供了服务发现（Eureka）、Circuit Breaker（Hystrix）、智能路由（Zuul）和客户端负载均衡（Ribbon）等功能。

Zuul 可以简单理解为一个类似于 Servlet 中的过滤器。和大部分基于 Java 的 Web 应用类似，Zuul 也采用了 Servlet 架构，因此 Zuul 处理每个请求的方式是针对每个请求用一个线程来处理。通常情况下，为了提高性能，所有请求会放到处理队列中，从线程池中选取空闲线程来处理该请求。这样的设计方式，足以应付一般的高并发场景。

Zuul 的核心组件是一系列的过滤器，它们可以完成以下功能：

- ❑ 身份认证和安全：识别每一个资源的验证要求，并拒绝那些不符的请求。
- ❑ 审计和监控：实现对 API 调用过程的审计和监控，追踪有意义数据及统计结果，从而为我们带来准确的生产状态数据。
- ❑ 动态路由：动态将请求路由到不同后端集群。
- ❑ 压力测试：逐渐增加指向集群的流量，以了解系统的性能。
- ❑ 负载分配：为每一种负载类型分配对应容量，并弃用超出限定值的请求。
- ❑ 静态响应处理：边缘位置进行响应，避免转发到内部集群。
- ❑ 多区域弹性：跨域 AWS Region 进行请求路由，旨在实现 ELB(ElasticLoad Balancing) 使用多样化。

Zuul 提供了四种过滤器的 API，分别为前置（pre）、后置（post）、路由（route）和异常

(error)，简单介绍如下，其生命周期如图 16-2 所示。

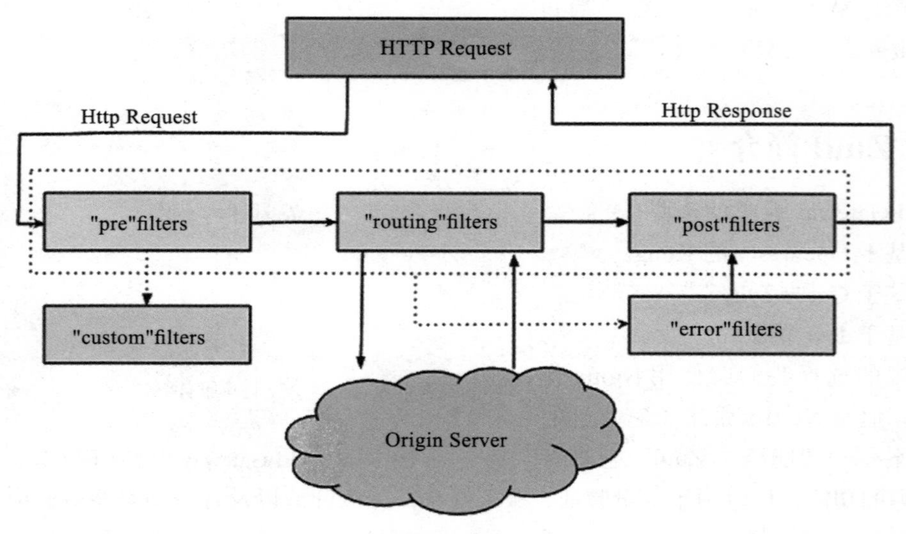

图 16-2 Zuul 四种过滤器的 API 生命周期

- pre：这种过滤器在请求到达 Origin Server 之前调用。比如身份验证，在集群中选择请求的 Origin Server，记 log 等。
- route：在这种过滤器中把用户请求发送给 Origin Server。发送给 Origin Server 的用户请求在这类过滤器中 build。并使用 Apache HttpClient 或者 Netfilx Ribbon 发送给 Origin Server。
- post：这种过滤器在用户请求从 Origin Server 返回以后执行。比如在返回的 response 上面加 response header，做各种统计等。并在该过滤器中把 response 返回给客户。
- error：在其他阶段发生错误时执行该过滤器。

这些过滤器的核心处理逻辑在 ZuulServlet 类中。关键代码说明如下：

```
public class ZuulServlet extends HttpServlet {
    ...
    @Override
    public void service(...) throws ServletException, IOException {
        try {
            // 初始化请求响应对象
            init((HttpServletRequest) servletRequest, (HttpServletResponse)
                servletResponse);
            ...
            RequestContext context = RequestContext.getCurrentContext();
            context.setZuulEngineRan();

            try {
                preRoute(); // "pre" 过滤器
            } catch (ZuulException e) {
```

```
                error(e);
                postRoute();
                return;
            }
            try {
                route();     //"route" 过滤器
            } catch (ZuulException e) {
                error(e);
                postRoute();
                return;
            }
            try {
                postRoute();//"post"过滤器
            } catch (ZuulException e) {
                error(e);
                return;
            }

        } catch (Throwable e) {
            //"error" 过滤器
            error(new ZuulException(e, 500, "UNHANDLED_EXCEPTION_" + e.getClass().
                getName()));
        } finally {
            RequestContext.getCurrentContext().unset();
        }
        ...
    }
}
```

另外，Zuul 还可以通过调用 FilterProcessor.runFilters（类型）来创建或添加并运行任何 filterType。关于自定义静态响应的“静态”类型，请参见 StaticResponseFilter 类。

一个请求会先按顺序通过所有的前置过滤器，之后在路由过滤器中转发给后端应用，得到响应后又会通过所有的后置过滤器，最后响应给客户端。在整个流程中如果发生了异常则会跳转到异常过滤器中。

一般来说，如果需要在请求到达后端应用前就进行处理的话，会选择前置过滤器，例如鉴权、请求转发、增加请求参数等行为。在请求完成后需要处理的操作放在后置过滤器中完成，例如统计返回值和调用时间、记录日志、增加跨域头等行为。路由过滤器一般只需要选择 Zuul 中内置的即可，异常过滤器一般只需要一个，这样可以在 Gateway 遇到错误逻辑时直接抛出异常中断流程，并直接统一处理返回结果。

Spring Cloud 对 Zuul 进行了整合和增强。目前，Zuul 默认使用的是 Apache 的 HTTP Client。也可以通过设置 ribbon.restclient.enabled=true 来使用 Rest Client。在 Zuul 中，每一个后端应用都称为一个 Route，为了避免一个 Route 抢占了太多资源影响到其他 Route 的情况出现，Zuul 使用 Hystrix 对每一个 Route 都做了隔离和限流。

更多关于 Zuul 的内容参考 https://github.com/Netflix/zuul。

16.3 项目实战

本节介绍如何使用 Spring Boot 集成 Zuul 来实现 API Gateway。

1. 创建项目

首先我们来创建基于 Kotlin、Gradle 的 Spring Boot 项目。使用的 Kotlin、Spring Boot、Spring Cloud 的版本号分别配置如下：

```
buildscript {
    ext {
        kotlinVersion = '1.2.20'
        springBootVersion = '2.0.1.RELEASE'
    }

    dependencies {
        classpath("org.springframework.boot:spring-boot-gradle-plugin:${springBoot
            Version}")
        classpath("org.jetbrains.kotlin:kotlin-gradle-plugin:${kotlinVersion}")
        classpath("org.jetbrains.kotlin:kotlin-allopen:${kotlinVersion}")
    }
}
...
ext {
    springCloudVersion = 'Finchley.M9'
}
...
dependencyManagement {
    imports {
        mavenBom "org.springframework.cloud:spring-cloud-dependencies:${springCloud
            Version}"
    }
}
```

2. 添加 Zuul 依赖

通常 Zuul 需要注册到 Eureka 上以实现高可用。这里我们为了简单演示，只实现一个单机版的 API Gateway。在 build.gradle 中添加 spring-cloud-starter-netflix-zuul 如下所示：

```
repositories {
    mavenCentral()
    maven { url "https://repo.spring.io/milestone" }
}

ext {
    springCloudVersion = 'Finchley.M9'
}

dependencies {
    compile('org.springframework.cloud:spring-cloud-starter-netflix-zuul')
    ...
}
```

```
dependencyManagement {
    imports {
        mavenBom "org.springframework.cloud:spring-cloud-dependencies:${spring
            CloudVersion}"
    }
}
```

3. 启用 Zuul 代理

在 Spring Boot 启动类上添加注解 @EnableZuulProxy，代码如下：

```
@SpringBootApplication
@EnableZuulProxy
open class DemoZuulApplication

fun main(args: Array<String>) {
    runApplication<DemoZuulApplication>(*args)
}
```

可将 @EnableZuulProxy 简单理解为 @EnableZuulServer 的增强版，当 Zuul 与 Eureka、Ribbon 等组件配合使用时，使用 @EnableZuulProxy。@EnableZuulProxy 注解默认加上了 @EnableCircuitBreaker，它的定义如下：

```
@EnableCircuitBreaker
@Target(ElementType.TYPE)
@Retention(RetentionPolicy.RUNTIME)
@Import(ZuulProxyMarkerConfiguration.class)
public @interface EnableZuulProxy
```

4. 配置 application.properties

在 application.properties 中添加 zuul.routes.*.url 配置如下：

```
zuul.routes.book_api.url=http://127.0.0.1:9000
zuul.routes.user_api.url=http://127.0.0.1:9001
server.port=8000
```

其中，book_api 是微服务 Book 的服务 API 地址标识，user_api 是微服务 User 的服务 API 地址标识。这个请求流程可以简单用图 16-3 来表示。

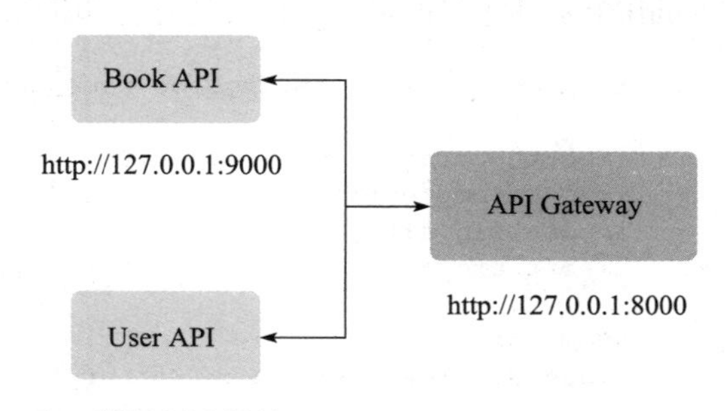

图 16-3　配置 application.properties 的请求流程

5. 启动测试微服务应用

分别启动我们的测试应用 demo_microservice_api_book 和 demo_microservice_api_user。先测试一下，保证服务自身可用。

访问 http://localhost:9000/book/1， 输 出

如下：

```
{
    "id": 1,
    "title": "Spring Boot 2.0 极简教程 ",
    "author": " 陈光剑 "
}
```

访问 http://localhost:9001/user/1，输出如下：

```
{
    "id": 1,
    "username": "user",
    "password": "123456"
}
```

6. 启动 API Gateway 服务

运行 demo_zuul 应用，访问 http://127.0.0.1:8000/user_api/user/1，可以得到输出如下：

```
{
    "id": 1,
    "username": "user",
    "password": "123456"
}
```

访问 http://127.0.0.1:8000/book_api/book/1，可以得到输出如下：

```
{
    "id": 1,
    "title": "Spring Boot 2.0 极简教程 ",
    "author": " 陈光剑 "
}
```

这样我们就实现了一个简单的 API Gateway。

7. 编写 Zuul 过滤器

下面我们在 API Gateway 实现一个请求跟踪过滤器 SimpleFilter。只需要继承抽象类 ZuulFilter 过滤器即可，让该过滤器打印请求日志。实现代码如下：

```
package com.easy.springboot.demo_zuul

import com.netflix.zuul.ZuulFilter
import com.netflix.zuul.context.RequestContext
import org.slf4j.LoggerFactory
import org.springframework.stereotype.Component

@Component
class SimpleFilter : ZuulFilter() {
    private val log = LoggerFactory.getLogger(SimpleFilter::class.java)
    override fun run(): Any? {
        val ctx = RequestContext.getCurrentContext()
```

```
        val request = ctx.request

        log.info(String.format("%s request to %s", request.method, request.requestURL.
            toString()))
        log.info(String.format("LocalAddr: %s", request.localAddr))
        log.info(String.format("LocalName: %s", request.localName))
        log.info(String.format("LocalPort: %s", request.localPort))

        log.info(String.format("RemoteAddr: %s", request.remoteAddr))
        log.info(String.format("RemoteHost: %s", request.remoteHost))
        log.info(String.format("RemotePort: %s", request.remotePort))

        return null
    }

    override fun shouldFilter(): Boolean {
        //判断是否需要过滤
        return true
    }

    override fun filterType(): String {
        //过滤器类型
        return "pre"
    }

    override fun filterOrder(): Int {
        //过滤器的优先级，越大越靠后执行
        return 1
    }

}
```

其中，fun filterType() 指定过滤器类型为 "pre"。

8. 测试 SimpleFilter 过滤器效果

重启应用，再次分别请求 http://127.0.0.1:8000/user_api/user/1 和 http://127.0.0.1:8000/book_api/book/1，我们可以在 API Gateway 应用的控制台后端看到类似下面的请求日志：

```
GET request to http://127.0.0.1:8000/user_api/user/1
LocalAddr: 127.0.0.1
LocalName: localhost
LocalPort: 8000
RemoteAddr: 127.0.0.1
RemoteHost: 127.0.0.1
RemotePort: 61747
GET request to http://127.0.0.1:8000/book_api/book/1
LocalAddr: 127.0.0.1
LocalName: localhost
LocalPort: 8000
RemoteAddr: 127.0.0.1
RemoteHost: 127.0.0.1
RemotePort: 61747
```

提示 本节涉及的三个示例工程源代码地址如下：

1）API Gateway 工程源代码：https://github.com/EasySpringBoot/demo_zuul

2）Book 微服务工程源代码：https://github.com/EasySpringBoot/demo_microservice_api_book

3）User 微服务工程源代码：https://github.com/EasySpringBoot/demo_microservice_api_user

16.4 本章小结

使用 API Gateway 可以将 "1 对 N" 问题转换成了 "1 对 1” 问题，同时在请求到达真正的服务之前，可以做一些预处理工作。API Gateway 可以完成诸如鉴权、流量控制、系统监控、页面缓存等功能，使用 Spring Boot 加上 Spring Cloud“全家桶”来实现微服务架构无疑是一种相当不错的选择。

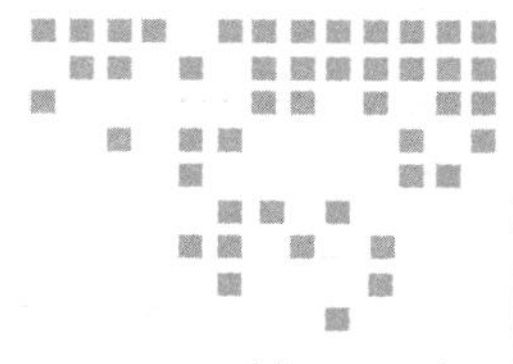

第 17 章 Chapter 17

Spring Boot 日志

在任何一个生产系统中，对日志的合理记录都是非常重要的。这对系统故障的定位处理极其关键。Spring Boot 支持 java.util.logging、Log4j、Logback 等作为日志框架，其中，Logback 作为默认日志框架。无论使用哪种日志框架，Spring Boot 都支持配置将日志输出到控制台或者文件中。

本章我们来详细介绍 Spring Boot 应用的日志的配置与使用。

17.1 Logback 简介

Java 日志框架众多，常用的有 java.util.logging、log4j、logback、commons-logging 等。本节简单介绍 Logback 日志框架。

SLF4J（Simple Logging Facade For Java）是一个针对于各类 Java 日志框架的统一 Facade 抽象。SLF4J 定义了统一的日志抽象接口，而真正的日志实现则是在运行时决定。

Logback 是由 SLF4J 作者开发的新一代日志框架，用于替代 log4j。它效率更高、能够适应诸多的运行环境。Logback 的架构设计足够通用，可适用于不同的环境。目前 Logback 分为三个模块：lobback-core、logback-classic 和 logback-access。core 模块是其他两个模块的基础，classic 是 core 的扩展，是 log4j 巨大改进的版本。Logback-classic 本身实现了 SLF4J 的 API，因此可以很容易在 logback 与其他日志系统之间转换，例如 log4j、JDK1.4 中的 java.util.logging（JUL）。access 模块集成了 Servlet 容器，提供了通过 HTTP 访问日志的功能，详细了解 access 可访问文档：http://logback.qos.ch/access.html。

Logback 的日志级别有 trace、debug、info、warn、error，级别排序为：trace < debug <

info < warn < error。

提示 关于日志级别详细信息，可参考官方文档：http://logback.qos.ch/manual/architecture.html。

一般情况下，我们不需要单独引入 spring-boot-starter-logging 这个起步依赖了，因为这是 spring-boot-starter 默认引入的依赖，其依赖树如下：

```
▼ org.springframework.boot:spring-boot-starter:2.0.1.RELEASE (Compile)
  ► org.springframework.boot:spring-boot-autoconfigure:2.0.1.RELEASE ((
  ► org.springframework.boot:spring-boot:2.0.1.RELEASE (Compile)
  ▼ org.springframework.boot:spring-boot-starter-logging:2.0.1.RELEASE
    ▼ ch.qos.logback:logback-classic:1.2.3 (Compile)
        ch.qos.logback:logback-core:1.2.3 (Compile)
        org.slf4j:slf4j-api:1.7.25 (Compile)
    ▼ org.apache.logging.log4j:log4j-to-slf4j:2.10.0 (Compile)
        org.slf4j:slf4j-api:1.7.25 (Compile)
        org.apache.logging.log4j:log4j-api:2.10.0 (Compile)
    ▼ org.slf4j:jul-to-slf4j:1.7.25 (Compile)
        org.slf4j:slf4j-api:1.7.25 (Compile)
    javax.annotation:javax.annotation-api:1.3.2 (Compile)
    org.springframework:spring-core:5.0.5.RELEASE (Compile)
    org.yaml:snakeyaml:1.19 (Compile)
```

从上面的依赖树，我们可以看出，spring-boot-starter-logging 依赖 logback-classic，logback-classic 依赖 logback-core、sl4j-api。

Spring Boot 为我们提供了功能齐全的默认日志配置，基本上就是“开箱即用”。

默认情况下，Spring Boot 的日志是输出到控制台的，不写入任何日志文件。

要让 Spring Boot 输出日志文件，最简单的方式是在 application.properties 配置文件中配置 logging.path 键值，如下：

```
logging.path=${user.home}/logs
```

这样在 ${user.home}/logs 目录下会生成默认的文件名命名的日志文件 spring.log。

我们可以在 application.properties 配置文件中配置 logging.file 键值，如下所示：

```
spring.application.name=lightsword
logging.file=${user.home}/logs/${spring.application.name}.log
```

这样日志文件的名字就是 lightsword.log 了。另外，二者不能同时使用，如同时使用，则只有 logging.file 生效。

17.2 配置 logback 日志

日志服务一般都在 ApplicationContext 创建前就初始化了，所以日志配置可以独立于 Spring 的配置。我们也可以通过系统属性和传统的 Spring Boot 外部配置文件，实现日志控制和管理。

根据不同的日志系统，Spring Boot 按如下“约定规则”组织配置文件名加载日志配置文件，参见表 17-1。

表 17-1　配置日志规则

日志框架	配置文件
Logback	logback-spring.xml，logback-spring.groovy，logback.xml，logback.groovy
Log4j	log4j-spring.properties，log4j-spring.xml，log4j.properties，log4j.xml
Log4j2	log4j2-spring.xml，log4j2.xml
Java Util Logging	logging.properties

Spring Boot 官方推荐优先使用带有 -spring 的文件名作为你的日志配置（如使用 logback-spring.xml，而不是 logback.xml），命名为 logback-spring.xml 的日志配置文件，Spring Boot 可以为它添加一些特有的配置项。

Logback 读取配置或属性文件的步骤是：

1）Logback 在类路径下尝试查找 logback.groovy 的文件。

2）如果没有找到 logback.groovy 文件，就在类路径下查找 logback-test.xml 文件。

3）若也没有找到 logback-test.xml 文件，就会在类路径下查找 logback.xml 文件。

我们也可以自定义 logback.xml 名称，然后在 application.properties 中指定它。例如：

```
#logging
logging.config=classpath:logback-dev.groovy
```

要把这个 logback-dev.groovy 配置文件放到类路径下，如下所示：

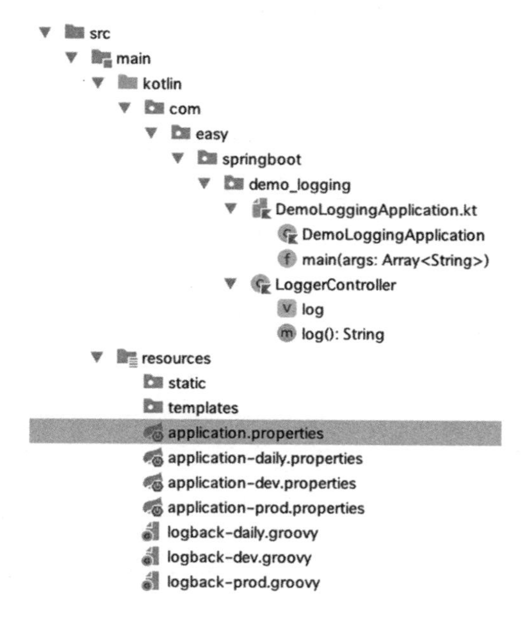

我们在 application.properties 指定环境：

```
spring.profiles.active=daily
```

对应的 application-daily.properties 指定日志的配置文件如下：

```
#logging
logging.config=classpath:logback-daily.groovy
```

另外，如果我们没有配置任何的 logback.xml、logback.groovy 等文件，Logback 就会使用 BasicConfigurator 启动默认配置，该配置会将日志输出到控制上。这样就意味着使用默认配置，它提供了默认的最基础的日志功能。

17.3 logback.groovy 配置文件

本节介绍 logback 配置文件的具体内容。

17.3.1 显示系统 Log 级别

我们首先编写一个 log 接口来展示当前系统的日志级别，代码如下：

```
package com.easy.springboot.demo_logging

import org.slf4j.LoggerFactory
import org.springframework.web.bind.annotation.GetMapping
import org.springframework.web.bind.annotation.RestController

@RestController
class LoggerController {
    val log = LoggerFactory.getLogger(LoggerController::class.java)

    @GetMapping("/log")
    fun log(): String {
        var logLevel = ""

        if (log.isTraceEnabled) {
            log.trace("5-TRACE")
            logLevel += "5-TRACE|"
        }

        if (log.isDebugEnabled) {
            log.debug("4-DEBUG")
            logLevel += "4-DEBUG|"
        }

        if (log.isInfoEnabled) {
            log.info("3-INFO")
            logLevel += "3-INFO|"
        }

        if (log.isWarnEnabled) {
```

```
            log.warn("2-WARN")
            logLevel += "2-WARN|"
        }

        if (log.isErrorEnabled) {
            log.error("1-ERROR")
            logLevel += "1-ERROR|"
        }
        return logLevel
    }
}
```

17.3.2 使用 logback.groovy 配置

logback.xml 配置文件烦琐而冗长。logback 框架支持 logback.groovy 简洁的 DLS 风格的配置。Groovy 是一门优秀的 DSL（领域特定语言，Domain-specific languages），用途非常广泛。

详细的 logback.groovy 配置语法可以参考 https://logback.qos.ch/manual/groovy.html。同时，logback 提供了直接把 logback.xml 转换成 logback.groovy 的工具：https://logback.qos.ch/translator/asGroovy.html（测试过，这个工具 include 标签暂时未作解析）。

推荐使用 Groovy DSL 作为 logback 日志配置文件的最佳实践。配置 logback-daily.groovy 如下所示：

```
import ch.qos.logback.classic.encoder.PatternLayoutEncoder
import ch.qos.logback.classic.filter.ThresholdFilter
import ch.qos.logback.core.ConsoleAppender
import ch.qos.logback.core.rolling.RollingFileAppender
import ch.qos.logback.core.rolling.TimeBasedRollingPolicy

import java.nio.charset.Charset
import static ch.qos.logback.classic.Level.INFO

def USER_HOME = System.getProperty("user.home")
def APP_NAME = "demo_logging"
def LOG_PATTERN = "%d{HH:mm:ss.SSS} [%thread] %-5level %logger{35} - %msg %n"
def LOG_FILE = "${USER_HOME}/logs/${APP_NAME}"
def FILE_NAME_PATTERN = "${APP_NAME}.%d{yyyy-MM-dd}.log"

scan("60 seconds")

context.name = "${APP_NAME}"
jmxConfigurator()

logger("org.springframework.web", INFO)
logger("com.easy.springboot.demo_logging", INFO)
```

```
appender("CONSOLE", ConsoleAppender) {
    encoder(PatternLayoutEncoder) {
        pattern = "${LOG_PATTERN}"
        charset = Charset.forName("utf8")
    }
}

appender("dailyRollingFileAppender", RollingFileAppender) {
    file = "${LOG_FILE}"
    rollingPolicy(TimeBasedRollingPolicy) {
        fileNamePattern = "${FILE_NAME_PATTERN}"
        maxHistory = 30
    }
    filter(ThresholdFilter) {
        level = INFO
    }
    encoder(PatternLayoutEncoder) {
        pattern = "${LOG_PATTERN}"
    }
}
root(INFO, ["CONSOLE", "dailyRollingFileAppender"])
```

上面的 Groovy 配置文件等价于如下的 logback-daily.xml 配置文件内容：

```
<?xml version="1.0" encoding="UTF-8"?>
<!--scan：当此属性设置为 true 时，配置文件如果发生改变，将会被重新加载，默认值为 true。-->
<!--scanPeriod：设置监测配置文件是否有修改的时间间隔，如果没有给出时间单位，默认单位是毫秒。
    当 scan 为 true 时，此属性生效。默认的时间间隔为 1 分钟。-->
<!--debug：当此属性设置为 true 时，将打印出 logback 内部日志信息，实时查看 logback 运行状态。
    默认值为 false。-->
<configuration scan="true" scanPeriod="60 seconds" debug="false">
    <property name="APP_NAME" value="lightsword"/>
    <contextName>${APP_NAME}</contextName>
    <include resource="org/springframework/boot/logging/logback/base.xml"/>
    <jmxConfigurator/>

    <logger name="org.springframework.web" level="INFO"/>
    <logger name=" com.easy.springboot.demo_logging " level="TRACE"/>

    <appender name="dailyRollingFileAppender" class="ch.qos.logback.core.rolling.
       RollingFileAppender">
        <File>${user.home}/logs/${APP_NAME}</File>
        <rollingPolicy class="ch.qos.logback.core.rolling.TimeBasedRollingPolicy">
            <!-- daily rolling over -->
            <FileNamePattern>${APP_NAME}.%d{yyyy-MM-dd}.log</FileNamePattern>
            <!-- keep 30 days' log history -->
            <maxHistory>30</maxHistory>
        </rollingPolicy>
        <filter class="ch.qos.logback.classic.filter.ThresholdFilter">
            <level>ERROR</level>
        </filter>
        <encoder>
            <Pattern>%d{HH:mm:ss.SSS} [%thread] %-5level %logger{35} - %msg
                %n</Pattern>
```

```
        </encoder>
    </appender>
    <!--TRACE, DEBUG, INFO, WARN, ERROR-->
    <root level="DEBUG">
        <appender-ref ref="CONSOLE"/>
        <!--<appender-ref ref="FILE"/>-->
        <appender-ref ref="dailyRollingFileAppender"/>
    </root>
</configuration>
```

我们可以看出，使用 groovy 表达的配置，更加简洁、富表现力。只需要在 application.properties 里面配置 logging.config 指定日志配置文件即可：

```
#logging
logging.config=classpath:logback-dev.groovy
```

另外，需要在 build.gradle 中添加 groovy 依赖：

```
dependencies {
    ...
    compile group: 'org.codehaus.groovy', name: 'groovy-all', version: '2.4.15'
}
```

完成上述配置，即可使用与 logback.xml 配置一样的日志功能了。

17.3.3　配置文件说明

在 logback.xml 形式配置文件内，总体结构是：最顶层是一个 <configuration> 标签，在 <configuration> 标签下可以有 0 到 n 个 <appender> 标签，0 到 n 个 <logger> 标签，最多只能有 1 个 <root> 标签，以及其他一些高级配置。下面我们针对上面的 logback.xml 配置文件作简要说明。

1. configuration 节点

配置文件中的根节点中的 <configuration scan="true" scanPeriod="60 seconds" debug="false"> 包含的属性简单说明如下：

- scan：当此属性设置为 true 时，配置文件如果发生改变，将会被重新加载，默认值为 true。
- scanPeriod: 设置监测配置文件是否有修改的时间间隔，如果没有给出时间单位，默认单位是毫秒。当 scan 为 true 时，此属性生效。默认的时间间隔为 1 分钟。
- debug: 当此属性设置为 true 时，将打印出 logback 内部日志信息，实时查看 logback 运行状态。默认值为 false。

2.jmxConfigurator

<jmxConfigurator/> 标签对应 Groovy 配置脚本中的 jmxConfigurator()。这个配置是开启 JMX 的功能。JMX（Java Management Extensions，即 Java 管理扩展）是一个为应用程序、设备、系统等植入管理功能的框架。JMX 可以跨越一系列异构操作系统平台、系统体系结构和网络传输协议，灵活的开发无缝集成的系统、网络和服务管理应用。

有了这个配置，我们可以直接在命令行输入：jconsole，这个命令会启动 JConsole 的

GUI 界面，选中我们的本地工程进程 com.easy.springboot.demo_logging.DemoLoggingApplication，点击“连接”进入主界面。如图 17-1 所示。

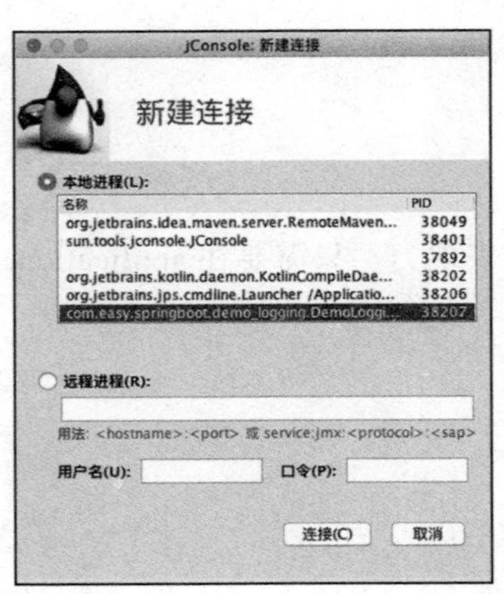

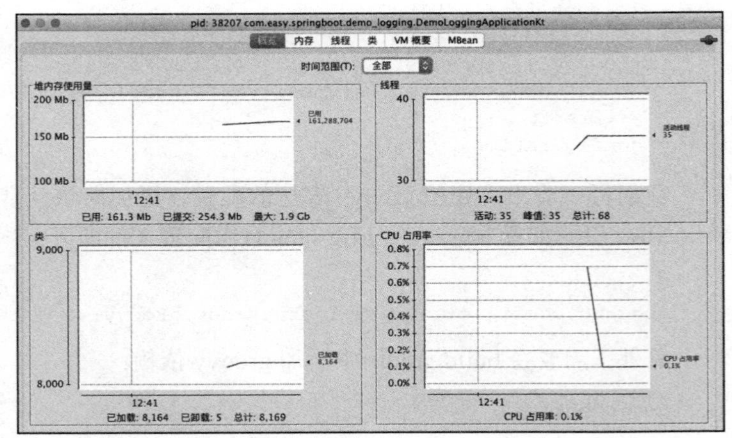

图 17-1 启动 JConsole 的 GUI 界面

点击 MBean 选项卡，我们可以看到关于 logback 的信息，如下：

```
▶ DefaultDomain
▶ JMImplementation
▶ Tomcat
▼ ch.qos.logback.classic
    ▼ demo_logging
        ▼ ch.qos.logback.classic.jmx.JMXConfigurator
            ▼ Attributes
                LoggerList
                Statuses
            ▼ Operations
                getLoggerLevel
                setLoggerLevel
                reloadByFileName
                reloadDefaultConfiguration
                reloadByURL
                getLoggerEffectiveLevel
▶ com.sun.management
▶ java.lang
▶ java.nio
▶ java.util.logging
▶ org.springframework.boot
```

启动系统，采用 spring.profiles.active=daily 配置，日志级别是：

```
logger("com.easy.springboot.demo_logging", INFO)
```

访问 http://127.0.0.1:8080/log，响应输出：“3-INFO|2-WARN|1-ERROR|”。

然后，点击“setLoggerLevel”，我们可以在设置界面动态修改系统的日志级别，如图 17-2 所示。

设置 p1 参数为：com.easy.springboot.demo_logging，设置 p2 参数为：TRACE，点击“setLoggerLevel”按钮，提示“Method successfully invoked”。

再次访问 http://127.0.0.1:8080/log，响应输出：“5-TRACE|4-DEBUG|3-INFO|2-WARN|1-ERROR|”。我们可以看出系统的日志级别已经变成了 TRACE。

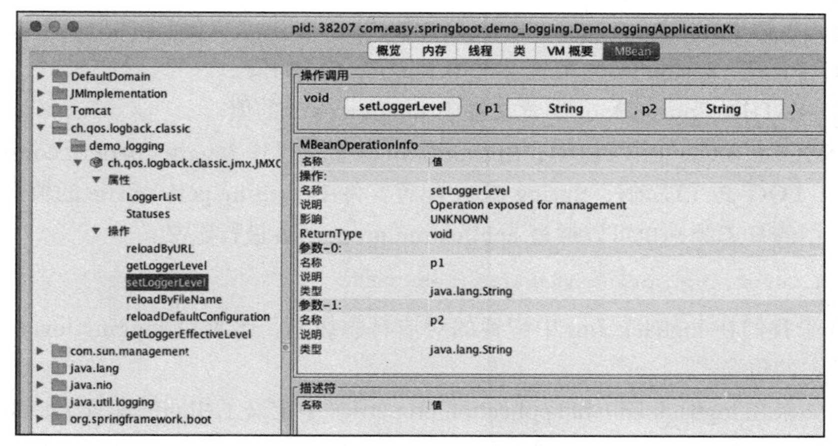

图 17-2　修改系统的日志级别

提示　关于 jconsole 的详细介绍，可以参考：https://docs.oracle.com/javase/8/docs/technotes/guides/management/jconsole.html。

3. Logger 节点

在配置 <logger name="org.springframework.web" level="INFO"/> 中，我们定义了捕获 org.springframework.web 的日志，日志级别是 DEBUG。捕获 com.springboot.in.action 的日志，日志级别是 TRACE。

上面引用的 org/springframework/boot/logging/logback/base.xml 文件是 Spring Boot 内置的，其内容为：

```
<?xml version="1.0" encoding="UTF-8"?>
<included>
    <include resource="org/springframework/boot/logging/logback/defaults.xml" />
    <property name="LOG_FILE" value="${LOG_FILE:-${LOG_PATH:-${LOG_TEMP:-${java.
        io.tmpdir:-/tmp}}/}spring.log}"/>
    <include resource="org/springframework/boot/logging/logback/console-appender.
        xml" />
    <include resource="org/springframework/boot/logging/logback/file-appender.xml" />
    <root level="INFO">
        <appender-ref ref="CONSOLE" />
        <appender-ref ref="FILE" />
    </root>
</included>
```

其中，引用的 defaults.xml、console-appender.xml、file-appender.xml 都在同一个目录下。默认情况下包含两个 appender：一个是控制台，一个是文件，分别定义在 console-appender.xml 和 file-appender.xml 中。这里面的内容就是 Spring Boot 默认实现的 logback 的日志配置。

Spring Boot 的日志模块里，预定义了一些系统变量：

- PID，当前进程 ID。
- LOG_FILE，Spring Boot 配置文件中 logging.file 的值。
- LOG_PATH，Spring Boot 配置文件中 logging.path 的值。
- CONSOLE_LOG_PATTERN，Spring Boot 配置文件中 logging.pattern.console 的值。
- FILE_LOG_PATTERN，Spring Boot 配置文件中 logging.pattern.file 的值。

对于应用的日志级别也可以通过 application.properties 进行定义：

```
logging.level.org.springframework.web=DEBUG
```

这相当于我们在 logback.xml 中配置的对应日志级别。名称以 logging.level 开头，后面跟要输入日志的包名。

另外，如果在 logback.xml 和 application.properties 中定义了相同的配置（如都配置了 org.springframework.web）但是输出级别不同，由于 application.properties 的优先级高于 logback.xml，所以会使用 application.properties 的配置。

4. ConsoleAppender

Logback 使用 appender 来定义日志输出，在开发过程中最常用的是将日志输出到控制台。我们直接使用 SpringBoot 内置的 ConsoleAppender 配置。这个配置的内容如下：

```
<?xml version="1.0" encoding="UTF-8"?>
<included>
    <conversionRule conversionWord="clr" converterClass="org.springframework.
        boot.logging.logback.ColorConverter" />
    <conversionRule conversionWord="wex" converterClass="org.springframework.
        boot.logging.logback.WhitespaceThrowableProxyConverter" />
    <conversionRule conversionWord="wEx" converterClass="org.springframework.
        boot.logging.logback.ExtendedWhitespaceThrowableProxyConverter" />
    <property name="CONSOLE_LOG_PATTERN" value="${CONSOLE_LOG_PATTERN:-
        %clr (%d{yyyy-MM-dd HH:mm:ss.SSS}){faint} %clr(${LOG_LEVEL_PATTERN:-
        %5p}) %clr(${PID:-}){magenta} %clr(---){faint} %clr([%15.15t]){faint}
        %clr(%-40.40logger{39}){cyan} %clr(:){faint} %m%n${LOG_EXCEPTION_
        CONVERSION_WORD:-%wEx}}"/>
    <property name="FILE_LOG_PATTERN" value="%d{yyyy-MM-dd HH:mm:ss.SSS}${LOG_LEVEL_
        PATTERN:-%5p} ${PID:- } --- [%t] %-40.40logger{39} : %m%n${LOG_EXCEPTION_
            CONVERSION_WORD:-%wEx}"/>

    <appender name="DEBUG_LEVEL_REMAPPER" class="org.springframework.boot.logging.
       logback.LevelRemappingAppender">
        <destinationLogger>org.springframework.boot</destinationLogger>
    </appender>

    <logger name="org.apache.catalina.startup.DigesterFactory" level="ERROR"/>
    <logger name="org.apache.catalina.util.LifecycleBase" level="ERROR"/>
    <logger name="org.apache.coyote.http11.Http11NioProtocol" level="WARN"/>
    <logger name="org.apache.sshd.common.util.SecurityUtils" level="WARN"/>
    <logger name="org.apache.tomcat.util.net.NioSelectorPool" level="WARN"/>
    <logger name="org.crsh.plugin" level="WARN"/>
    <logger name="org.crsh.ssh" level="WARN"/>
```

```
    <logger name="org.eclipse.jetty.util.component.AbstractLifeCycle" level="ERROR"/>
    <logger name="org.hibernate.validator.internal.util.Version" level="WARN"/>
    <logger name="org.springframework.boot.actuate.autoconfigure.CrshAutoConfiguration"
        level="WARN"/>
    <logger name="org.springframework.boot.actuate.endpoint.jmx" additivity="false">
        <appender-ref ref="DEBUG_LEVEL_REMAPPER"/>
    </logger>
    <logger name="org.thymeleaf" additivity="false">
        <appender-ref ref="DEBUG_LEVEL_REMAPPER"/>
    </logger>
</included>
<included>
    <appender name="CONSOLE" class="ch.qos.logback.core.ConsoleAppender">
        <encoder>
            <pattern>${CONSOLE_LOG_PATTERN}</pattern>
            <charset>utf8</charset>
        </encoder>
    </appender>
</included>
```

其中，

charset 表示对日志进行编码。

pattern 简单说明如下：

- %d{HH:mm:ss.SSS}——日志输出时间。
- %thread——输出日志的进程名字，用方括号括起来。这个信息在 Web 应用以及异步任务处理中很有用。
- %-5level——日志级别，并且使用 5 个字符靠左对齐。
- %logger{36}——日志输出者的名字。
- %msg——日志消息。
- %n——平台的换行符。

在这种格式下一条日志的输出内容格式如下：

```
02:37:22.752 [http-nio-8888-exec-1] DEBUG o.s.s.w.a.AnonymousAuthenticationFilter
    - Populated SecurityContextHolder with anonymous token: 'org.springframework.
    security.authentication.AnonymousAuthenticationToken@6fab4e5e: Principal:
    anonymousUser; Credentials: [PROTECTED]; Authenticated: true; Details:
    org.springframework.security.web.authentication.WebAuthenticationDetails@
    fffe3f86: RemoteIpAddress: 127.0.0.1; SessionId: E30F2AF513F94C7FC7611353B6
    1A26C6; Granted Authorities: ROLE_ANONYMOUS'
```

5. RollingFileAppender

另一种通用功能是将日志输出到文件。同时，随着应用的运行时间越来越长，日志也会增长得越来越多，将其输出到同一个文件并非一个好办法。我们有 RollingFileAppender 用于切分文件日志：

```
<appender name="dailyRollingFileAppender" class="ch.qos.logback.core.rolling.
    RollingFileAppender">
```

```
        <File>${user.home}/logs/${APP_NAME}</File>
        <rollingPolicy class="ch.qos.logback.core.rolling.TimeBasedRollingPolicy">
            <!-- daily rolling over -->
            <FileNamePattern>${APP_NAME}.%d{yyyy-MM-dd}.log</FileNamePattern>
            <!-- keep 30 days' log history -->
            <maxHistory>30</maxHistory>
        </rollingPolicy>
        <encoder>
            <Pattern>%d{HH:mm:ss.SSS} [%thread] %-5level %logger{35} - %msg %n</Pattern>
            </encoder>
    </appender>
```

其中，核心的配置部分是 rollingPolicy 的定义：

- FileNamePattern，定义了日志的切分方式——把每一天的日志归档到一个文件中。
- maxHistory，表示只保留最近几天的日志，以防止日志填满整个磁盘空间，代码中保留了30天的日志。我们也可以使用 %d{yyyy-MM-dd_HH-mm} 来定义精确到的日志切分方式。

6. Threshold filter

ThresholdFilter 是 Logback 定义的日志打印级别的过滤器。例如配置 ThresholdFilter 来过滤掉 ERROR 级别以下的日志不输出到文件中：

```
<filter class="ch.qos.logback.classic.filter.ThresholdFilter">
    <level>ERROR</level>
</filter>
```

7. root 节点

关于 root 节点的配置如下：

```
<root level="DEBUG">
    <appender-ref ref="CONSOLE"/>
    <!--<appender-ref ref="FILE"/>-->
    <appender-ref ref="dailyRollingFileAppender"/>
</root>
```

root 节点是必选节点，用来指定最基础的日志输出级别，只有一个 level 属性：用来设置打印级别，大小写无关：TRACE、DEBUG、INFO、WARN、ERROR、ALL 和 OFF，不能设置为 INHERITED 或者同义词 NULL，默认值是 DEBUG。

17.4 本章小结

Spring Boot 集成 logback 日志框架非常简单。同时，使用基于 Groovy DSL 的 logback.groovy 配置文件，风格简洁优雅。使用 spring.profile 配置多环境（dev、daily、prod 等）的日志配置文件也非常简单方便。通过配置 jmxConfigurator 可以在 jconsole 管理后台动态修改系统的日志级别。

提示 本章实例工程源代码地址为 https://github.com/EasySpringBoot/demo_logging。

第Ⅲ部分 Part 3

Spring Boot 系统监控、测试与运维

Chapter 18 第 18 章

Spring Boot 应用的监控：Actuator 与 Admin

在企业级应用中，对系统进行运行状态监控通常是必不可少的。Spring Boot 提供了 Actuator 模块实现应用的监控与管理，对应的起步依赖是 spring-boot-starter-actuator。

spring-boot-actuator 模块提供了一个监控和管理生产环境的模块，可以使用 http、jmx、ssh、telnet 等管理和监控应用，提供了应用的审计（Auditing）、健康（Health）状态信息、数据采集（Metrics Gathering）统计等监控运维的功能。同时，我们可以扩展 Actuator 端点自定义监控指标。这些指标都是以 JSON 接口数据的方式呈现。而使用 Spring Boot Admin 可以实现这些 JSON 接口数据的界面展现。

本章介绍 Spring Boot Actuator 和 Admin 实现对 Spring Boot 应用的监控与管理。

18.1 Actuator 简介

在实际的生产系统中，怎样知道应用运行良好呢？往往需要对系统实际运行的情况（例如 cpu、io、disk、db、业务功能等指标）进行监控运维，这需要耗费不少精力。

在 SpringBoot 中，完全不需要面对这样的难题。Spring Boot Actuator 提供了众多 HTTP 接口端点（Endpoint），其中包含了丰富的 Spring Boot 应用程序运行时的内部状态信息。同时，我们还可以自定义监控端点，实现灵活定制。

Actuator 是 spring boot 提供的对应用系统的自省和监控功能，Actuator 对应用系统本身的自省功能，可以让我们方便快捷地实现线上运维监控的工作。这有点儿像 DevOps。通过 Actuator，可以使用数据化的指标去度量应用的运行情况。比如查看服务器的磁盘、内存、CPU 等信息，系统运行了多少线程、gc 的情况、运行状态等。

spring-boot-actuator 模块提供了一个监控和管理生产环境的模块，可以使用 http、jmx、ssh、telnet 等管理和监控应用。

随着 DevOps 的兴起，以及 docker 技术的普及，微服务在一定场合会越来越受欢迎。即使不说微服务，springboot 这种可以直接内嵌 Web 服务器打成一个 jar 包的方式，也更符合 DevOps 的趋势：打成 jar 包，往服务器上一扔，十分方便，自带 Actuator，也省了大半监控工作，真正做到可以把精力花在刀刃上。

18.2 启用 Actuator

在 Spring Boot 项目中添加 Actuator 起步依赖即可启用 Actuator 功能。在 Gradle 项目配置文件 build.gradle 中添加如下代码：

```
dependencies {
    compile('org.springframework.boot:spring-boot-starter-actuator')
    ...
}
```

为了看到 Spring Boot 中提供的全部的端点信息，在 Spring Boot 1.5.x 版本中默认启用所有 Endpoint，这些端点如下：

```
{
    "links" : [ {
        "rel" : "self",
        "href" : "http://127.0.0.1:8010/actuator"
    }, {
        "rel" : "metrics",
        "href" : "http://127.0.0.1:8010/metrics"
    }, {
        "rel" : "autoconfig",
        "href" : "http://127.0.0.1:8010/autoconfig"
    }, {
        "rel" : "configprops",
        "href" : "http://127.0.0.1:8010/configprops"
    }, {
        "rel" : "dump",
        "href" : "http://127.0.0.1:8010/dump"
    }, {
        "rel" : "trace",
        "href" : "http://127.0.0.1:8010/trace"
    }, {
        "rel" : "logfile",
        "href" : "http://127.0.0.1:8010/logfile"
    }, {
        "rel" : "beans",
        "href" : "http://127.0.0.1:8010/beans"
    }, {
        "rel" : "env",
```

```
        "href" : "http://127.0.0.1:8010/env"
    }, {
        "rel" : "heapdump",
        "href" : "http://127.0.0.1:8010/heapdump"
    }, {
        "rel" : "serverEndpoint",
        "href" : "http://127.0.0.1:8010/serverEndpoint"
    }, {
        "rel" : "jolokia",
        "href" : "http://127.0.0.1:8010/jolokia"
    }, {
        "rel" : "info",
        "href" : "http://127.0.0.1:8010/info"
    }, {
        "rel" : "loggers",
        "href" : "http://127.0.0.1:8010/loggers"
    }, {
        "rel" : "showEndpoints",
        "href" : "http://127.0.0.1:8010/showEndpoints"
    }, {
        "rel" : "auditevents",
        "href" : "http://127.0.0.1:8010/auditevents"
    }, {
        "rel" : "health",
        "href" : "http://127.0.0.1:8010/health"
    }, {
        "rel" : "docs",
        "href" : "http://127.0.0.1:8010/docs"
    }, {
        "rel" : "mappings",
        "href" : "http://127.0.0.1:8010/mappings"
    } ]
}
```

在Spring Boot 2.0中，Actuator模块做了较大更新，默认启用的端点如下：

```
{
    _links: {
        self: {
        href: "http://127.0.0.1:8008/actuator",
        templated: false
        },
        health: {
        href: "http://127.0.0.1:8008/actuator/health",
        templated: false
            },
        info: {
        href: "http://127.0.0.1:8008/actuator/info",
        templated: false
        }
    }
}
```

如果想启用所有端点，在 application.properties 中按如下配置：

```
#endpoints in Spring Boot 2.0
#http://127.0.0.1:8008/actuator
management.endpoints.enabled-by-default=true
management.endpoints.web.expose=*
```

重新启动应用，将看到一个信息更全的 Actuator 端点列表，这个列表将在下面小节中介绍。

18.3　揭秘端点

本节详细介绍 Actuator 提供的端点内容。

18.3.1　常用的 Actuator 端点

使用 /actuator 端点可以获取应用启用的所有端点列表，其中常用的 Actuator 端点如表 18-1 所示。

表 18-1　常用的 Actuator 端点

HTTP 方法	路径	描述	是否敏感信息
GET	/autoconfig	查看自动配置的使用情况，显示一个 auto-configuration 的报告，该报告展示所有 auto-configuration 候选者及其情况	true
GET	/conditions	获取自动配置条件信息。访问 /actuator/conditions 接口获取 Spring Boot 自动配置的条件信息，这些信息反映了 Spring Boot 在实现自动配置过程中的条件信息	true
GET	/configprops	查看配置属性，包括默认配置，显示一个所有 @Configuration-Properties 的整理列表	true
GET	/beans	获取 bean 及其关系列表，显示一个应用中所有 Spring Beans 的完整列表。访问 /actuator/beans 可以获取当前 Spring Boot 应用 Spring 容器中所有的 Bean 信息及其依赖关系	true
GET	/env	查看所有环境变量	true
GET	/env/{name}	查看具体变量值	true
GET	/health	查看应用健康指标	false
GET	/info	查看应用信息	false
GET	/loggers	获取系统日志信息	true
GET	/mappings	查看所有 URL 映射	true
GET	/metrics	获取系统度量指标信息	true
GET	/metrics/{name}	获取系统度量指标信息时，查看具体指标	true
GET	/scheduledtasks	获取系统定时任务信息	true
POST	/shutdown	关闭应用，允许应用以优雅的方式关闭（默认情况下不启用）	true
GET	/threaddump	获取系统线程转储信息	true
GET	/trace	查看基本追踪信息，默认为最新的一些 HTTP 请求	true

下面举例介绍其中的几个。

1. /configprops

很多时候，我们可能不知道在 application.properties 配置文件中到底有哪些默认的配置。现在可以直接访问 /actuator/configprops 这个接口获取当前 Spring Boot 应用的所有配置信息。这里面有常见的 Web Server、Spring Data JPA 的数据源等配置信息。有了这些信息，就再也不用担心 application.properties 中属性配置的问题了。

2. /env

访问 /actuator/env 可以获取到几乎你能想到的所有关于当前 Spring Boot 应用程序的运行环境信息，例如：操作系统信息（systemProperties）、环境变量信息、JDK 版本及 ClassPath 信息、当前启用的配置文件（activeProfiles）、propertySources、应用程序配置信息（applicationConfig）等。有了这些信息，可以帮助我们更好地了解 Spring Boot 应用的运行状况。

3. /health

当使用一个未认证连接访问时显示一个简单的状态，使用认证连接访问则显示全部信息详情。示例数据如下：

```
// 20180207010657
// http://127.0.0.1:8008/actuator/health

{
    "status": "UP",
    "details": {
        "diskSpace": {
            "status": "UP",
            "details": {
                "total": 120108089344,
                "free": 2241015808,
                "threshold": 10485760
            }
        },
        "db": {
            "status": "UP",
            "details": {
                "database": "MySQL",
                "hello": 1
            }
        }
    }
}
```

其中，details 信息的展示需要在 application.properties 文件中配置：

```
management.endpoint.health.show-details=true
```

否则，只能看到一个“UP”的信息输出，如下所示：

```
{
    "status": "UP"
}
```

4. /info

查看应用信息是否在 application.properties 中配置的，例如：

```
info.app.name=KSB Security
info.app.version=v1.0.0
info.app.description=Spring Boot With Security
```

5. /loggers

访问 /actuator/loggers 获取系统的日志信息，示例数据如下：

```
{
    "levels": [
        "OFF",
        "ERROR",
        "WARN",
        "INFO",
        "DEBUG",
        "TRACE"
    ],
    "loggers": {
        "ROOT": {
            "configuredLevel": "INFO",
            "effectiveLevel": "INFO"
        },
        ......
    }
}
```

6. /mappings

查看所有 URL 映射，即所有 @RequestMapping 路径的整理列表。访问 /actuator/mappings 获取系统中所有 URL 映射信息。示例数据如下：

```
{
    "/webjars/**": {
        "bean": "resourceHandlerMapping"
    },
    "/**": {
        "bean": "resourceHandlerMapping"
    },
    "/**/favicon.ico": {
        "bean": "faviconHandlerMapping"
    },
    "{[/httpTest],methods=[POST]}": {
        "bean": "requestMappingHandlerMapping",
```

```
        "method": "public com.ksb.ksb_with_security.controller.HttpTestResult
            com.ksb.ksb_ with_security.controller.HttpTestController.doTest(com.
            ksb.ksb_with_ security.controller.HttpTestRequest,org.springframework.
          validation. BindingResult)"
    },
    ...
}
```

7. /metrics

访问 /actuator/metrics 获取系统度量指标信息，示例数据如下：

```
{
    "names": [
        "data.source.active.connections",
        "jvm.buffer.memory.used",
        "jvm.memory.used",
        "jvm.buffer.count",
        "http.server.requests",
        "logback.events",
        "process.uptime",
        "jvm.memory.committed",
        "data.source.max.connections",
        "system.load.average.1m",
        "jvm.buffer.total.capacity",
        "jvm.memory.max",
        "system.cpu.count",
        "process.start.time",
        "data.source.min.connections"
    ]
}
```

8. /metrics/{name}

对应访问 names 中指标，可以获取具体的指标信息。例如要获取 jvm.memory.used 信息，访问 /actuator/metrics/jvm.memory.used 即可获取。输出的数据结构如下：

```
{
    "name": "jvm.memory.used",
    "measurements": [
        {
            "statistic": "Value",
            "value": 413923152
        }
    ],
    "availableTags": [
        {
            "tag": "area",
            "values": [
                "heap",
                "heap",
                "heap",
                "nonheap",
                "nonheap",
```

```
                "nonheap"
            ]
        },
        {
            "tag": "id",
            "values": [
                "PS Old Gen",
                "PS Survivor Space",
                "PS Eden Space",
                "Code Cache",
                "Compressed Class Space",
                "Metaspace"
          ]
        }
    ]
}
```

9. /scheduledtasks

访问 /actuator/scheduledtasks 获取系统中定时任务的信息，示例数据如下：

```
{
    "cron": [
        {
            "runnable": {
                "target": "com.ksb.ksb_with_security.schedule.DemoSchedule.job1"
            },
            "expression": "0/10 * * * * *"
        }
    ],
    "fixedDelay": [

    ],
    "fixedRate": [
        {
            "runnable": {
                "target": "com.ksb.ksb_with_security.schedule.DemoSchedule.job2"
            },
            "initialDelay": 0,
            "interval": 3000
                }
    ]
}
```

上面的输出数据对应的定时任务的代码如下：

```
@Component
@EnableScheduling
class DemoSchedule {


    @Scheduled(cron = "0/10 * * * * *")
    fun job1() {
```

```
        ...
    }

    @Scheduled(fixedRate = 3000)
    fun job2() {
        ...
    }

}
```

10. /threaddump

访问 /actuator/threaddump 可以获取系统线程转储信息，示例数据如下：

```
{
    "threads": [
        {
            "threadName": "SimplePauseDetectorThread_2",
            "threadId": 113,
            "blockedTime": -1,
            "blockedCount": 2,
            "waitedTime": -1,
            "waitedCount": 424974,
            "lockName": null,
            "lockOwnerId": -1,
            "lockOwnerName": null,
            "inNative": false,
            "suspended": false,
            "threadState": "TIMED_WAITING",
            "stackTrace": [
                {
                    "methodName": "sleep",
                    "fileName": "Thread.java",
                    "lineNumber": -2,
                    "className": "java.lang.Thread",
                    "nativeMethod": true
                },
                ...]
}
```

11. /trace

访问 http://127.0.0.1:8008/actuator/trace 可以获取系统请求跟踪信息，例如下面的数据就是上面小节中我们访问 http://127.0.0.1:8008/actuator/scheduledtasks 获取系统中定时任务信息请求的跟踪记录：

```
{
    "traces": [
        ...
        {
            "timestamp": "2018-02-06T16:54:38.860+0000",
            "info": {
```

```
                "method": "GET",
                "path": "/actuator/scheduledtasks",
                "headers": {
                    "request": {
                        "host": "127.0.0.1:8008",
                        "connection": "keep-alive",
                        "cache-control": "max-age=0",
                        "user-agent": "Mozilla/5.0 (Macintosh; Intel Mac OS X 10
                            _12_1) AppleWebKit/537.36 (KHTML, like Gecko) Chrome/
                            63.0.3239.132 Safari/537.36",
                        "upgrade-insecure-requests": "1",
                        "accept": "text/html,application/xhtml+xml,application/
                            xml;q=0.9,image/webp,image/apng,*/*;q=0.8",
                        "referer": "http://127.0.0.1:8008/login",
                        "accept-encoding": "gzip, deflate, br",
                        "accept-language": "zh-CN,zh;q=0.9,en;q=0.8",
                        "cookie": "cna=dHZeEm9PhWkCAbR/ddciOmfw; ctoken=9AZWbDTWL
                            BdhfrUHF99zwelare;…
                    },
                    "response": {
                        "X-Content-Type-Options": "nosniff",
                        "X-XSS-Protection": "1; mode=block",
                        "Cache-Control": "no-cache, no-store, max-age=0, must-
                            revalidate",
                        "Pragma": "no-cache",
                        "Expires": "0",
                        "X-Frame-Options": "DENY",
                        "Content-Type": "application/vnd.spring-boot.actuator.v2+json;
                            charset=UTF-8",
                        "Transfer-Encoding": "chunked",
                        "Date": "Tue, 06 Feb 2018 16:54:38 GMT",
                        "status": "200"
                    }
                },
                "timeTaken": "4"
            }
        },
        ...
        }
    ]
}
```

可以看出，上面的数据中包含了完整的请求头与响应头等信息。

18.3.2　启用和禁用端点

在 Spring Boot 1.5.x 中默认情况下，所有端点（除了 /shutdown）都是启用的。禁用所有端点的配置如下：

```
endpoints.enabled=false
```

禁用某个特定的端点的配置如下：

```
endpoints.endpoint-id.enabled=false
```

禁用后，再次访问 URL 时，会出现 404 错误。

在 Spring Boot 2.0 中，通过如下配置：

```
management.endpoints.enabled-by-default=false
```

禁用所有 Actuator 端点。禁用后，访问 http://127.0.0.1:8008/actuator，只有如下信息：

```
{
    "_links": {
        "self": {
            "href": "http:// 127.0.0.1:8008/actuator",
            "templated": false
        }
    }
}
```

如果是默认启用所有 Actuator 端点，但是想要禁用某些端点信息，可以配置如下：

```
management.endpoint.beans.enabled=false
management.endpoint.info.enabled=false
management.endpoint.health.enabled=false
```

这样再访问 http://127.0.0.1:8008/actuator 时，将不会看到 /beans、/info、/health 的端点信息。另外，可以通过 application.properties 中的属性定制 Actuator 的使用。完整的 Actuator 配置属性列表参考 application.properties 中的如下部分：

```
# ----------------------------------------
# ACTUATOR PROPERTIES
# ----------------------------------------
```

这些配置属性都在 management.* 命名空间下。

提示 本节的实例工程代码地址为 https://github.com/KotlinSpringBoot/ksb_with_security/tree/front_back_end_2018.2.2。

更多关于 Spring Boot 2.0 中的 Actuator 的介绍请参考：https://docs.spring.io/spring-boot/docs/2.0.x/actuator-api/html/。

18.4 自定义 Actuator 端点

Spring Boot Actuator 模块提供了灵活的接口，方便我们自己定制监控端点。例如 Endpoint、PublicMetrics、HealthIndicator、CounterService、GaugeService 接口等。为了跟下一小节的 Spring Boot Admin（目前仅支持 Spring Boot >=1.5.9.RELEASE and <2.0.0.M1 版本）衔接，本节基于 Spring Boot 1.5.10 中的 Actuator 模块展开。

18.4.1　Endpoint 接口

SpringBoot 的 Endpoint 主要是用来监控应用服务的运行状况，并在 MVC 中集成以提供 HTTP 接口。内置的 Endpoint（比如 HealthEndpoint）会监控 disk 和 db 的状况，如图 18-1 所示。

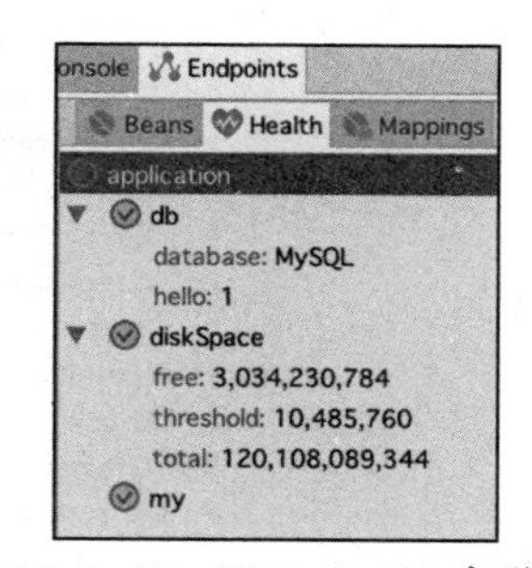

图 18-1　HealthEndpoint 会监控 disk 和 db 的状况

MetricsEndpoint 则会监控内存和 GC 等指标的状况：

```
{
    "mem": 1089990,
    "mem.free": 86536,
    "processors": 4,
    "instance.uptime": 796368,
    "uptime": 829333,
    "systemload.average": 5.74365234375,
    "heap.committed": 968704,
    "heap.init": 131072,
    "heap.used": 881143,
    "heap": 2097152,
    ...
}
```

Endpoint 的接口协议如下：

```
public interface Endpoint<T> {
    String getId();
    boolean isEnabled();
    boolean isSensitive();
    T invoke();
}
```

其中的方法说明如下：

- getId()——返回端点 ID，名称类似于 healthEndpoint、beansEndpoint、metricsEndpoint、traceEndpoint 等。
- isEnabled()——是否启用该端点。
- isSensitive()——用于权限控制。
- invoke()——调用端点核心逻辑，返回监控的数据内容。

Endpoint 对象 Bean 的加载是依靠 spring.factories 实现的，spring-boot-actuator 包下的 META-INF/spring.factories 配置了 EndpointAutoConfiguration，如下所示：

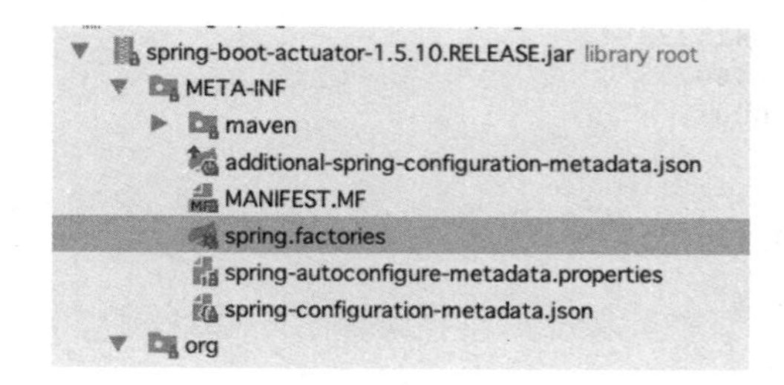

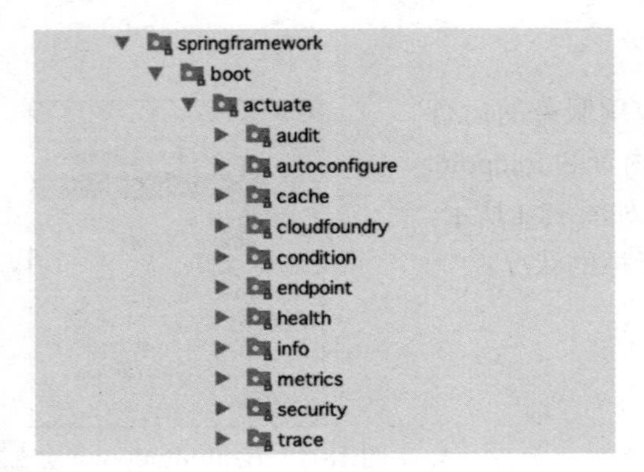

spring.factories 文件中的内容参考框架源码。其中的 EndpointAutoConfiguration 会自动注入必要的 Endpoint。下面来具体介绍如何自定义实现 Actuator 监控端点。

18.4.2 实现 Endpoint 接口

下面来自定义一个显示当前 Spring Boot 应用程序运行机器的信息。这个端点 ID 是 /serverEndpoint，输出的数据结构是 Map<String，Map<String，String>>，示例数据如下：

```
// 20180207113759
// http:// 127.0.0.1:8010/serverEndpoint

{
    "ServerInfo": {
        "hostAddress": "127.0.0.1",
        "hostName": "jacks-MacBook-Air.local",
        "OS": "Mac OS X"
    },
    "DiskInfo": {
        "freeSpace": "3143M",
        "usableSpace": "2893M",
        "totalSpace": "114544M"
    },
    "MemInfo": {
        "totalPhysicalMemorySize": "8192M",
        "freePhysicalMemorySize": "84M",
        "committedVirtualMemorySize": "6225M",
        "freeSwapSpaceSize": "1314M",
        "totalSwapSpaceSize": "7168M",
        "processCpuLoad": "0.1716696728153335",
        "systemCpuLoad": "0.8331151832460733",
        "processCpuTime": "1401557475000",
        "arch": "x86_64",
        "availableProcessors": "4",
        "systemLoadAverage": "5.63134765625",
        "version": "10.12.1"
    }
}
```

完整的实现代码在 ServerEndpoint.kt 中，可参考示例工程源代码。

启动应用，通过 http://127.0.0.1:8010/actuator 可以看到 /serverEndpoint 端点的信息：

```
{
    "links": [
        {
            "rel": "self",
            "href": "http:// 127.0.0.1:8010/actuator"
        },
        {
            "rel": "metrics",
            "href": "http:// 127.0.0.1:8010/metrics"
        },
        …
        {
            "rel": "serverEndpoint",
            "href": "http:// 127.0.0.1:8010/serverEndpoint"
        },
        …
    ]
}
```

访问其中的 http://127.0.0.1:8010/serverEndpoint，可以看到输出数据，如下：

```
// 20180207114245
// http://127.0.0.1:8010/serverEndpoint

{
  "ServerInfo": {
    "hostAddress": "127.0.0.1",
    "hostName": "jacks-MacBook-Air.local",
    "OS": "Mac OS X"
  },
  "DiskInfo": {
    "freeSpace": "2121M",
    "usableSpace": "1871M",
    "totalSpace": "114544M"
  },
  "MemInfo": {
    "totalPhysicalMemorySize": "8192M",
    "freePhysicalMemorySize": "61M",
    "committedVirtualMemorySize": "6226M",
    "freeSwapSpaceSize": "1566M",
    "totalSwapSpaceSize": "8192M",
    "processCpuLoad": "0.19684252441458763",
    "systemCpuLoad": "0.7204906900328587",
    "processCpuTime": "1626221354000",
    "arch": "x86_64",
    "availableProcessors": "4",
    "systemLoadAverage": "5.2431640625",
    "version": "10.12.1"
  }
}
```

18.4.3 继承 AbstractEndpoint 抽象类

Spring Boot Actuator 内置的 /env 端点实现代码在 EnvironmentEndpoint 中。其中的关键

代码如下：

```
@ConfigurationProperties(prefix = "endpoints.env")
public class EnvironmentEndpoint extends AbstractEndpoint<Map<String, Object>> {
    …
        @Override
        public Map<String, Object> invoke() {
            Map<String, Object> result = new LinkedHashMap<String, Object>();
        …
        }
    ...
}
```

可以看出，EnvironmentEndpoint 是继承 AbstractEndpoint 抽象类，重写 invoke() 方法来实现的。所以，我们也可以通过这种方法来自定义端点。

下面实现一个显示 Spring Boot 应用中所有端点信息（类似 /actuator 功能）的 /showEndpoints。这个端点输出的数据结构如下：

```
{
    "serverEndpoint": {
        "id": "serverEndpoint",
        "enabled": true,
        "sensitive": false
    },
    "showEndpoints": {
        "id": "showEndpoints",
        "enabled": true,
        "sensitive": false
    },
    ..
}
```

首先，声明一个 ShowEndpoints 类，它继承 AbstractEndpoint 抽象类并实现 ApplicationContextAware 接口：

```
@Component
class ShowEndpoints :
        AbstractEndpoint<MutableMap<String, MyEndpoint>?>("showEndpoints"),
        ApplicationContextAware {

    val log = LoggerFactory.getLogger(ShowEndpoints::class.java)
    private var applicationContext: ApplicationContext? = null

    @Throws(BeansException::class)
    override fun setApplicationContext(applicationContext: ApplicationContext) {
        this.applicationContext = applicationContext
    }
    ...
}
```

Spring 容器会检测容器中的所有 Bean，如果发现某个 Bean 实现了 ApplicationContext-

Aware 接口，Spring 容器会在创建该 Bean 之后，自动调用该 Bean 的 setApplicationContext-Aware() 方法，将容器对象本身作为 applicationContext 实例变量参数传给该方法。

接下来可以通过该 applicationContext 实例变量来访问容器本身。使用 Application-Context 对象调用 getBeansOfType（Class<T>type）获取当前 Spring Boot 应用程序中所有的 Endpoint 接口类型和 MvcEndpoint 接口类型的 Bean。相关代码是：

```
val endpoints = this.applicationContext?.getBeansOfType(Endpoint::class.java)
val mvcEndpoints = applicationContext?.getBeansOfType(MvcEndpoint::class.java)
```

完整的实现代码可以参考示例工程源代码 ShowEndpoints.kt。

18.4.4　实现健康指标接口 HealthIndicator

下面在 /health 端点中自定义健康信息 myCustome，输出的数据如下：

```
// 20180207120100
// http://127.0.0.1:8010/health
{
  "status": "UP",
  "myCustom": {
    "status": "UP",
    "imageRepository.selectTest": 0
  },
  "diskSpace": {
    "status": "UP",
    "total": 120108089344,
    "free": 3034247168,
    "threshold": 10485760
  },
  "db": {
    "status": "UP",
    "database": "MySQL",
    "hello": 1
  }
}
```

只需要实现 HealthIndicator 接口即可。HealthIndicator 接口协议如下：

```
package org.springframework.boot.actuate.health;
public interface HealthIndicator {
    Health health();
}
```

具体的实现代码如下：

```
@Component
class MyCustomHealthIndicator : HealthIndicator {

    override fun health(): Health {
        val errorCode = check()          // 健康检查方法示例
        return if (errorCode != 0) {
            Health.down().withDetail("Error Code", errorCode).build()
        } else Health
                .up()
                .withDetail("imageRepository.selectTest", errorCode)
```

```
            .build()                       // 返回 selectTest 健康信息
    }

    @Autowired lateinit var imageRepository: ImageRepository
    // 健康检查方法逻辑
    private fun check(): Int {
        return imageRepository.selectTest()
    }
}
```

其中，imageRepository.selectTest() 方法代码是

```
@Query(value = "select 0", nativeQuery = true)
fun selectTest():Int
```

18.4.5 实现度量指标接口 PublicMetrics

Actuator 中提供的 PublicMetrics 接口实现类有：

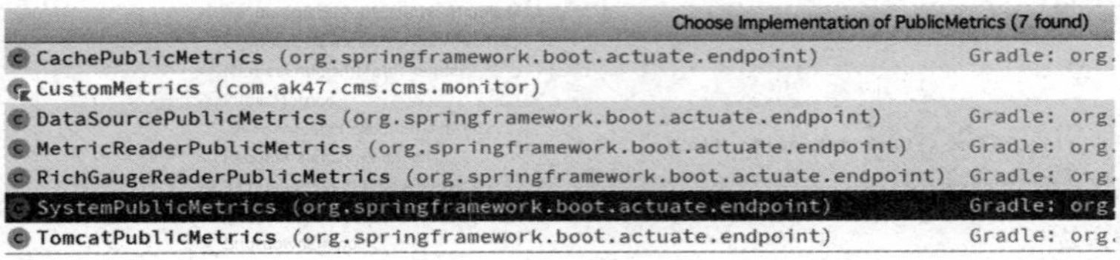

Choose Implementation of PublicMetrics (7 found)

- CachePublicMetrics (org.springframework.boot.actuate.endpoint) Gradle: org.
- CustomMetrics (com.ak47.cms.cms.monitor)
- DataSourcePublicMetrics (org.springframework.boot.actuate.endpoint) Gradle: org.
- MetricReaderPublicMetrics (org.springframework.boot.actuate.endpoint) Gradle: org.
- RichGaugeReaderPublicMetrics (org.springframework.boot.actuate.endpoint) Gradle: org.
- SystemPublicMetrics (org.springframework.boot.actuate.endpoint) Gradle: org.
- TomcatPublicMetrics (org.springframework.boot.actuate.endpoint) Gradle: org.

TomcatPublicMetrics 提供了 Tomcat 的数据统计的 PublicMetrics 的实现 . 该类实现了 PublicMetrics, ApplicationContextAware 接口。DataSourcePublicMetrics 则提供了数据源状态的监控统计。其中, Actuator 内置提供的 /metrics 端点中跟系统 System 指标相关的实现类是 SystemPublicMetrics。SystemPublicMetrics 类实现了 PublicMetrics 接口。PublicMetrics 接口协议如下：

```
import org.springframework.boot.actuate.metrics.Metric;
public interface PublicMetrics {
    Collection<Metric<?>> metrics();
}
```

SystemPublicMetrics 的成员方法如下：

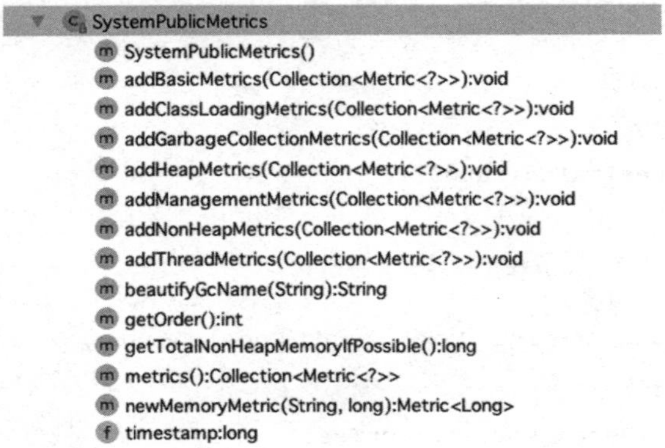

SystemPublicMetrics

- SystemPublicMetrics()
- addBasicMetrics(Collection<Metric<?>>):void
- addClassLoadingMetrics(Collection<Metric<?>>):void
- addGarbageCollectionMetrics(Collection<Metric<?>>):void
- addHeapMetrics(Collection<Metric<?>>):void
- addManagementMetrics(Collection<Metric<?>>):void
- addNonHeapMetrics(Collection<Metric<?>>):void
- addThreadMetrics(Collection<Metric<?>>):void
- beautifyGcName(String):String
- getOrder():int
- getTotalNonHeapMemoryIfPossible():long
- metrics():Collection<Metric<?>>
- newMemoryMetric(String, long):Metric<Long>
- timestamp:long

下面分别对上述 SystemPublicMetrics 里的列表中的成员方法进行简介。

1. addBasicMetrics()

代码说明如下：

```
protected void addBasicMetrics(Collection<Metric<?>> result) {
    //NOTE: ManagementFactory must not be used here since it fails on GAE
    Runtime runtime = Runtime.getRuntime();
    //1. 内存使用统计：mem = 总内存 + 堆外内存使用量
    result.add(newMemoryMetric("mem",
            runtime.totalMemory() + getTotalNonHeapMemoryIfPossible()));
    //2. 可用内存统计 mem.free  = 可用内存
    result.add(newMemoryMetric("mem.free", runtime.freeMemory()));
    //3. 处理器核数 processors
    result.add(new Metric<Integer>("processors", runtime.availableProcessors()));
    //4. 运行时间统计 : instance.uptime = 当前时间 - 启动时间
    result.add(new Metric<Long>("instance.uptime",
            System.currentTimeMillis() - this.timestamp));
}
```

在内存使用情况统计中，堆外内存使用 getTotalNonHeapMemoryIfPossible() 方法的代码说明如下：

```
private long getTotalNonHeapMemoryIfPossible() {
    try {
        return ManagementFactory.getMemoryMXBean().getNonHeapMemoryUsage().getUsed();
    }
    catch (Throwable ex) {
        return 0;
    }
}
```

2. addManagementMetrics()

代码说明如下：

```
private void addManagementMetrics(Collection<Metric<?>> result) {
    try {
        //1. jvm 启动时间：uptime = 启动时间，单位 ms
        result.add(new Metric<Long>("uptime",
                ManagementFactory.getRuntimeMXBean().getUptime()));
        //2. 系统负载：systemload.average = 启动负载
        result.add(new Metric<Double>("systemload.average",
                ManagementFactory.getOperatingSystemMXBean().getSystemLoadAverage()));
        //3. jvm 的 Heap 堆内存监控统计
        addHeapMetrics(result);
        //4. 堆外内存的统计
        addNonHeapMetrics(result);
        //5. 线程的统计
        addThreadMetrics(result);
        //6. 类加载相关的统计
        addClassLoadingMetrics(result);
        //7. 垃圾回收的统计
```

```
        addGarbageCollectionMetrics(result);
    }
    catch (NoClassDefFoundError ex) {

    }
}
```

其中，jvm 的 Heap 堆内存的监控统计的代码说明如下：

```
protected void addHeapMetrics(Collection<Metric<?>> result) {
    MemoryUsage memoryUsage = ManagementFactory.getMemoryMXBean()
        .getHeapMemoryUsage();
    //1. 获得所提交的字节内存量 --> 这个内存量是保证 java 虚拟机使用的
    result.add(newMemoryMetric("heap.committed", memoryUsage.getCommitted()));
    //2. 获得 jvm 的初始化内存数，单位：字节 . 如果初始内存大小未定义，则此方法返回 -1
    result.add(newMemoryMetric("heap.init", memoryUsage.getInit()));
    //3. 获得内存的使用量
    result.add(newMemoryMetric("heap.used", memoryUsage.getUsed()));
    //4. 获得内存的最大值，返回 -1，如果为指定
    result.add(newMemoryMetric("heap", memoryUsage.getMax()));
}
```

堆外内存统计的代码说明如下：

```
private void addNonHeapMetrics(Collection<Metric<?>> result) {
    MemoryUsage memoryUsage = ManagementFactory.getMemoryMXBean()
        .getNonHeapMemoryUsage();
    result.add(newMemoryMetric("nonheap.committed", memoryUsage.getCommitted()));
    result.add(newMemoryMetric("nonheap.init", memoryUsage.getInit()));
    result.add(newMemoryMetric("nonheap.used", memoryUsage.getUsed()));
    result.add(newMemoryMetric("nonheap", memoryUsage.getMax()));
}
```

线程的统计代码说明如下：

```
protected void addThreadMetrics(Collection<Metric<?>> result) {
    ThreadMXBean threadMxBean = ManagementFactory.getThreadMXBean();
    //1. 获得 jvm 启动以来或者统计重置以来的最大值
    result.add(new Metric<Long>("threads.peak",
        (long) threadMxBean.getPeakThreadCount()));
    //2. 获得 daemon 线程的数量
    result.add(new Metric<Long>("threads.daemon",
        (long) threadMxBean.getDaemonThreadCount()));
    //3. 获得 jvm 启动以来被创建并且启动的线程数
    result.add(new Metric<Long>("threads.totalStarted",
        threadMxBean.getTotalStartedThreadCount()));
    //4. 获得当前存活的线程数包括 daemon，非 daemon 的
    result.add(new Metric<Long>("threads", (long) threadMxBean.getThreadCount()));
}
```

类加载相关的统计，代码说明如下：

```
protected void addClassLoadingMetrics(Collection<Metric<?>> result) {
    ClassLoadingMXBean classLoadingMxBean = ManagementFactory.getClassLoading
```

```
        MXBean();
    //1. 获得 jvm 目前加载的 class 数量
    result.add(new Metric<Long>("classes",
        (long) classLoadingMxBean.getLoadedClassCount()));
    //2. 获得 jvm 启动以来加载 class 的所有数量
    result.add(new Metric<Long>("classes.loaded",
        classLoadingMxBean.getTotalLoadedClassCount()));
    //3. 获得 jvm 卸载 class 的数量
    result.add(new Metric<Long>("classes.unloaded",
        classLoadingMxBean.getUnloadedClassCount()));
}
```

垃圾回收的统计，代码说明如下：

```
protected void addGarbageCollectionMetrics(Collection<Metric<?>> result) {
    //1. 获得 GarbageCollectorMXBean
    List<GarbageCollectorMXBean> garbageCollectorMxBeans = ManagementFactory
            .getGarbageCollectorMXBeans();
    //2. 遍历
    for (GarbageCollectorMXBean garbageCollectorMXBean : garbageCollectorMxBeans) {
        String name = beautifyGcName(garbageCollectorMXBean.getName());
        //2.1 获得 gc 的次数
        result.add(new Metric<Long>("gc." + name + ".count",
                garbageCollectorMXBean.getCollectionCount()));
        //2.2 获得 gc 的时间
        result.add(new Metric<Long>("gc." + name + ".time",
                garbageCollectorMXBean.getCollectionTime()));
    }
}
```

下面来自定义 CustomMetrics 实现，提供的功能在 metrics 度量指标中添加：

1）Spring 容器中注册 Bean 的数量 beanDefinitionCount。

2）使用 getBeanNamesForType 函数获取 Any 类型及其子类型的所有 Bean 的数量。

3）获取 @Controller 注解标注的 Bean 的数量。

实现代码如下：

```
@Component
class CustomMetrics
@Autowired
constructor(private val applicationContext: ApplicationContext) : PublicMetrics {
    override fun metrics(): Collection<Metric<*>> {
        val metrics = ArrayList<Metric<*>>()
        metrics.add(Metric<Int>("spring.bean.definitions", applicationContext.
            beanDefinitionCount))
        metrics.add(Metric<Int>("spring.beans", applicationContext.getBeanNames
            ForType(Any::class.java).size))
        metrics.add(Metric<Int>("spring.controllers", applicationContext.getBean
            NamesForAnnotation(Controller::class.java).size))
        return metrics
    }
}
```

启动应用，访问 http://127.0.0.1:8010/metrics，可以在输出结果中看到 spring.bean.definitions、spring.beans、spring.controllers 的数据：

```
{
    "mem": 1204876,
    "mem.free": 372465,
    "processors": 4,
    "instance.uptime": 4528996,
    "uptime": 4566507,
    "systemload.average": 8.40869140625,
    …
    "spring.bean.definitions": 298,
    "spring.beans": 312,
    "spring.controllers": 7,
    …
    "counter.status.200.sotuSearchByTypeJson": 1,
    "counter.status.200.sotu_gank_view": 1,
    "counter.status.200.sotuGankSearchJson": 1,
    "counter.status.200.sotuSearchJson": 8
}
```

18.4.6 统计方法执行数据

可以通过使用 AOP 的技术，实现 CounterService、GaugeService 接口中的方法完成对方法执行数据的监控统计。

CounterService 和 GaugeService 是在 org.springframework.boot.actuate.metrics 包下面定义的两个接口：

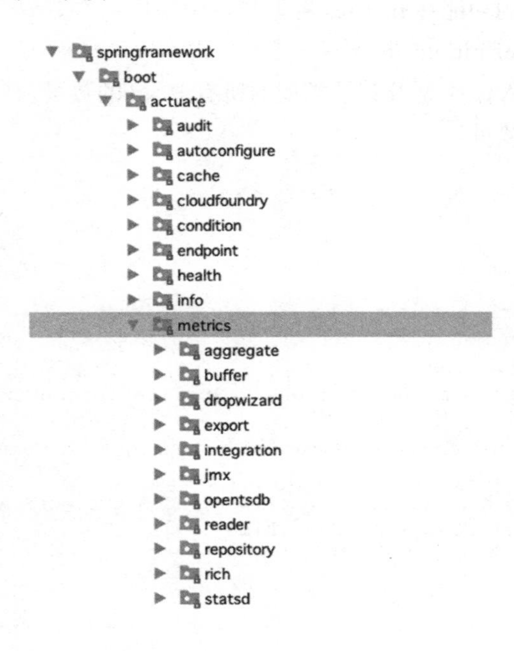

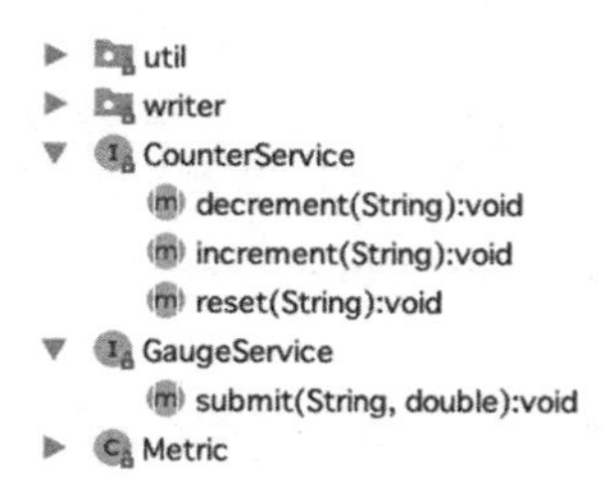

顾名思义，CounterService 用于计数（Counter）统计，GaugeService 用于具体指标数据的测量（Gauge）。

1. 统计方法的调用次数

统计 controller 包下面所有方法的调用次数，可以通过 AOP 实现：在调用指定的接口之前，首先调用 counterService.increment()，某个方法被调用之后，则对它的统计值 +1。具体的实现代码如下：

```
@Aspect
@Component
class ServiceMonitor {
    @Autowired
    lateinit var counterService: CounterService;

    @Before("execution(* com.ak47.cms.cms.controller..*.*(..))")
    fun countServiceInvoke(joinPoint: JoinPoint) {
        counterService.increment("${joinPoint.getSignature()}");
    }

}
```

上面的代码中，我们添加了 Aspect 组件 ServiceMonitor，表示在每个 Controller 的方法调用之前（通过 @Before 注解中配置的 execution 表达式实现），首先增加调用次数。

在 application.properties 中设置打开 AOP 功能：

```
spring.aop.auto=true
```

然后启动应用，通过浏览器访问 http://127.0.0.1:8010/metrics，可以发现 com.ak47.cms.cms.controller 包下面方法的调用次数已经被统计好了：

```
{
    ...
    "counter.Page com.ak47.cms.cms.controller.ImageController.sotuSearchFavoriteJson
        (int,int,String)": 1,
    "counter.ModelAndView com.ak47.cms.cms.controller.RouterController.tech_
        article_view(Model,HttpServletRequest)": 2,
    "counter.status.304.star-star": 1,
    "counter.ModelAndView com.ak47.cms.cms.controller.SearchKeyWordController.
        sotuView(Model,HttpServletRequest)": 2,
```

```
    ...
    "counter.boolean com.ak47.cms.cms.controller.ImageController.delete(long)": 10,
    "counter.ModelAndView com.ak47.cms.cms.controller.RouterController.sotu_
        view(Model,HttpServletRequest)": 3,
    "counter.Page com.ak47.cms.cms.controller.ImageController.sotuGankSearchJson
        (int,int,String)": 6,
    ...
    "counter.Page com.ak47.cms.cms.controller.SearchKeyWordController.sotuSearchJson
        (int,int,String)": 2

}
```

2. 统计方法的执行时长

如果希望统计每个接口的调用时长，则需要通过实现 GagueService 接口来完成。同样使用 AOP 实现——添加 @Around 环绕通知，步骤具体说明如下：

1) 在接口调用之前，记录开始时间戳 long start = System.currentTimeMillis()；

2) 在接口调用之后，计算结束时间戳 long end = System.currentTimeMillis()，然后相减（end-start）计算出耗费的时间，单位是 ms。

3) 最后，调用 GaugeService 的 submit 方法，提交统计数据：

```
GaugeService.submit(String metricName, double value)
```

在 ServiceMonitor 类中添加对应的监控代码如下：

```
@Aspect
@Component
class ServiceMonitor {

    ...

    @Autowired
    lateinit var gaugeService: GaugeService;

    @Around("execution(* com.ak47.cms.cms.service..*.*(..))")
    fun latencyService(pjp: ProceedingJoinPoint) {
        val start = System.currentTimeMillis()
        pjp.proceed()
        val end = System.currentTimeMillis()
        gaugeService.submit(pjp.getSignature().toString(), (end - start).toDouble())
    }

}
```

重启应用，浏览器访问 http://127.0.0.1:8010/metrics，可以看出 metrics 中已经统计了 com.ak47.cms.cms.service 包下面的类中方法调用执行的时间性能数据，如下所示：

```
{
    ...
    "gauge.response.loggers.name": 57.0,
    "gauge.response.dump": 630.0,
```

```
    "gauge.response.sotuGankSearchJson": 4580.0,
    "gauge.response.delete": 12.0,
    "gauge.response.searchKeyWordJson": 84.0,
    "gauge.response.listTechArticle": 98.0,
    "gauge.response.info": 7.0,
    "gauge.response.heapdump": 5.0,
    "gauge.response.sotuSearchJson": 1655.0,
    "gauge.response.star-star.favicon.ico": 9.0,
    "gauge.response.env": 5.0,
    "gauge.response.sotuSearchByTypeJson": 132.0,
    "gauge.response.logfile": 29.0,
    "gauge.response.doHuaBanImageCrawJob": 12890.0,
    "gauge.void com.ak47.cms.cms.service.CrawTechArticleService.doCrawITEye
        TechArticle()": 5.0,
    "gauge.void com.ak47.cms.cms.service.CrawTechArticleService.doCrawJianShuTech
        Article()": 1.0,
    "gauge.response.tech_article_view": 16.0,
    "gauge.response.search_keyword_view": 56.0,
    "gauge.void com.ak47.cms.cms.service.CrawImageService.doCrawHuaBanImages()":
        12872.0,
    ...
}
```

CounterService 和 GaugeService 这两个 Service 基本覆盖了大多数应用需求，如果需要监控其他的度量信息，则可以定制我们自己的 Metrics。

18.5 使用 Admin

在前面的小节中，我们看到的 Actuator 的端点数据信息都是 JSON 格式的数据，没有可视化的界面。本节介绍如何使用使用 Spring Boot Admin 把这些 Actuator 监控数据以可视化的界面展示出来。

18.5.1 Admin 简介

Spring Boot Admin 是 Spring Boot 应用程序运行状态监控和管理的后台界面。这个管理后台中提供了丰富的应用指标的监控、运行环境和运行状态的数据，还有动态修改系统日志等级的功能。Spring Boot Admin 是 codecentric 出品的社区开源项目。前端 UI 使用 AngularJs 开发。

Spring Boot Admin 为已注册的应用程序提供丰富的监控运维功能。这个功能列表如下：

- 显示健康状况。
- 显示应用运行时详细信息，如：JVM 和内存指标等。
- 计数器和测量指标。
- 数据源度量。
- 缓存度量。

- ❑ 跟踪和下载日志文件。
- ❑ 查看 jvm 系统和环境属性。

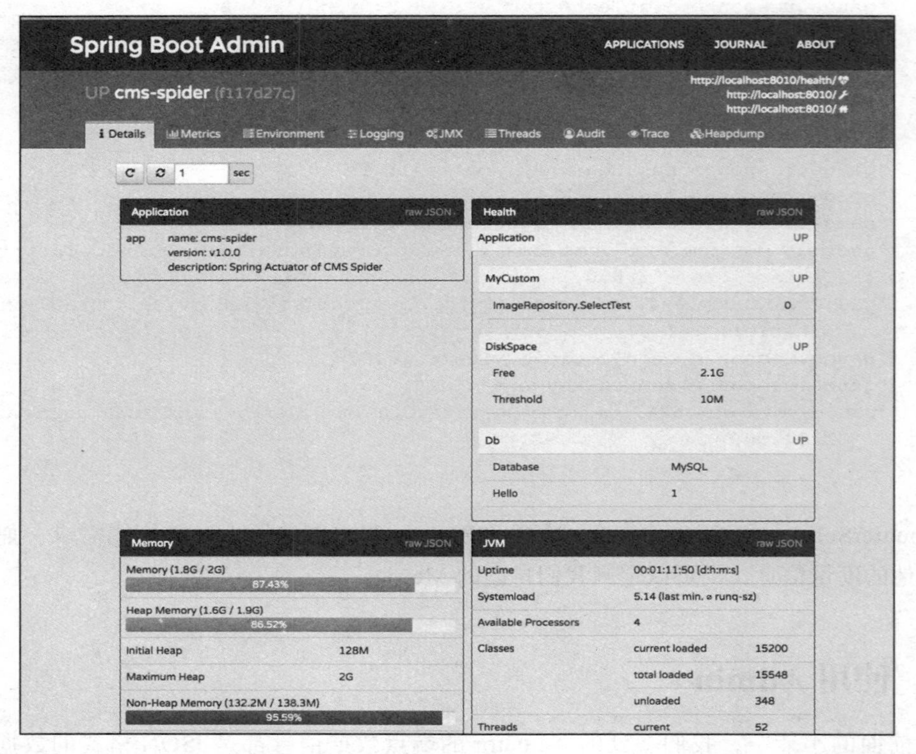

图 18-2 Details 下面提供了 Application、Health、Memory、JVM 等指标监控数据

- ❑ 一键管理 loglevel（目前仅用于 Logback）。
- ❑ 管理执行 JMX-beans。
- ❑ 查看线程转储。
- ❑ 查看跟踪信息。
- ❑ Hystrix-Dashboard 集成。
- ❑ 下载 heapdump。
- ❑ 状态更改通知（支持：电子邮件、Slack、Hipchat 等）。
- ❑ 状态更改事件日志（非永久性）。

例如，在 Details 下面提供了 Application、Health、Memory、JVM 等指标监控数据，如图 18-2 所示。

在 Metrics 下面则通过柱状图可视化展示度量指标的数据，如图 18-3 所示。

在 Logging 下面可以看到系统中日志的统一管理列表，支持动态修改类中的日志等级，如图 18-4 所示。

Spring Boot Admin　APPLICATIONS　JOURNAL　ABOUT

UP cms-spider (f117d27c)

http://localhost:8010/health/
http://localhost:8010/
http://localhost:8010/

Details　Metrics　Environment　Logging　JMX　Threads　Audit　Trace　Heapdump

Filter

classes　raw JSON

classes
15202

classes.loaded
15550

classes.unloaded
348

counter　raw JSON

counter.status.200.sotu_huaban_view
5

counter.status.200.heapdump
3

counter.Page com.ak47.cms.cms.controller.ImageController.sotuSearchByTypeJson(int,int,String,int)
5

counter.status.404.star-star
7

counter.Page com.ak47.cms.cms.controller.ImageController.sotuSearchFavoriteJson(int,int,String)
1

counter.String com.ak47.cms.cms.controller.CrawController.doGankImageCrawJob()
2

counter.ModelAndView com.ak47.cms.cms.controller.RouterController.sotu_huaban_view(Model,HttpServletRequest)
5

counter.status.304.star-star
1

counter.status.200.searchKeyWordJson
2

图 18-3　Metrics 下面则通过柱状图可视化展示度量指标的数据

Spring Boot Admin　APPLICATIONS　JOURNAL　ABOUT

UP cms-spider (f117d27c)

http://localhost:8010/health/
http://localhost:8010/
http://localhost:8010/

Details　Metrics　Environment　Logging　JMX　Threads　Audit　Trace　Heapdump

Filter by name ...　620/1012

ROOT	TRACE DEBUG INFO WARN ERROR OFF
com.ak47.cms.cms.CmsApplicationKt	TRACE DEBUG INFO WARN ERROR OFF
com.ak47.cms.cms.controller.ApiController	TRACE DEBUG INFO WARN ERROR OFF
com.ak47.cms.cms.monitor.ShowEndpoints	TRACE DEBUG INFO WARN ERROR OFF
com.ak47.cms.cms.schedule.CustomScheduleConfig	TRACE DEBUG INFO WARN ERROR OFF
com.ak47.cms.cms.service.CrawImageService	TRACE DEBUG INFO WARN ERROR OFF
com.ak47.cms.cms.service.CrawTechArticleService	TRACE DEBUG INFO WARN ERROR OFF
com.ak47.cms.cms.task.CenterBankRateTask	TRACE DEBUG INFO WARN ERROR OFF
com.ak47.cms.cms.task.FinanceInfoCalendarTask	TRACE DEBUG INFO WARN ERROR OFF
com.ak47.cms.cms.task.FocusLiveNewsTask	TRACE DEBUG INFO WARN ERROR OFF

show more

show all

Reference Guide - Sources - Code licensed under Apache License 2.0

图 18-4　Logging 下面可以看到系统中日志的统一管理列表

> 提示 更多关于 Spring Boot Admin 的介绍，请参考：https://codecentric.github.io/springboot-admin/1.5.7/。
> 源代码工程地址为 https://github.com/codecentric/spring-boot-admin。

使用 Spring Boot Admin 非常简单，只需要很少的步骤，即可拥有一个美观功能丰富的 Spring Boot 应用程序的监控管理后台。Spring Boot Admin 的配置和使用分为监控端（Server）和被监控端（Client）。目前，Spring Boot Admin 支持 Spring Boot 的版本是 Spring Boot>=1.5.9.RELEASE and <2.0.0.M1。本节实例项目中 Spring Boot 版本采用 1.5.10。

Spring Boot Admin 的工作原理是客户端 client 向服务端 server 注册数据：客户端启动后会实例化一个 RegistrationApplicationListener，该 Listener 默认会每隔 10s 到服务端去注册数据，如果数据已经存在，会执行更新 refresh。客户端 client 需要知道去哪里注册数据，所以要配置服务端 server 的地址。

18.5.2 创建 Admin Server 项目

创建 Admin Server 端项目如下：

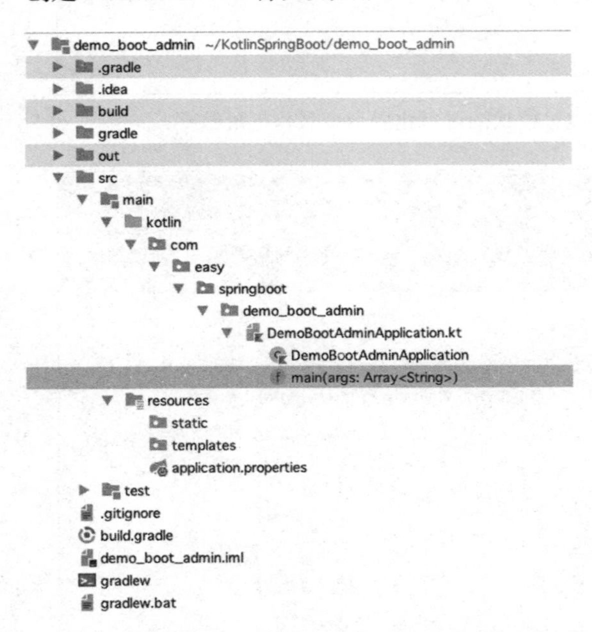

Admin Server 端项目没有任何逻辑代码，只需要完成下面三个简单的步骤。

1) 引入依赖：

```
compile('de.codecentric:spring-boot-admin-starter-server')
```

2) 在 Spring Boot 启动类上添加注解 @EnableAdminServer：

```
@SpringBootApplication
@EnableAdminServer
open class DemoBootAdminApplication

fun main(args: Array<String>) {
    SpringApplication.run(DemoBootAdminApplication::class.java, *args)
}
```

3)application.properties 配置如下：

```
server.port=9000
spring.application.name=Spring Boot Admin Web
spring.boot.admin.routes.endpoints=env,metrics,dump,jolokia,info,configprops,
    trace,logfile,refresh,flyway,liquibase,heapdump,loggers,auditevents,hystrix.
    stream,showEndpoints,serverEndpoint
```

18.5.3　在客户端使用 Admin Server

客户端只需要两个步骤使用 Admin Server。

1) 在 build.gradle 中添加 client 起步依赖：

```
compile group: 'de.codecentric', name: 'spring-boot-admin-starter-client', version:'1.5.7'
```

2) 然后，配置开放的 Actuator 端点和 Admin Server 的 URL 即可：

```
endpoints.enabled=true
endpoints.sensitive=false
management.security.enabled=false
#Config the Spring Boot Admin URL
spring.boot.admin.url=http:// localhost:9000
```

启动 Admin Server 应用和 Client 端应用，访问 http://localhost:9000，我们将看到 Spring Boot Admin 的管理界面，如图 18-5 所示。

点击 Details 进入详情页面，我们可以看到关于 cms-spider 应用的丰富指标数据和运行状态信息等。

例如，Health 健康状态数据如图 18-6 所示。

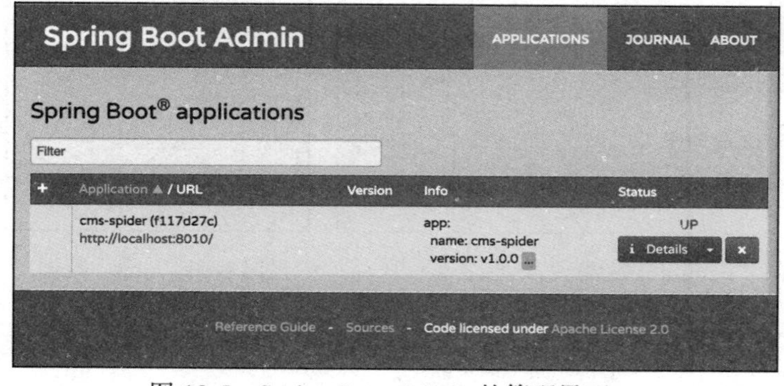

图 18-5　Spring Boot Admin 的管理界面

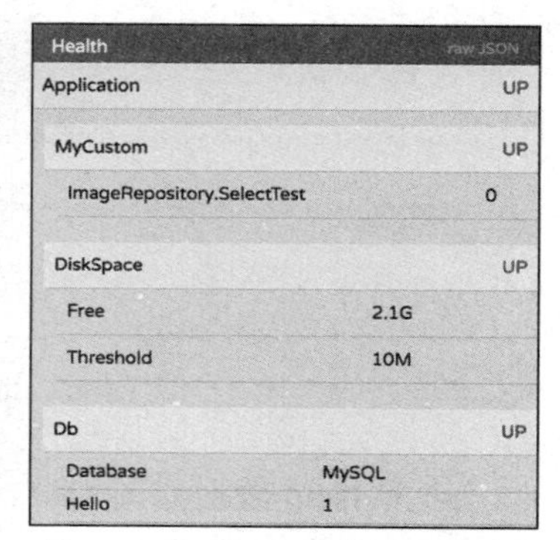

图 18-6　Health 健康状态数据

其中，MyCustom 是我们自定义的健康指标数据。

Memory 内存信息如图 18-7 所示。JVM 状态信息如图 18-8 所示。

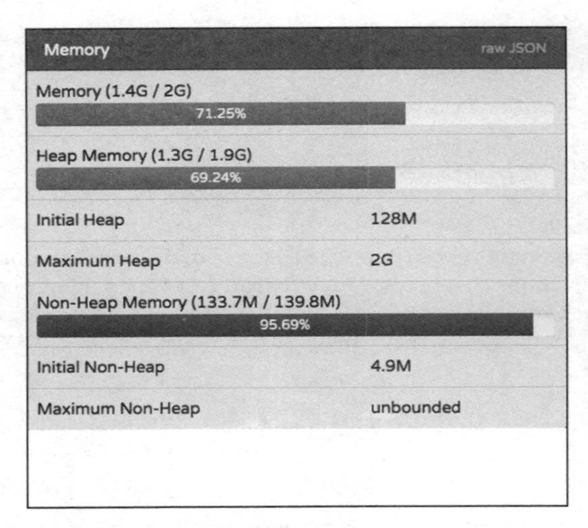

图 18-7　Memory 内存信息　　　　图 18-8　JVM 状态信息

class 与 counter 数据如图 18-9 所示。gauge 数据如图 18-10 所示。

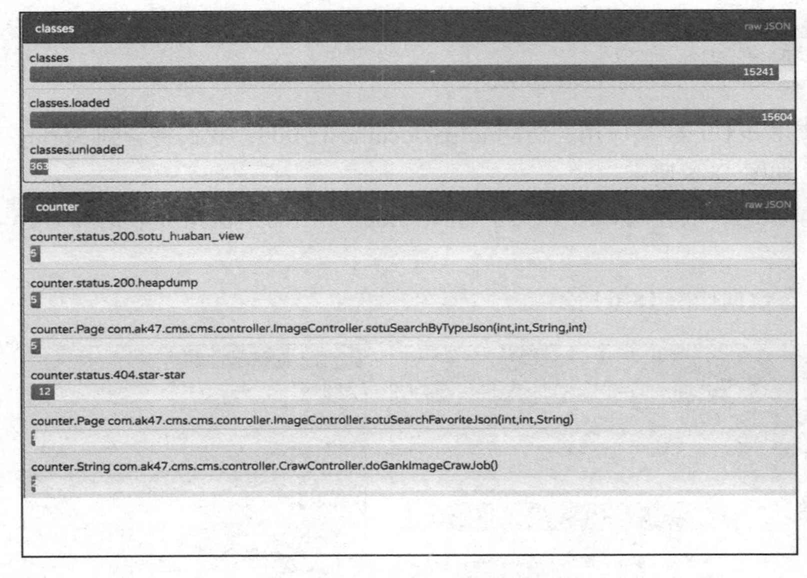

图 18-9　class 与 counter 数据

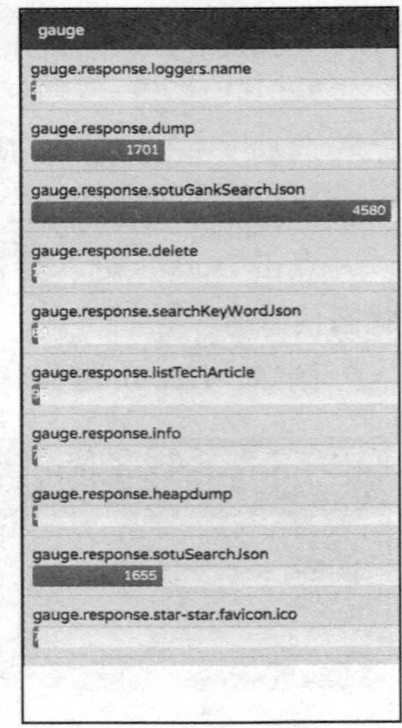

图 18-10　gauge 数据

gc 和 heap 数据如图 18-11 所示。系统负载、线程、uptime 数据如图 18-12 所示。自定义的 metrics 指标数据如图 18-13 所示。

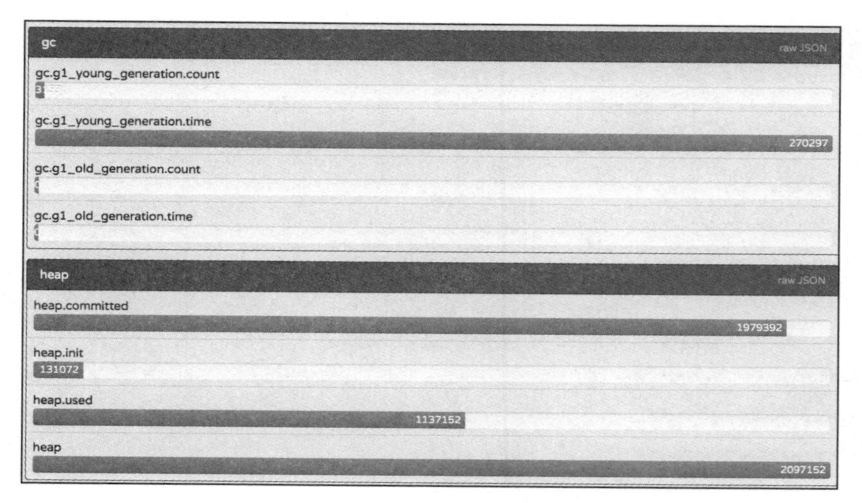

图 18-11　gc 和 heap 数据

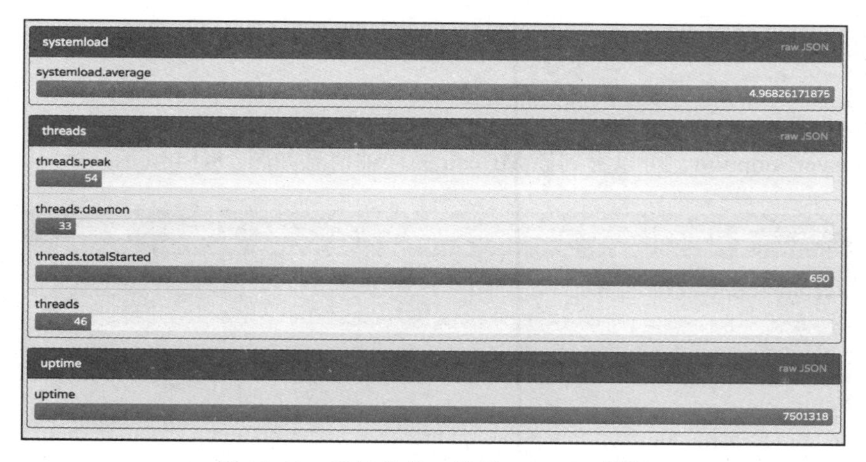

图 18-12　系统负载、线程、uptime 数据

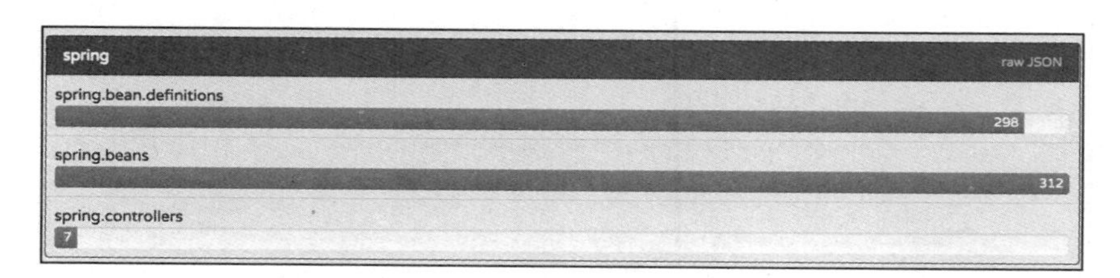

图 18-13　自定义的 metrics 指标数据

JMX 界面管理 MBeans，支持直接调用接口执行。例如，在 org.springframework.boot 域下的 MBeans 列表如下：

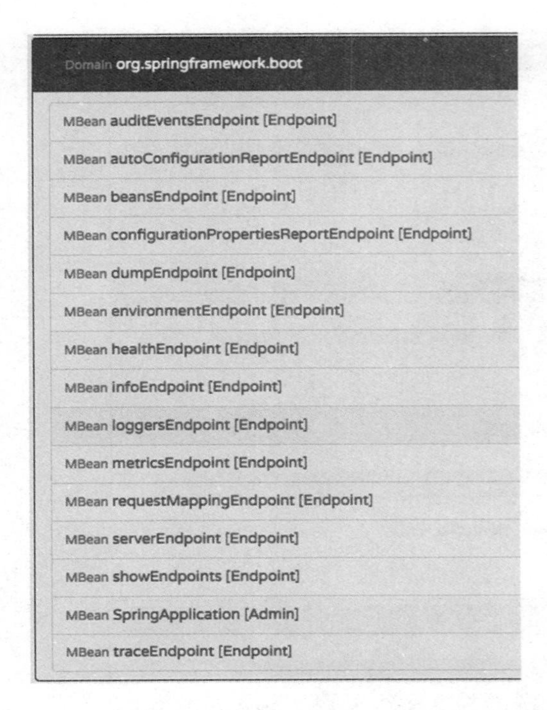

点开 serverEndpoint，可以看到该 MBean 的详细信息如图 18-14 所示。

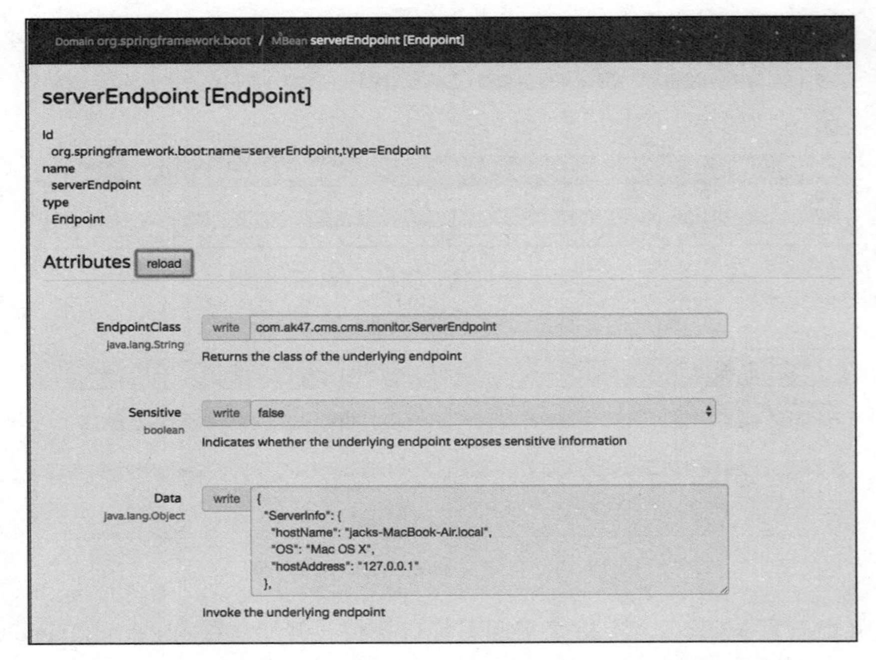

图 18-14 点开 serverEndpoint，可以看到该 MBean 的详细信息

而且可以点击“reload”实时更新端口输出数据，如图 18-15 所示。

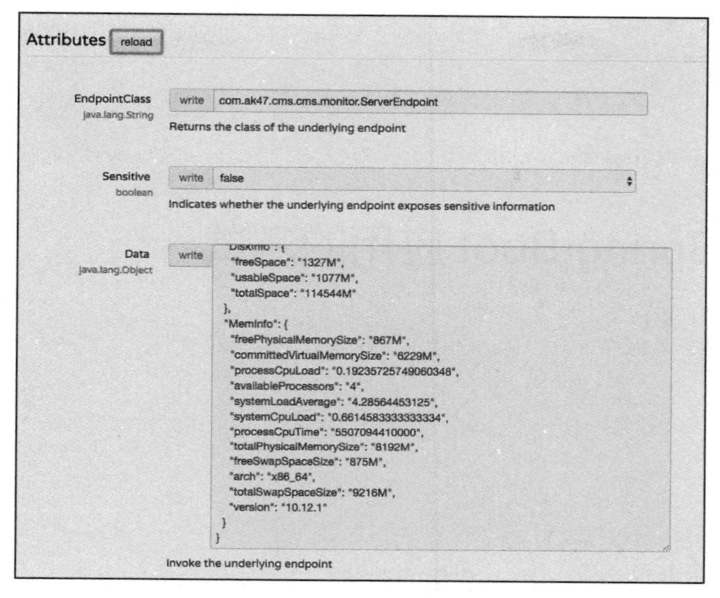

图 18-15　点击“reload”实时更新端口输出数据

还有 Threads、Trace 等详细信息，此处不一一介绍。请下载源码工程，部署查看效果。

提示　本节 Spring Boot 客户端应用源代码地址为 https://github.com/AK-47-D/cms-spider/tree/boot_admin_2018.2.4。

Spring Boot Admin Server 工程源代码地址为 https://github.com/KotlinSpringBoot/demo_boot_admin。

18.6　本章小结

Spring Boot Actuator 提供了强大的应用自省功能，提供了丰富的端点信息，覆盖 Spring Boot 应用程序运行的方方面面。同时，结合可视化的 Spring Boot Admin 管理界面，一切显得如此“高大上”。而在此过程中，我们只需要极简的几步配置即可完成这些事情。这正是 Spring Boot 的“初心”所在。

下章介绍 Spring Boot 应用的测试与部署。

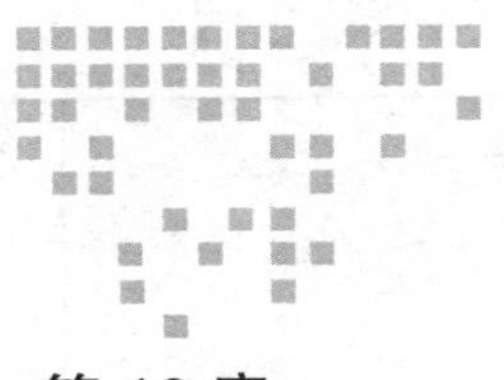

Chapter 19 第 19 章

Spring Boot 应用的测试

本章介绍如何进行 Spring Boot 应用的测试，这是质量保障的重点工作。我们在项目开发中使用分层架构，在测试中也进行分层测试。

19.1 准备工作

本节先来创建一个基于 Spring MVC、Spring Data JPA 的 Spring Boot，完成 Dao 层、Service 层、Controller 层代码的编写，为后面的测试代码的编写做准备。

使用 http://start.spring.io/ 创建项目、导入此 Gradle 项目到 IDEA 中。配置 Kotlin Compiler 版本与 Target JVM 版本。最后等待项目构建完毕。我们将得到一个初始 Spring Boot 工程。详细的代码参考本章给出的示例工程源码。

下面我们来详细讲解怎样针对 Spring Boot 项目进行分层测试。

19.2 分层测试

我们在开发阶段过程中，单元测试通常是必要的。Spring Boot 提供的 spring-boot-test 模块基于 spring-test 模块和 junit 框架，封装集成了功能强大的结果匹配校验器 assertj、hamcrest Matcher、Web 请求 Mock 对象、httpclient、JsonPath（测试 JSON 数据）、mockito、selenium 等。

测试代码通常放在 src/test 目录下，包目录规范是跟 src/main 目录保持一致。测试代码目录结构设计如下：

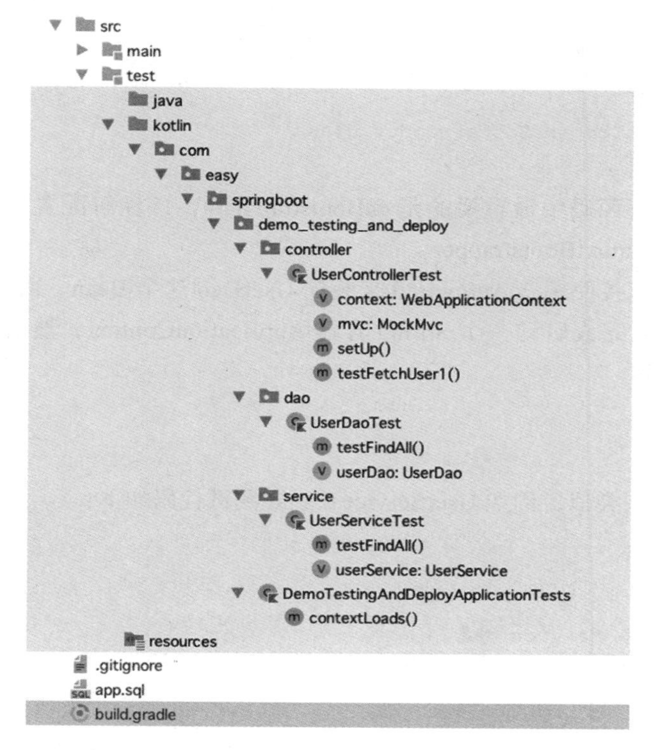

测试代码的分层逻辑与项目源代码中的 dao 层、service 层、controller 层各自对应。

下面我们来开发具体的测试类。

19.2.1 dao 层测试

在包 com.easy.springboot.demo_testing_and_deploy.dao 下面添加 UserDaoTest.kt 测试类，代码如下：

```
@RunWith(SpringRunner::class)
@SpringBootTest
class UserDaoTest {
    @Autowired lateinit var userDao: UserDao

    @Test
    fun testFindAll() {
        Assert.assertTrue(userDao.findAll().size == 2)
    }
}
```

其中，需要测试类上需要添加 @RunWith（SpringRunner.class）和 @SpringBootTest 注解。这里的 @RunWith 这里就不多做解释了，在 JUnit 中这个是最常用的注解。

@SpringBootTest 这个注解是 SpringBoot 项目测试的核心注解，标识该测试类以 SpringBoot 方式运行，该注解的定义如下：

```
@Target(ElementType.TYPE)
@Retention(RetentionPolicy.RUNTIME)
@Documented
@Inherited
@BootstrapWith(SpringBootTestContextBootstrapper.class)
public @interface SpringBootTest{}
```

在上面的 @SpringBootTest 注解源码中最重要的是 @BootstrapWith，该注解配置了测试类的启动核心类 SpringBootTestContextBootstrapper。

在 UserDaoTest 测试类中可以直接使用 @Autowired 来装配 UserDao 这个 Bean。而且，@SpringBootTest 注解会自动帮我们完成启动一个 Spring 容器 ApplicationContext，然后连接数据库，执行一套完整的业务逻辑。

19.2.2 service 层测试

service 层的代码测试类跟 dao 层类似，例如 UserServiceTest.kt 测试代码如下：

```
@RunWith(SpringRunner::class)
@SpringBootTest
class UserServiceTest {
    // 直接使用 @Autowired 注解注入 Service 对象
    @Autowired lateinit var userService: UserService

    @Test
    fun testFindAll() {
        Assert.assertTrue(userService.findAll().size == 2)
    }
}
```

19.2.3 使用 Mockito 测试 service 层代码

上面的测试代码是连接真实的数据库来执行真实的 dao 层数据库查询逻辑。而在实际开发的场景中，我们有时候需要独立于数据库进行 service 层逻辑的开发。这个时候就可以直接把数据库 dao 层代码 Mock 掉。例如在 UserService 中有一个 getOne() 方法，具体的实现代码如下：

```
interface UserService {
    ...
    fun getOne(id:Long):User?
}

@Service
class UserServiceImpl : UserService {
    @Autowired lateinit var userDao: UserDao
    ...

    override fun getOne(id: Long): User? {
        return userDao.getOne(id)
    }
}
```

下面，我们就使用 Mockito 来把 UserDao 层代码 Mock 掉。Mockito 主要用于 service 层的 Mock 测试。Mock 的对象一般是对 dao 层的依赖；另外就是别人的 Service 实现类。

新建测试类 MockUserServiceTest.kt 代码如下：

```
@RunWith(MockitoJUnitRunner::class)
class MockUserServiceTest {
    @Mock
    lateinit var mockUserDao: UserDao        //Mock 一个 DAO 层的接口
    @InjectMocks
    lateinit var userService: UserServiceImpl //Mock 一个 Service 的实现类，用
        @InjectMocks。注意这里是实现类 UserServiceImpl

    @Before
    fun setUp() {
        //initMocks 必须，否则 @Mock 注解无效
        MockitoAnnotations.initMocks(this)
    }

    @Test
    fun testGetOne() {
        val mockUser = User()
        mockUser.id = 101
        mockUser.username = "mockUser"
        mockUser.password = "123456"

        val roles = mutableSetOf<Role>()
        val r1 = Role()
        r1.role = "ROLE_USER"
        val r2 = Role()
        r1.role = "ROLE_ADMIN"
        roles.add(r1)
        roles.add(r2)
        mockUser.roles = roles
        //模拟 UserDao 对象
        `when`(mockUserDao.getOne(1)).thenReturn(mockUser)

        val u = userService.getOne(1)
        println(ObjectMapper().writeValueAsString(u))
        Assert.assertTrue(u?.password == "123456")
    }
}
```

该测试的执行 Runner 是 @RunWith（MockitoJUnitRunner::class），有两点需注意：

- 使用 @Mock 注解标记这个对象是被 Mock 的。
- 使用 @InjectMocks 注解标注一个实现类 UserServiceImpl，Mockito 会自动把 @Spy 或 @Mock 标注的 Mock 对象注入到实现类 UserServiceImpl 的方法执行中，相当于把实现类中的 UserDao 对象使用 mockUserDao 对象给“偷梁换柱”了。

运行上面的测试类，可以发现测试成功，如图 19-1 所示。

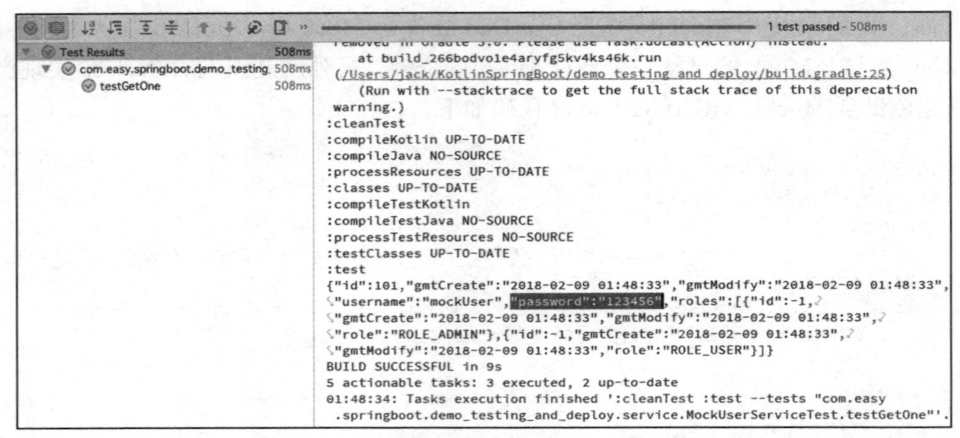

图 19-1 MockUserServiceTest 测试成功

在测试代码的打印日志中，输出的 getOne(1) 方法的返回对象是我们 Mock 的对象 mockUser：

```
{"id":101,"gmtCreate":"2018-02-09 01:48:33","gmtModify":"2018-02-09 01:48:33",
"username":"mockUser","password":"123456","roles":[{"id":-1,"gmtCreate":"
2018-02-09 01:48:33","gmtModify":"2018-02-09 01:48:33","role":"ROLE_ADMIN"},
{"id":-1,"gmtCreate":"2018-02-09 01:48:33","gmtModify":"2018-02-09 01:48:33",
"role":"ROLE_USER"}]}
```

 提示 更多关于 Mockito 的使用请参考官网文档：http://site.mockito.org/

19.2.4 controller 层测试

通过上面的实例，我们已经了解了在实际项目开发测试中对 dao 层代码和 service 层代码的测试，还学习了 Mockito 技术的相关内容。spring-boot-starter-test 中提供了对项目测试功能的强大支持，更难得的是其中增加了对 controller 层测试的支持。

下面我们来测试接口 http://127.0.0.1:8012/user/1。该接口的输出的 JSON 数据如下：

```
{
    "id": 1,
    "gmtCreate": "2018-02-08 12:58:14",
    "gmtModify": "2018-02-08 12:58:14",
    "username": "user",
    "password": "user",
    "roles": [
        {
          "id": 1,
          "gmtCreate": "2018-02-08 12:58:14",
          "gmtModify": "2018-02-08 12:58:14",
          "role": "ROLE_USER"
        }
    ]
}
```

UserControllerTest 测试代码如下：

```
@RunWith(SpringJUnit4ClassRunner::class)
@SpringBootTest
class UserControllerTest {

    @Autowired
    lateinit var context: WebApplicationContext
    lateinit var mvc: MockMvc
    @Before
    fun setUp() {
        mvc = MockMvcBuilders.webAppContextSetup(context).build()
    }

    @Test
    fun testFetchUser1() {
        mvc.perform(MockMvcRequestBuilders.get("/user/1")
                .contentType(MediaType.APPLICATION_JSON_UTF8)
                .accept(MediaType.APPLICATION_JSON))
                .andExpect(MockMvcResultMatchers.status().isOk)
                .andDo(MockMvcResultHandlers.print())
                .andExpect(MockMvcResultMatchers.content().string(Matchers.
                    containsString("""
                    "username":"user"
                """.trimIndent())))
                .andDo {
                    println("it.request.method=${it.request.method}")
                    println("it.response.contentAsString=${it.response.content
                        AsString}")
                }
                .andExpect(MockMvcResultMatchers.jsonPath("$.id", Matchers.equal
                    To(1)))
                .andExpect(MockMvcResultMatchers.jsonPath("$.roles[0].role",
                    Matchers.equalTo("ROLE_USER")))

    }
}
```

其中，MockMvc 是一个被 final 修饰的类型，该类无法被继承使用。这个类在包 org.springframework.test.web.servlet 下面，是 Spring 提供的模拟 SpringMVC 请求的实例类，该类由 MockMvcBuilders 通过 WebApplicationContext 实例进行创建。MockMvcBuilder 接口签名如下：

```
package org.springframework.test.web.servlet;
public interface MockMvcBuilder {
    MockMvc build();
}
```

上面的代码简单说明如下：

- Perform() 方法其实只是为了构建一个请求，并且返回 ResultActions 实例，使用该实例可以获取到请求的返回内容。
- MockMvcRequestBuilders 支持构建多种请求，如：Post、Get、Put、Delete 等常用的

请求方式，其中的参数 "/user/1" 是我们需要请求的本项目的相对路径，/ 是项目请求的根路径。另外，还可以调用 param() 方法用于在发送请求时携带参数。

- andExpect() 是 ResultActions 中成员，入参是 ResultMatcher 类型：ResultActions andExpect（ResultMatcher matcher）。在发送请求后对响应结果进行匹配校验时调用。其中 MockMvcResultMatchers 抽象类是一个静态工厂，用于生产 ResultMatcher 对象。MockMvcResultMatchers 中提供了丰富的匹配器。

19.2.5 JSON 接口测试

使用 JsonPath 我们可以像 JavaScript 语法一样方便地进行 JSON 数据返回的访问操作。例如下面的这两行代码：

```
.andExpect(MockMvcResultMatchers.jsonPath("$.id", Matchers.equalTo(1)))
.andExpect(MockMvcResultMatchers.jsonPath("$.roles[0].role", Matchers.equalTo
    ("ROLE_USER")))
```

这里的 Matchers 类是 org.hamcrest 包下面的类。org.hamcrest.Matchers 类提供了丰富的断言方法，这些方法的具体使用可以阅读 Matchers 类的源码深入了解。

其中，"$.id" 和 "$.roles[0].role" 就是 JsonPath 的表达式语法。

更多关于 JsonPath 的内容可以参考：https://github.com/json-path/JsonPath。

运行上面的测试代码，测试成功，如图 19-2 所示。

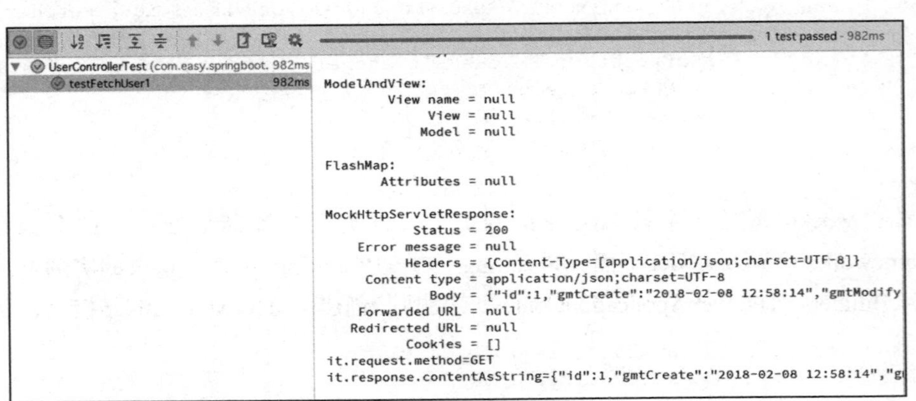

图 19-2 UserControllerTest 测试成功

使用命令 $ gradle test 可以一次性全部执行 src/test 目录下面的测试类。在 IDEA 中可以直接邮寄 src/test 目录，选择 Run > All Tests 执行所有测试类，如图 19-3 所示。

另外，Gradle Test 生成的测试报告在 build/reports/tests/test/index.html 中。测试报告的部分内容截图如图 19-4 所示。

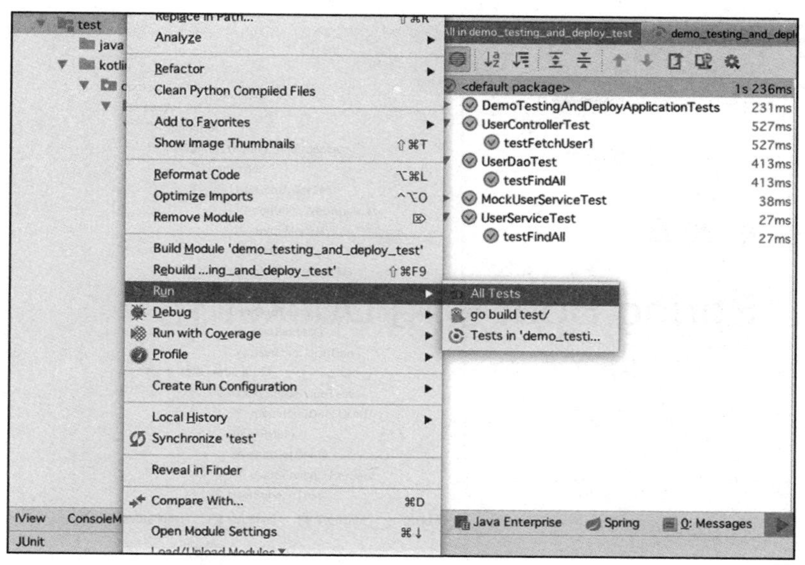

图 19-3　选择 Run > All Tests 执行所有测试类

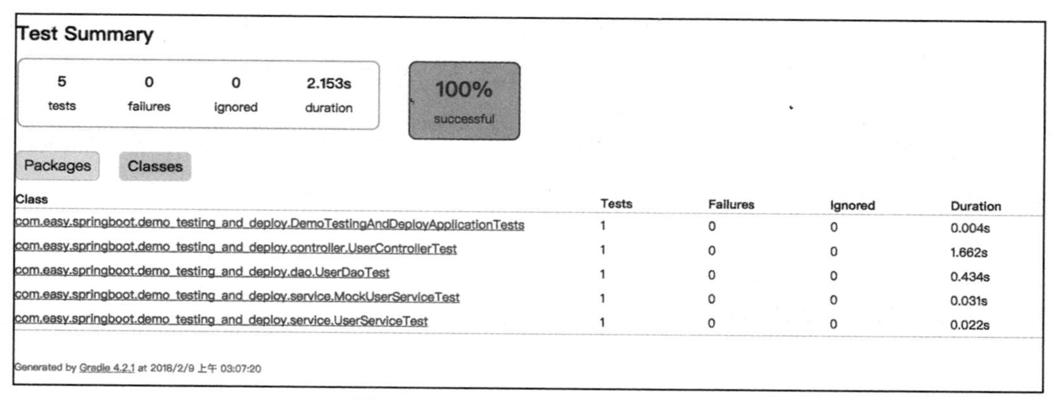
Test Summary

5	0	0	2.153s
tests	failures	ignored	duration

100%
successful

Packages　Classes

Class	Tests	Failures	Ignored	Duration
com.easy.springboot.demo_testing_and_deploy.DemoTestingAndDeployApplicationTests	1	0	0	0.004s
com.easy.springboot.demo_testing_and_deploy.controller.UserControllerTest	1	0	0	1.662s
com.easy.springboot.demo_testing_and_deploy.dao.UserDaoTest	1	0	0	0.434s
com.easy.springboot.demo_testing_and_deploy.service.MockUserServiceTest	1	0	0	0.031s
com.easy.springboot.demo_testing_and_deploy.service.UserServiceTest	1	0	0	0.022s

Generated by Gradle 4.2.1 at 2018/2/9 上午 03:07:20

图 19-4　测试报告中的 Summary

19.3　本章小结

本章介绍了 Spring Boot 项目如何测试。Spring Boot 应用对 Web 层测试提供强大的支持：采用 MockMvc 方式测试 Web 请求，根据传递的不同参数以及请求返回对象反馈信息进行验证测试。另外，针对 JSON 数据接口，使用 JsonPath 可以方便地进行 JSON 数据结果的校验。

提示　本章项目工程源代码：https://github.com/KotlinSpringBoot/demo_testing_and_deploy

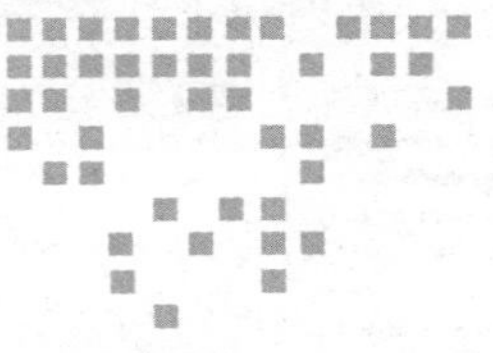

Chapter 20 第 20 章

Spring Boot 应用 Docker 化

前面的章节中，我们都是在 IDE 环境中开发运行测试 Spring Boot 应用程序。在开发测试发布整个软件生命周期的过程中，我们通常需要完成打包部署发布到日常、预发、线上机器运行等运维相关工作。

本章前半部分介绍 Spring Boot 应用的打包和部署，后半部分重点介绍如何使用 Docker 来构建部署运行 Spring Boot 应用。

20.1 Spring Boot 应用打包

1. 准备工作

首先，使用 http://start.spring.io/ 创建一个打包方式为 war 的 Spring Boot Kotlin 应用，采用 Gradle 构建。点击 Generate Project 等待创建完毕，下载 zip 包，导入 IDEA 中。可以看到，相比于项目打成 jar 包方式，打成 war 包的项目中多了一个用于初始化 Servlet 的 ServletInitializer 类。代码如下：

```
class ServletInitializer : SpringBootServletInitializer() {

    override fun configure(application: SpringApplicationBuilder) : SpringApplication
        Builder {
        return application.sources(DemoPackageAndDeployApplication::class.java)
    }

}
```

我们知道 Spring Boot 默认集成了内嵌 Web 容器（例如 Tomcat、Jetty 等），这个时候，Spring Boot 应用支持"一键启动"，像一个普通 Java 程序一样，从 main 函数入口开始启动。现在，我们将项目打包成 war 包，放到独立的 Web 容器中。

而如果我们这个 war 包中没有配置 Spring MVC 的 DispatcherServlet 的 web.xml 文件或者初始化 Servlet 的类，那么 Tomcat 就无法识别这个 war 包。这个时候，我们需要告诉 Tomcat 这个 war 包的启动入口。而 SpringBootServletInitializer 就是来完成这件事情的。

通过重写 configure（SpringApplicationBuilder）方法，使用 SpringApplicationBuilder 来配置应用程序的 sources 类。为了测试应用运行的效果，我们在 DemoPackageAndDeploy-Application.kt 中添加 HelloWorld REST 接口方便测试：

```
@SpringBootApplication
open class DemoPackageAndDeployApplication

fun main(args: Array<String>) {
    runApplication<DemoPackageAndDeployApplication>(*args)
}

@RestController
class HelloWorld {
    @GetMapping(value = ["", "/"])
    fun hello(): Map<String, Any> {
        val result = mutableMapOf<String, Any>()
        result["msg"] = "Hello,World"
        result["time"] = Date()
        return result
    }
}
```

2. 项目打包成可执行 jar

在 IDEA 的右边的 Gradle 工具栏中列出了 Gradle 构建项目的命令，如下：

```
▼ demo_package_and_deploy (auto-import enabled)
    ▶ Source Sets
    ▼ Tasks
        ▼ application
            bootRun
        ▼ build
            assemble
            bootJar
            bootWar
            build
            buildDependents
            buildNeeded
            classes
            clean
            jar
            testClasses
            war
```

我们可以直接点击 bootJar 把项目打成 jar 包。当然，在运维部署脚本中通常使用命令行：gradle bootJar。执行日志如下：

```
17:44:21: Executing task 'bootJar'...

:compileKotlin UP-TO-DATE
:compileJava NO-SOURCE
:processResources UP-TO-DATE
:classes UP-TO-DATE
:bootJar UP-TO-DATE

BUILD SUCCESSFUL in 1s
3 actionable tasks: 3 up-to-date
17:44:22: Task execution finished 'bootJar'.
```

执行完毕，我们可以在项目的 build/libs 目录下看到打好的 jar 包，如下所示：

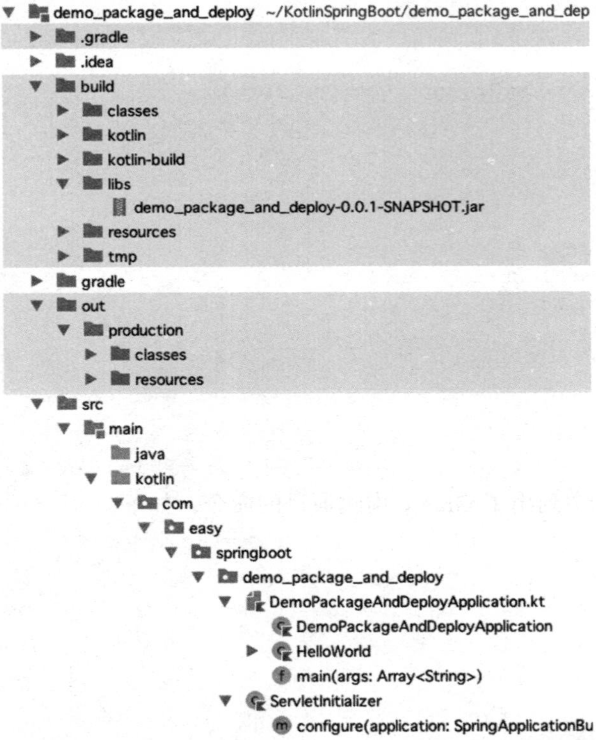

```
▼ demo_package_and_deploy ~/KotlinSpringBoot/demo_package_and_dep
  ▶ .gradle
  ▶ .idea
  ▼ build
    ▶ classes
    ▶ kotlin
    ▶ kotlin-build
    ▼ libs
        demo_package_and_deploy-0.0.1-SNAPSHOT.jar
    ▶ resources
    ▶ tmp
  ▶ gradle
  ▼ out
    ▼ production
      ▶ classes
      ▶ resources
  ▼ src
    ▼ main
        java
      ▼ kotlin
        ▼ com
          ▼ easy
            ▼ springboot
              ▼ demo_package_and_deploy
                ▼ DemoPackageAndDeployApplication.kt
                    DemoPackageAndDeployApplication
                  ▶ HelloWorld
                    main(args: Array<String>)
                ▼ ServletInitializer
                    configure(application: SpringApplicationBu
```

然后，我们就可以直接使用 java–jar 命令执行该 jar 包了

```
$ java -jar build/libs/demo_package_and_deploy-0.0.1-SNAPSHOT.jar
```

此时，我们浏览器访问 http://127.0.0.1:8080/，可以看到输出

```
{
    "msg": "Hello,World",
    "time": "2018-02-09T09:38:31.933+0000"
}
```

不过，使用 java –jar 命令行来启动系统的这种方式：

```
java -jar build/libs/demo_package_and_deploy-0.0.1-SNAPSHOT.jar
```

只要控制台关闭，服务就不能访问了。我们可以使用 nohup 与 & 命令让进程在后台运行：

```
nohup java -jar build/libs/demo_package_and_deploy-0.0.1-SNAPSHOT.jar &
```

3. 定制配置文件启动应用

我们也可以在启动的时候选择读取不同的配置文件。例如，在项目 src/main/resources 目录下面有不同环境的配置文件。如下所示：

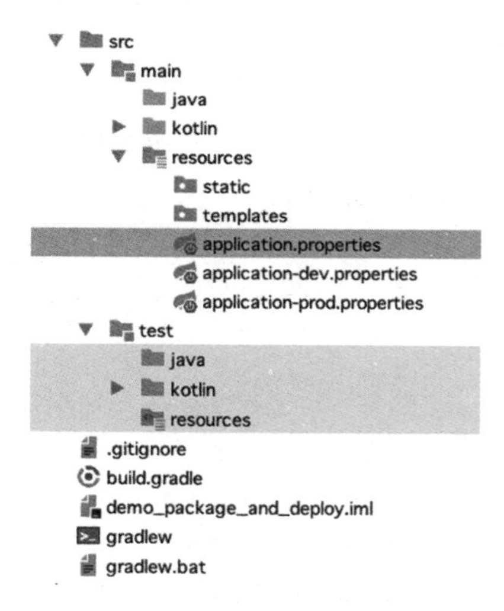

其中，application-dev.properties 中配置服务器端口号为 9000：

```
server.port=9000
```

执行 bootJar 重新打 jar 包，执行下面的命令：

```
java -jar
build/libs/demo_package_and_deploy-0.0.1-SNAPSHOT.jar
--spring.profiles.active=dev
```

可以看到应用成功启动，并监听 9000 端口：

```
...
o.s.b.w.embedded.tomcat.TomcatWebServer  : Tomcat started on port(s): 9000 (http)
    with context path ''
2018-02-09 18:18:47.336  INFO 69156 ---
[           main] .e.s.d.DemoPackageAndDeploy ApplicationKt : Started
DemoPackageAndDeployApplicationKtin 6.493 seconds (JVM running for 7.589)
```

4. 项目打包成 war 包

在上面创建的项目中，Gradle 构建配置文件 build.gradle 内容如下：

```
buildscript {
    ...
}
...
apply plugin: 'war'
...
configurations {
    providedRuntime
}
dependencies {
    ...
    providedRuntime('org.springframework.boot:spring-boot-starter-tomcat')
}
```

其中，apply plugin: 'war' 使用 war 插件来完成项目的打包工作。

直接使用 gradle bootWar，即可把项目打成 war 包。然后，就可以像普通 J2EE 项目一样部署到 Web 容器。同样，war 包的路径默认也是放在 build/libs 下面。

另外，如果下面这行代码还在：

```
@SpringBootApplication
open class DemoPackageAndDeployApplication

fun main(args: Array<String>) {
    runApplication<DemoPackageAndDeployApplication>(*args)
}
```

项目打成的 war 包，依然支持 java-jar 运行：

```
$ java -jar build/libs/demo_package_and_deploy-0.0.1-SNAPSHOT.war
```

这个 war 包很不错，既可以直接扔到 Tomcat 容器中执行，也可以直接在命令行启动运行。

提示 项目打 war 包的示例项目源代码：https://github.com/EasySpringBoot/demo_package_and_deploy

20.2 Spring Boot 应用运维

本节简单介绍 Spring Boot 应用的生产运维相关内容。

20.2.1 查看 JVM 参数的值

使用如下命令：

```
ps -ef|grep java
```

拿到对于 Java 程序的 pid（第 2 列）：

```
501 69156 68678    0  6:18PM ttys002    0:21.59 /usr/bin/java -jar
    build/libs/demo_package_and_deploy-0.0.1-SNAPSHOT.jar --spring.profiles.active=dev
```

可以根据 Java 自带的 jinfo 命令：

```
jinfo -flags 69156
```

来查看 jar 启动后使用的是什么 GC、新生代、老年代，以及分批的内存都是多少，示例如下：

```
$ jinfo -flags 69156
Attaching to process ID 69156, please wait...
Debugger attached successfully.
Server compiler detected.
JVM version is 25.40-b25
Non-default VM flags: -XX:CICompilerCount=3 -XX:InitialHeapSize=134217728
    -XX:MaxHeapSize=2147483648 -XX:MaxNewSize=715653120 -XX:MinHeapDeltaBytes=524288
    -XX:NewSize=44564480 -XX:OldSize=89653248 -XX:+UseCompressedClassPointers
    -XX:+UseCompressedOops -XX:+UseFastUnorderedTimeStamps -XX:+UseParallelGC
```

其中的参数简单说明如下：

- -XX：CICompilerCount：最大的并行编译数。
- -XX：InitialHeapSize 和 -XX：MaxHeapSize：指定 JVM 的初始堆内存和最大堆内存大小。
- -XX：MaxNewSize：JVM 堆区域新生代内存的最大可分配大小。
- -XX：+UseParallelGC：垃圾回收使用 Parallel 收集器。

我们可以在 Java 命令行中配置需要的 JVM 参数指标。

提示　更多关于 JVM 选项参数配置参考：http://www.oracle.com/technetwork/java/javase/tech/vmoptions-jsp-140102.html。

20.2.2　应用重启

要想重启应用，要首先找到该应用的 Java 进程，然后 kill 掉 Java 进程。完成这个逻辑的 shell 脚本如下：

```
kill -9 $(ps -ef|grep java|awk '{print $2}')
```

然后，再使用命令行重新启动应用即可。

20.3　使用 Docker 构建部署运行 Spring Boot 应用

本节介绍如何使用 Docker 来构建部署 Spring Boot 应用。

20.3.1 Docker 简介

Docker 是一个 Go 语言开发的开源的轻量级应用容器引擎，诞生与 2013 年。Docker 的核心概念是：镜像、容器、仓库。关键字是：分布式应用（distributed applications），微服务（microservices），容器（containers），虚拟化（docker virtualization）。

Docker 容器“轻量级”的含义主要是跟传统的虚拟机方式的对比而言。如图 20-1 所示。

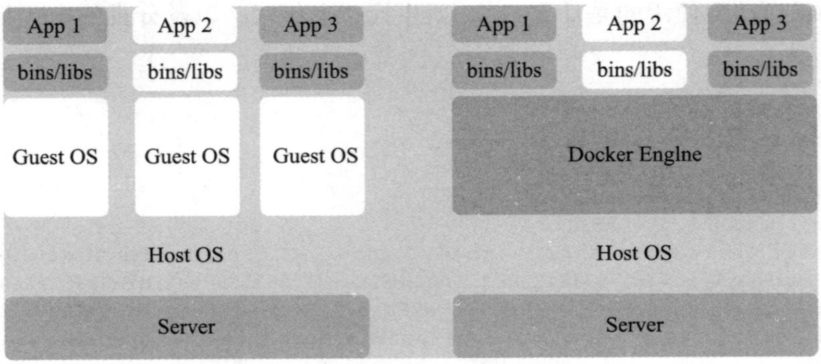

图 20-1 Docker“轻量级”容器 VS. 传统的虚拟机方式

传统的虚拟机技术是在硬件层面实现虚拟化，需要额外的虚拟机管理软件跟虚拟机操作系统这层。而 Docker 是在操作系统层面上的虚拟化，直接使用的是本地操作系统资源，因此更加轻量级。

Docker 的主要目标是通过对应用组件的封装、分发、部署、运行等生命周期的管理，做到“一次封装，到处运行”。

Docker 是实现微服务（microservices）应用程序开发的理想选择。开发、部署和回滚都将变成“一键操作”。传统的在服务器上进行各种软件包的安装、环境配置、应用程序的打包部署、启动进程等零散的运维操作——被更高层次的“抽象”，放到了一个“集装箱”中，我们只是“开箱即用”。Docker 把交付运行环境比作“海运”：OS 如同一个货轮，每一个在 OS 上运行的软件都如同一个集装箱，用户可以通过标准化手段自由组装运行环境，同时集装箱的内容可以由用户自定义，也可以由专业人员制造——这样交付一个软件，就是一系列标准化组件集的交付，如同乐高积木，用户只需要选择合适的积木组合，最后个标准化组件就是给用户的应用程序。这就是基于 docker 的 PaaS 产品的原型。

一个完整的 Docker 有以下几个部分组成：

- DockerClient 客户端
- Docker Daemon 守护进程
- Docker Image 镜像
- DockerContainer 容器

在 Docker 的网站上介绍了使用 Docker 的典型场景：

- 应用打包部署自动化。
- 创建轻量、私有的 PaaS 环境。
- 实现自动化测试和持续的集成 / 部署。
- 部署与扩展 web app、数据库和后端服务。

由于 Docker 基于 LXC 的轻量级虚拟化的特点，相比 KVM 之类虚拟机而言，最明显的特点就是启动快，资源占用小（轻量级），这正是构建隔离的标准化运行环境、轻量级的 PaaS、自动化测试和持续集成环境、以及一切可以横向扩展的应用等场景的最佳选择。

提示　更多关于 Docker 的介绍参考：https://docs.docker.com。Dockers 项目空间是：https://github.com/docker。入门书籍推荐《Docker 技术入门与实战》(杨保华)。

20.3.2　环境搭建

本小节介绍如何搭建 Docker 环境。

1. 安装 Docker

去 docker 官网 https://docs.docker.com/install/ 下载对应的操作系统上的安装包。安装完毕，打开 Docker 运行，可以看到 Mac 系统菜单栏上的显示的 Docker 应用信息如图 20-2 所示。

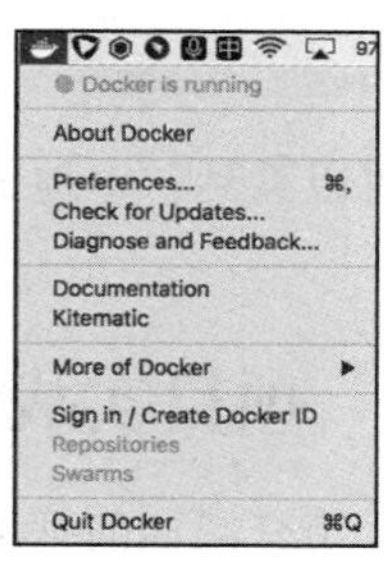

图 20-2　Mac 系统菜单栏上的 Docker 图标

想知道 docker 提供了哪些命令行操作吗？执行 docker help 即可看到一个详细的命令说明。例如，在命令行查看 Docker 版本信息：

```
$ docker version
Client:
    Version: 17.12.0-ce
    API version: 1.35
    Go version:  go1.9.2
    Git commit:  c97c6d6
    Built: Wed Dec 27 20:03:51 2017
    OS/Arch: darwin/amd64

Server:
    Engine:
        Version:  17.12.0-ce
        API version:  1.35 (minimum version 1.12)
        Go version: go1.9.2
        Git commit: c97c6d6
        Built:  Wed Dec 27 20:12:29 2017
        OS/Arch:  linux/amd64
        Experimental: false
```

查看详细的 docker 信息：

```
$ docker info
Containers: 0
    Running: 0
    Paused: 0
    Stopped: 0
Images: 1
Server Version: 17.12.0-ce
…
```

2. 从仓库 pull Java 环境镜像

使用 sudo docker pull java 命令从 Docker 官方仓库获取 Java 运行环境镜像：

```
$ sudo docker pull java
Password:
Using default tag: latest
latest: Pulling from library/java
...
bb9cdec9c7f3: Pull complete
Digest: sha256:c1ff613e8ba25833d2e1940da0940c3824f03f802c449f3d1815a66b7f8c0e9d
Status: Downloaded newer image for java:latest
```

下载完毕之后，可以通过 docker images 命令查看镜像列表：

```
$ docker images
REPOSITORY TAG     IMAGE ID           CREATED         SIZE
Java        latest  d23bdf5b1b1b       12 months ago   643MB
```

可以看到，本地镜像中已经有了 Java 运行环境。

20.4 项目实战

本节介绍如何把上面的 Spring Boot 项目 Docker 容器化，主要步骤为如下 3 步：

1）添加 Docker 构建插件。

2）配置 Dockerfile 文件创建自定义的镜像。

3）构建 Docker 镜像。

下面就来详细介绍。

20.4.1 添加 Docker 构建插件

在 Gradle 项目构建配置文件 build.gradle 中添加 com.palantir.docker 插件：

```
buildscript {
    ext {
        kotlinVersion = '1.2.20'
        springBootVersion = '2.0.0.RC1'
    }
    repositories {
```

```
        //gradle-docker plugin repo
        maven { url "https://plugins.gradle.org/m2/" }
        ...
    }
    dependencies {
        ...
        classpath('gradle.plugin.com.palantir.gradle.docker:gradle-docker:0.17.2')
    }
}


apply plugin: 'com.palantir.docker'

...

docker {
    name "${project.group}/${jar.baseName}"
    files jar.archivePath
    buildArgs(['JAR_FILE': "${jar.archiveName}"])
}
```

其中，buildArgs（['JAR_FILE': "${jar.archiveName}"]）中配置的'JAR_FILE': "${jar.archiveName}"是我们的 Spring Boot 项目打成 jar 包的名称，会传递到 Dockerfile 文件中使用（下一步骤中将会看到）。

提示　关于 Docker 插件 com.palantir.docker 的介绍参考文档：https://github.com/palantir/gradle-docker

这个插件发布在 https://plugins.gradle.org/m2/ 仓库中，所以我们添加 maven 仓库的依赖：

```
repositories {
    //gradle-docker plugin repo
    maven { url "https://plugins.gradle.org/m2/" }
    ...
}
```

gradle-docker 提供的版本有：https://plugins.gradle.org/m2/com/palantir/docker/com.palantir.docker.gradle.plugin/

20.4.2　配置 Dockerfile 文件创建自定义的镜像

Dockerfile 文件放置在项目根目录：

```
▼ demo_package_and_deploy ~/KotlinSpringBoot/demo_package_and_
  ▶ .gradle
  ▶ .idea
  ▶ build
  ▶ gradle
```

```
▶ out
▼ src
  ▼ main
      java
    ▼ kotlin
      ▼ com
        ▼ easy
          ▼ springboot
            ▼ demo_package_and_deploy
              ▶ DemoPackageAndDeployApplication.kt
              ▶ ServletInitializer
    ▼ resources
        static
        templates
        application.properties
        application-dev.properties
        application-prod.properties
  ▶ test
  .gitignore
  build.gradle
  demo_package_and_deploy.iml
  Dockerfile
  gradlew
  gradlew.bat
  LICENSE
  README.md
▶ External Libraries
```

Dockerfile 文件内容如下：

```
FROM java:latest
VOLUME /tmp
ARG JAR_FILE
ADD ${JAR_FILE} app.jar
ENTRYPOINT ["java","-Djava.security.egd=file:/dev/./urandom","-jar","/app.jar"]
```

20.4.3 Dockerfile 配置说明

Dockerfile 配置说明如表 20-1 所示。

表 20-1 Dockerfile 配置说明

配置项	说 明
FROM java:latest	使用本地 Docker 仓库中的 Java 镜像
VOLUME /tmp	挂载 /tmp 目录
ARG JAR_FILE	配置构建参数 JAR_FILE，这里的 JAR_FILE 是在 build.gradle 中 buildArgs 中配置的
ADD ${JAR_FILE} app.jar	将文件 ${JAR_FILE} 拷贝到 docker container 的文件系统对应的路径 app.jar
ENTRYPOINT ["java", "-Djava.security.egd= file:/dev/./urandom", "-jar", "/app.jar"]	Docker container 启动时执行的命令。注意：一个 Dockerfile 中只能有一条 ENTRYPOINT 命令；如果多条，则只执行最后一条

（续）

配置项	说　　明
-Djava.security.egd=file:/dev/./urandom	配置 JRE 使用非阻塞的 Entropy Source。SecureRandom generateSeed 使用 /dev/random 生成种子。但是 /dev/random 是一个阻塞数字生成器，如果它没有足够的随机数据提供，它就一直等，这迫使 JVM 等待。通过在 JVM 启动参数中配置这么一行：-Djava.security.egd=file:/dev/./urandom 解决这个阻塞问题

Dockerfile 是一个文本格式的配置文件，我们可以使用 Dockerfile 文件快速创建自定义的镜像。Dockerfile 支持的丰富的运维指令。这些指令分为 4 部分：

- 基础镜像信息
- 维护者信息
- 镜像操作指令
- 容器启动时的执行指令

Dockerfile 中使用以 # 开头的注释行。Dockerfile 中指令的一半格式是 INSTRUCTION arguments。下面介绍 Dockerfile 指令。

1. FROM

Dockerfile 文件第一行必须为 FROM 指令。如果在同一个 Dockerfile 中创建多个镜像时，可以使用多个 FROM 指令（每个镜像 FROM 一次）。

语法格式：

```
FROM <REPOSITORY>
```

或者

```
FROM <REPOSITORY>:<TAG>
```

2. MAINTAINER

指定维护者信息。

语法格式：

```
MAINTAINER <name>
```

3. RUN

语法格式：

```
RUN <command>
```

或者

```
RUN [ “executable” ,"param1","param2"]
```

前者将在 shell 终端中运行命令，即 /bin/sh -c；后者使用 exec 执行。

第二种方式可以指定其他终端实现，如：RUN[“/bin/bash”， "-c","echo hello"]。

当命令较长时，可以使用 \ 来换行。

4. CMD

指定启动容器时执行的命令，每个 Dockerfile 只能有一条 CMD 命令。如果指定了多个该命令，只有最后一条会被执行。

CMD 命令支持三种格式：

```
CMD ["executable","param1","param2"]: 使用 exec 执行
CMD command param1 param2: 在 /bin/sh 中执行，提供给需要交互的应用
CMD ["param1","param2"]: 提供给 ENTRYPOINT 的默认参数
```

5. EXPOSE

告诉 docker 服务端容器要暴露的端口号，供互联系统使用。

语法格式：

```
EXPOSE <port> [<port>...]
```

在容器启动时，需要通过 -P 参数选项让 Docker 主机自动分配一个端口转发到指定的端口；使用 -p 可以具体指定哪个本地端口会映射过来。

6. ENV

指定环境变量，在执行 RUN 指令的时候使用，在容器运行时都有效。

语法格式：

```
ENV <key> <value>
```

7. ADD

向容器中复制文件。

语法格式：

```
ADD <src> <dest>
```

指定的 <src> 会被复制到 <dest>。其中的 <src> 可以是 Dockerfile 所在目录的一个相对路径（文件或者目录），也可以是一个 URL，还可以是一个 tar 文件（会自动解压为一个目录）。

8. COPY

语法格式：

```
COPY <src> <dest>
```

复制主机的 <src>（Dockerfile 所在目录的相对路径，文件或目录）为容器的 <dest>，目标路径不存在时会自动创建。

9. ENTRYPOINT

配置容器启动后执行的命令，并且不可以被 docker run 提供过的参数覆盖。

语法格式：

```
ENTRYPOINT ["executable","param1","param2"]
```

同样的，每个 Dockerfile 中只能有一个该指令，指定多个时，只有最后一个生效。

10. VOLUME

创建一个可以从本地主机或者其他容器挂载的挂载点，一般用来存放数据库和需要保持的数据等。

语法格式：

```
VOLUME ["/data"]
```

11. USER

指定运行容器时的用户名或 UID，后续的 RUN 也会使用指定用户。

语法格式：

```
USER daemon
```

12. WORKDIR

为后续的 RUN、CMD、ENTRYPOINT 指令配置工作路径。

语法格式：

```
WORKDIR /path/to/workdir
```

可以使用多个该指令，后续指令如果是相对路径，则会基于之前的命令指定路径。比如：

```
WORKDIR /A
WORKDIR B
RUN X
```

则最终路径为：/A/B。

13. ONBUILD

配置当前所创建的镜像作为其他新创建镜像的基础镜像时，所执行的操作指令。

语法格式：

```
ONBUILD [INSTRUCTION]
```

比如镜像 A 中的 Dockerfile：

```
...
ONBUILD ADD  .   /app/src
ONBUILD RUN /usr/local/... --dir  /app/src
...
```

如果基于镜像 A 创建新镜像 B 时，在 B 的 Dockerfile 中使用 FROM[镜像 A] 时，会自动执行镜像 A 中的 Dockerfile 中所配置的 ONBUILD 指令内容。

> 提示 更多关于 Dockerfile 配置文件的介绍参考官方文档：https://docs.docker.com/engine/reference/builder/#usage

20.4.4 构建镜像

最后一个步骤就是创建自定义应用的镜像。编写好 Dockerfile 后，执行 gradle bootJar docker 命令创建镜像：

```
$ gradle bootJar docker

BUILD SUCCESSFUL in 6s
6 actionable tasks: 3 executed, 3 up-to-date
```

构建完毕。命令行执行 docker images 可以查看我们创建的 com.easy.springboot:demo_package_and_deploy 应用的 Docker 镜像：

```
jack@jacks-MacBook-Air:~$ docker images
REPOSITORY                                  TAG       IMAGE ID       CREATED           SIZE
com.easy.springboot/demo_package_and_deploy latest    94dcdae5dc3e   2 seconds ago     663MB
<none>                                      <none>    3ceff440b061   32 minutes ago    663MB
java                                        latest    d23bdf5b1b1b   12 months ago     643MB
jack@jacks-MacBook-Air:~$
```

直接在命令行执行：

```
$ docker run -p 8080:9000 -t com.easy.springboot/demo_package_and_deploy
```

即可启动我们构建发布在 Docker 镜像仓库中的 Spring Boot 应用镜像了。

我们的 Spring Boot 应用镜像运行在 Docker 容器沙箱环境中，端口号是 9000，作为外部 Host OS 环境要访问这个服务，需要添加 TCP 端口映射：把本机 8080 端口映射到 Docker 容器端口 9000，如图 20-3 所示。

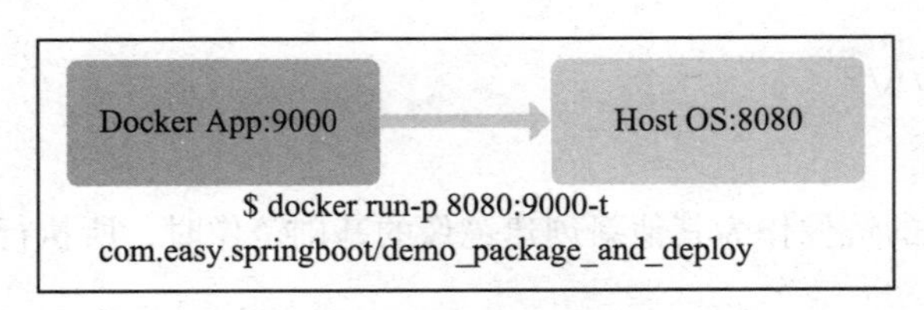

图 20-3 把本机 8080 端口映射到 Docker 容器端口 9000

其中：

- -p 是将容器的端口 9000 映射到 docker 所在操作系统的端口 8080；
- -t 是打开一个伪终端，以便后续可以进入查看控制台 log。

使用 docker ps 命令查看运行中的容器：

```
$ docker ps
CONTAINER ID          IMAGE                                            COMMAND
                   CREATED              STATUS                PORTS
                     NAMES
```

```
36fbfaf05359        com.easy.springboot/demo_package_and_deploy   "java -Djava.
    securit…"   25 minutes ago      Up 25 minutes       0.0.0.0:8080->9000/tcp
    infallible_kare
```

上面的运行镜像进程简单说明如表20-2所示。

表20-2 镜像进程简单说明

列项	值	说明
CONTAINER ID	36fbfaf05359	镜像进程ID，可以使用 `$ docker kill 36fbfaf05359` 命令来杀掉改进程
IMAGE	com.easy.springboot/ demo_package_and_deploy	镜像名称
COMMAND	"java -Djava.securit…"	进程的命令
CREATED	25 minutes ago	进程创建时间
STATUS	Up 25 minutes	进程状态
PORTS	0.0.0.0:8080->9000/tcp	端口映射：本机8080映射到容器内的9000端口
NAMES	infallible_kare	容器NAMES

其中，Docker列出容器命令简介如下。

语法：

```
docker ps [OPTIONS]
```

OPTIONS说明：

- -a：显示所有的容器，包括未运行的。
- -f：根据条件过滤显示的内容。
- --format：指定返回值的模板文件。
- -l：显示最近创建的容器。
- -n：列出最近创建的n个容器。
- --no-trunc：不截断输出。
- -q：静默模式，只显示容器编号。
- -s：显示总的文件大小。

20.4.5 运行测试

本机浏览器打开 http://127.0.0.1:8080/，可以看到成功输出，如图20-4所示。

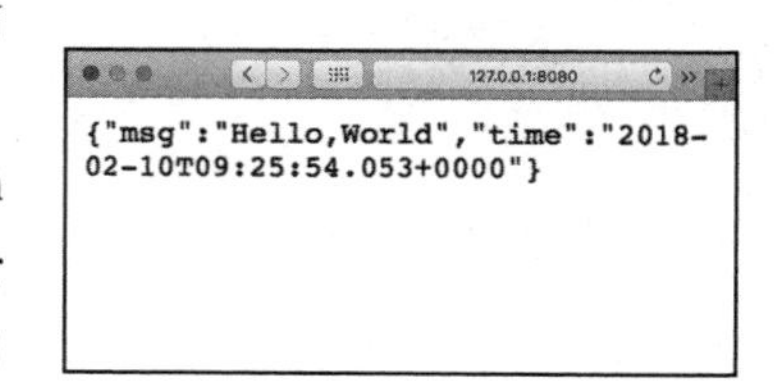

图20-4 输出结果

我们还可以使用docker push命令把Docker服务镜像push到云端，例如Docker Hub（https://hub.docker.com）。Docker Hub在国外，有时候拉取Image极其缓慢，可以使用国内的镜像来实现加速。

首先，你在 Docker Hub 要有注册账号，且创建了相应的库。其次，docker 推送前，先要执行 docker login 登录：

```
$ docker login
Login with your Docker ID to push and pull images from Docker Hub. If you don't
    have a Docker ID, head over to https://hub.docker.com to create one.
Username:
......
```

然后，执行 push 命令即可：

```
$ docker push com.easy.springboot/demo_package_and_deploy
```

提示 本节项目源代码：https://github.com/EasySpringBoot/demo_package_and_deploy/tree/spring_boot_with_docker_2018.2.10

20.5 本章小结

本章简单介绍了 Spring Boot 项目的打包、分环境运行、生产运维等操作。通常，在企业项目实践中，会实现一套 Spring Boot 应用部署发布的自动化运维平台工具。本章还给出了一个完整的 Spring Boot 项目 Docker 化的实战案例。

经过前面的学习，相信你已经对如何使用基于 Kotlin 编程语言的 Spring Boot 项目开发有了一个比较好的掌握。